T0143201

The Stockholm Paradigm

The Stockholm Paradigm

Climate Change and Emerging Disease

DANIEL R. BROOKS, ERIC P. HOBERG, AND WALTER A. BOEGER

The University of Chicago Press   Chicago and London

The University of Chicago Press, Chicago 60637
The University of Chicago Press, Ltd., London
© 2019 by The University of Chicago
All rights reserved. No part of this book may be used or reproduced
in any manner whatsoever without written permission, except
in the case of brief quotations in critical articles and reviews.
For more information, contact the University of Chicago Press,
1427 E. 60th St., Chicago, IL 60637.
Published 2019
Printed in the United States of America

28 27 26 25 24 23 22 21 20 19   1 2 3 4 5

ISBN-13: 978-0-226-63230-8 (cloth)
ISBN-13: 978-0-226-63244-5 (paper)
ISBN-13: 978-0-226-63258-2 (e-book)
DOI: https://doi.org/10.7208/chicago/9780226632582.001.0001

Library of Congress Cataloging-in-Publication Data

Names: Brooks, D. R. (Daniel R.), 1951– author. | Hoberg, Eric P.,
    1953– author. | Boeger, Walter A., author.
Title: The Stockholm paradigm : climate change and emerging
    disease / Daniel R. Brooks, Eric P. Hoberg, and Walter A. Boeger.
Description: Chicago : The University of Chicago Press, 2019. |
    Includes bibliographical references and index.
Identifiers: LCCN 2018055503 | ISBN 9780226632308
    (cloth : alk. paper) | ISBN 9780226632445 (pbk. : alk. paper) |
    ISBN 9780226632582 (e-book)
Subjects: LCSH: Bioclimatology. | Climatic changes. |
    Parasitic diseases.
Classification: LCC QH543 .B76 2019 | DDC 577.2/2—dc23
LC record available at https://lccn.loc.gov/2018055503

♾ This paper meets the requirements of ANSI/NISO Z39.48–1992
(Permanence of Paper).

Contents

Preface

Each decade of our careers has had a different theme. First was the feeling that we had our entire lives to study the amazing world of biological diversity, its evolution and ecology. Next, we began to realize that the biosphere was changing, and it was becoming more difficult to find some of the species that had become our cherished friends; nonetheless, we still could find isolated pockets of the way it used to be. Third, we saw that the biosphere was changing so rapidly that we needed to document what existed, paying special attention to what might be useful for sustainable human socioeconomic development. Fourth, and ongoing, our minds are filled with one word: *triage*.

We are convinced that humanity faces a grave imminent challenge, and the responses to that challenge have not been sufficient. We live in a world of climate change, and previous episodes of climate change have had profound impacts on the biosphere and the evolution of life on this planet. The crisis of emerging diseases forces us to confront the interface between a landscape engineered and maintained by humanity at increasing cost, and the largely "neglected other" that we sometimes call the biosphere. As we push into that neglected other and as elements of it push into "us," predictable things are happening. We therefore need a more robust understanding of the evolutionary biology of pathogens to explain why we have a crisis of emerging diseases, one that points the way to reducing their impact on us. This book pulls no punches, but that is not because we are committed misanthropes. It is because we believe that while the crisis we face is nobody's fault, we are all to blame. We are going to try to make every reader feel ambivalent about some of their most cherished scientific and socioeconomic notions, to push people out of their comfort zones so they can consider a new perspective. We are not alone. The scientific community has become galvanized in the early twenty-first century and is enthusiastically investigating links between climate change, disease, and humanity, resulting in an explosion of scientific literature matching the explosion of emerging disease. In notes at the end of the text and in the bibliography, we refer to about 1000 published works and supplementary data for those who want to begin to delve more deeply into particular topics. But that is only a fraction of the 10,000 or so published articles about climate change and disease we have read in the past few years. During the time this book was in review, news reports pertaining to emerging diseases appeared daily, and newly published scientific articles supporting our perspective appeared about once a week. This book confronts some painful realities and provides suggestions for how we might better cope with what is on our front doorstep and threatening to overwhelm us. We hope our words will be persuasive, especially to those caring and committed people who have been doing their best following traditional yet faulty principles. We must begin taking collective personal responsibility for our own survival. We are fond of this world and this existence; we would like it to continue, and we would like a larger proportion of humanity to share its benefits.

Acknowledgments

Our joint thanks to: Scott Gardner, University of Nebraska, who hosted a workshop at the University of Nebraska field station in 2012 that was the basis for the DAMA protocol (chapter 9); Sören Nylin, Stockholm University, who hosted a workshop held at the Tovetorp Research Station of Stockholm University, made possible by the Marcus Wallenberg Foundation ("Changing Species Associations in a Changing World: A Marcus Wallenberg Symposium" [MWS 2015.0009]). We learned a lot from the participants: Sal Agosta, Staffan Bensch, Mariana Braga, Matthew Forister, Peter Hambäck, Niklas Janz, Tommi Nyman, Martin Österling, Alexander Schäpers, Alycia Stigall, Niklas Wahlberg, and Christopher Wheat.

In 2017 two of us (Brooks and Hoberg) attended a workshop in Balatonfüred, Hungary, that focused on refining the scope of DAMA and enacting DAMA-based projects. Our deepest appreciation to Eörs Szathmáry and to András Báldi, director of the Institute of Ecology of the Hungarian Academy of Science, who hosted and sponsored the workshop. Special thanks to the participants—Alicia Juarrero, Vitalyi Karchenko, Lajos Rózsa, Viktor Mueller, Dávid Herczeg, and Gábor Földvári (who also made very helpful comments on the manuscript of this book).

David Chitwood and Lisa Castlebury from the U.S. Agricultural Research Service, Mycology and Nematology Genetic Diversity and Biology Laboratory, kindly provided us access to mycological literature from the early twentieth century. Kirill Galaktionov in St. Petersburg, Russia and Vitus Kontrimavichus in Vilnius, Lithuania provided insights into the often tragic history of Russian parasitology and its substantial and underappreciated contributions to ecology and evolution.

Deborah McLennan, University of Toronto, provided two detailed and exceptionally helpful critiques of the manuscript, along with substantial intellectual support. Alicia Juarrero's foundational work on the importance of context in understanding complex systems was critical for our thinking. As well, in her role as CEO of Vector Analytica, Alicia provided substantial insight particularly about community-based vector- and disease-monitoring systems.

Finally, we would like to thank the anonymous readers for taking the time to give the manuscript a thorough going-over.

My most heartfelt thanks go to my coauthors, who have been personal friends and invaluable colleagues for more than 30 years. Thanks in particular to Walter Boerger, who was awarded a Ciencias sem Fronteiras grant that allowed me to come to Curitiba for three months each year from 2014 to 2016. That fellowship program made it possible for us to hire Sabrina Araujo as a postdoctoral fellow; it was she who produced the first modeling of the Stockholm Paradigm in collaboration with Mariana Braga. Eörs Szathmáry, who invited me to be a senior visiting fellow at the Collegium Budapest in 2010–11, encouraged me strongly to pursue this work at the Institute for Advanced Study Kőszeg (iASK) in 2016, where I met Robert Manchin and András Szöllösi-Nagy. During my time there, iASK invited Walter to give a presentation of the modeling discussed in chapter 5 that is still remembered there. My thanks to Ferenc Miszlevitz, director of iASK, and the institute's kind and generous staff. I express my deepest thanks to the Stellenbosch Institute for Advanced Study (STIAS), where I was a senior visiting fellow from February to June 2017. That provided the opportunity for me to develop my contributions to this book in the company of a diverse group of highly accomplished people. I particularly thank Hendrik Geyer, STIAS director; Christoff Pauw, program manager; and the incredible STIAS staff. William Kunin, Leeds University, in particular provided enormous insights and critical feedback while sharing some memorable weekend road trips. Special thanks to other friends and colleagues who provided foundations and insights: Sven Jakobsson, Sören Nylin, Niklas Janz, Hans-Erik Wanntorp, and Christer Wiklund, Stockhholm University; Mike Singer, Plymouth University; Tim Benton, Leeds University; Ian Goldin, Oxford University; Mike Stuart, University of North Carolina, Asheville; Peter Watts, Toronto; and Sean Cleary, South Africa. Finally, I thank my wife, Zsuzsanna, without whom I would have been lost.

—D. R. B.

Mine has been a uniquely personal journey, now over 40 years and more, having seen much of a world in accelerating change, affording insights about the pace and scope of disturbances in our climate, landscapes, and seascapes around the globe. Traversing Arctic and Antarctic latitudes in the course of a single year was not unlike the peregrinations of wayward seabirds that I followed early in those years, with days in storm, wind, and sun. I owe considerable gratitude to those who made so much possible, from the earliest sojourns in Alaska and Siberia for field biology on land, across the northern seas, to more recent explorations across northwestern North America, Fennoscandia, and the Arctic. Explorations of biodiversity and climate change in northern systems emerged from the Beringian Coevolution Project and the Research Group for Arctic Parasitology in the late 1990s, including more than 20 years collaborating with Joe Cook at the Museum of Southwestern Biology. In my former position as the curator of the U.S. National Parasite Collection, support from the U.S. Department of Agriculture, Agricultural Research Service, since the 1990s facilitated broad-based studies of parasite diversity, biogeography, and the impact of pathogens among ungulates. It has been now more than 30 years that I have known Dan and Walter, and there are always surprises, a rich mixture of ideas and insights from which I am always learning. I am grateful most recently for the introductions to Tovetorp Field Station of Stockholm University and opportunities for rich discussions with Sören Nylin, Niklas Janz, and their colleagues at the interface of a synthesis for plant-insect systems and parasitology. Special thanks to the people on the path—Dan and Walter in particular, and the brilliance of Sal Agosta, Mariana Braga, and Sabrina Araujo. Through these integrated approaches we started to define the links of a larger picture of perturbation as a unifying theme in faunal assembly, host associations, and global diversity—all of which are among the foundations of the Stockholm Paradigm. Finally, thanks to Jerry Swartz and Bob Rausch, who understood that parasitology was the integrator across the biosphere.

—E.P.H.

I have to begin by thanking five people who influenced my thoughts in zoology. My first professor in zoology, Joaber Pereira, Jr., was fundamental in creating the basic understanding that a scientist's life may lack many things, but not new ideas. He is the one who introduced me to the fantastic world of biological associations through parasitology. Fortunately, my subsequent mentors, Vernon E. Thatcher and Delane C. Kritsky, have not attempted to fix my biological delusions. Early in my professional life Dan implanted the not-so-obvious-at-that-time idea that all biological processes must be understood through a historical and evolutionary perspective. We had the opportunity to work more closely these last few years, and the experience has rebooted my interest in critical evolutionary thinking. Finally, my wife and colleague biologist, Maria Regina T. Boeger, has always supported my professional decisions and has been—poor woman—the first person to actually hear and discuss with me almost all of my insights (in biology or other fields). She and my son, Bruno, have always been my loves and my direct links to the rest of the world. My present understanding of biological association also results from my past history as a professional scientist and professor. The places I worked, the focus of my research, my collaborators, my graduate students have all contributed to the way I understand the biological world today. I will not cite any, in order to avoid errors of omission. They know who they are. And a final word of gratitude to the thousands of undergraduate students that I have the pleasure to teach for more than 28 years. Classrooms and fresh minds are great sources of new insights.

—W.A.B.

ONE

How Bad Is It, Anyway?

The world's a little bit under the weather and I'm not feeling too good myself.
—Sign at the Rally and March for Science, Washington, D.C., April 22, 2017

This planet's climate is changing. Inexorably. We now live in a "second-order world" in which change is not simply accumulating, it is accelerating. Projections for the year 2100 made around the turn of the century already seem woefully inadequate.[1] Despite incrementally decreasing birthrates, the human population grows daily; it's on the move, and deep technological footprints mark its passage. We alter landscapes and perturb ecosystems, inserting ourselves and other species into novel regions of the world, contributing to permanent changes in the biosphere. This should surprise no one. More than a century ago the Swedish Nobel Laureate Svante Arrhenius showed that carbon dioxide emissions from industrial sources would increase atmospheric temperatures.[2] Three generations ago Charles Elton, a founder of modern ecology and a bellwether for the disruptive effects of climate change, wrote: "We must make no mistake; we are seeing one of the greatest historical convulsions in the world's fauna and flora."[3] Elton concluded that human civilization was about to experience severe socioeconomic disruptions. Global changes in temperature and rainfall would lead to drought, famine, and disease. These would in turn lead to conflict and disease. And conflict would lead inevitably to migration and more disease. Now each day is a testimony to Elton's insights. Climate change alters the movements and geographic distributions of myriad species. Transporting people and goods brings previously isolated species into sudden contact with one another. Drought and flood create famine and conflict. Two-thirds of humanity risk water shortages, and the rest face the threat of floods.[4] Paradoxically, both drought and flood carry increased risk of disease transmission. Furthermore, they all lead to migration, and not just of humans.[5] Diseases emerge everywhere on a daily basis, affecting us and the species upon which we depend for our survival. Elton would surely not have been surprised that we now find ourselves deeply embedded in the midst of an epidemiological crisis. In these arenas of movement and new contact, a world of pathogens continually encounters suitable hosts with no resistance and no time to evolve any.

This is also not news—maladies rare or unknown half a century ago, like HIV and Ebola, West Nile virus, avian influenza, Zika, dengue, monkeypox, hanta, and Lyme disease—are increasingly commonplace. Some old associates, such as malaria and yellow fever, have resurfaced with a vengeance. We are even seeing a resurgence of polio and tuberculosis, two diseases we thought we had largely vanquished. Yet others lurk in places becoming more accessible to humans. Some, such as bubonic plague in the United States, are expanding their geographic distributions as a result of successful efforts by conservation biologists to reintroduce reservoir hosts into areas with high human traffic.[6] In such a world—this world—events like these are ongoing. The entire range of species upon which humanity depends for socioeconomic reasons is affected.

Scarcely a week passes without news of some freshly discovered strain of pathogen trading up to a human host . . . or European cattle . . . or Nigerian tomatoes . . . or critical pollinator insects . . . or sea urchins. News items flashing across our monitors since we began writing this book in mid-2016, in no particular order: temporary cessation of blood transfusions in Greece for fear of malaria; closing of the Yellowstone River to fishing for fears of whirling disease and viral pathogens infecting trout; tourism losses in Florida amid Zika fears; shortages of the yellow fever vaccine in Brazil and in Central African countries amid outbreaks on both their continents; appearance of the South American tomato leaf miner in Nepal; an invasive insect killing trees on which street lights are hung, leading to crime increases associated with the loss of night lighting; an army worm invasion from North America to Africa, putting 10–40 percent of Africa's corn crop at risk; red squirrels in the United Kingdom harboring two different variants of the bacterium that causes leprosy; sudden oak death disease in Hawaii; a rust fungus that is resistant to every known fungicide and is infecting wheat in Eurasia; spruce beetle outbreaks in North America; plagues of locusts in southern Brazil and northern Argentina; filarioid worms in reindeer in Fenno-Scandinavia; blue-tongue virus and anthrax returning to Europe; an anthrax outbreak in eastern Siberia that may have come from anthrax bacteria frozen in permafrost for the past century until it began melting; a yellow fever outbreak giving way to a monkey-pox outbreak, giving way to an Ebola outbreak, giving way to a polio outbreak in the Democratic Republic of Congo in a span of four months; a massive cholera outbreak in Yemen associated with ongoing conflict and a breakdown in the water and sewage treatment infrastructure; plague in Madagascar; malaria in Tuscany, southern France, and southern Switzerland; leptospirosis breaking out in Puerto Rico following Hurricane Maria; a bacterium imported to Europe that threatens garden plants and fruit trees; a large influenza outbreak in North America on the centenary of the Spanish flu pandemic; the threat of measles at Chicago's O'Hare Airport; a Lassa fever outbreak in Nigeria; city parks in São Paulo closed for fears of yellow fever and malaria; Ebola reaching an urban center of two million in Congo; ongoing outbreaks of Crimean-Congo hemorrhagic fever, Zika, Ebola, dengue, chikungunya, Marburg, norovirus, arenavirus, rhinovirus; and on and on . . .

This is the new normal, the crisis of emerging infectious disease (EID), caused by pathogens fitting one of the following criteria: (1) something "new," that is, a pathogen previously unknown to science; (2) a known pathogen becoming more pathogenic than before; (3) a known pathogen becoming more pathogenic at the edges of its geographic range; (4) a known pathogen appearing in a geographic location where it had never been reported before; and (5) a known pathogen infecting a host it had never been reported infecting before. This is *not* a crisis of a few headline-worthy viruses infecting humans in tropical countries. Emerging diseases are caused by viruses, bacteria, fungi, protists, and multicellular animals, all exacting a socioeconomic cost from humans and the plants and animals upon which we depend. They can be found in developed as well as developing countries, and they occur in the wild lands, agricultural lands, and rural, suburban, and urban regions. They are transmitted in soil, water, and food; by eating, drinking, and casual contact with other organisms or surfaces that they touch; and by a constellation of biting, piercing, and sucking arthropods that feed on us and our livestock.

So here we are, plodding prey for a microcosmos of hunters that are finding us with—it seems—increasing ease. Some are ancient associates, but something new pops up on a daily basis. And we are afraid of them—starting in the past century they have been portrayed as the ultimate boogeymen intent on doing us in: "Searchers had not yet, like Pasteur . . . begun to challenge God, to shake their fists at the meaningless cruelties of nature toward mankind, her children."[7] Even these insights are not new—such classic books as *Microbe Hunters* (1926) and *Rats, Lice, and History* (1935) drew attention to this phenomenon,[8] so humanity should have been ready to accept the same message when it was delivered 60 years later by the microbiologist Laurie Garrett and the geographer Jared Diamond in their seminal books.[9]

If None of This Is New, Why Is It "News"?

You would think that the public health response would be a well-oiled machine by now. But yet again we have been blindsided.[10]

Scientists have been warning for at least a generation about the potential impact of EID in a world experiencing climate change.[11] The global response to those warnings, however, encourages the public to think of EID as isolated events, mostly affecting humans, that can only be dealt with by reacting after the fact and that have nothing to do with climate change. We allocate massive resources to pathogens that have already made themselves known while ignoring the far greater threat posed by those waiting in the wings. The ones we know are the tip of the iceberg—most of the world's pathogens haven't been discovered yet. They're discovering us easily, however; those weekly outbreaks and the endlessly variable strains of recent years are ample evidence of that. The new status quo is a succession of outbreaks battering an increasingly fragile public and agricultural health infrastructure that cannot meet the challenges of global climate change.[12] We rarely have an idea where the next one will pop up because we have done such a poor job of cataloguing the world's pathogens. All we know is that they're far better at finding us than we have been at finding them. This known issue is likely to continue to be reported as "surprising news" lasting only a few days when the real news story, as Laurie Garrett pointed out in 2001, is our continuing failure to cope with the crisis by anticipating and preparing for more outbreaks.

Elements of the Problem

The emerging disease crisis can be best understood in the context of three fundamental realities that will be recurring themes throughout this book.

It's about Us and Climate Change

About three million years ago our African ancestors moved from the forest to the savannah. Gradually changing from scavengers to predators, sharing prey with grassland carnivores, those ancient humans acquired pathogens that had first been found in hyenas, large cats, and African hunting dogs. They carried those pathogens out of Africa, helping them move into native hosts in new environments, while native pathogens residing in nonhumans returned the favor, infecting the newly arrived humans. Over time, agriculture, domestication, and urbanization brought people and animals into even closer contact, expanding the menus of yet more pathogens and making transmission easier than ever.

In the past 100,000 years, human geographic expansion disseminated EID risk on a global scale. If doctors had existed then, they would have remarked on a worrisome surge in the number of EIDs, responding to the crisis as best they could, after the fact, and wondered how to manage the outbreaks. The human ecological footprint has had a disproportionate influence on the distribution of diversity on the planet for the past 12,000 years.[13] By that time at least some humans had adopted sedentary lifestyles, accompanied by the rapid emergence of animal and plant domestication (corn was being domesticated in Mexico between 11,000 and 9,000 years ago)[14] and new diseases as humans inserted themselves more intimately into the lives of their newfound domestic companions. That time period also witnessed an improvement in the physical conditions of life, especially for women, who, narrowing the size gap with males, began having bigger, healthier babies. This was a mixed blessing. The human population began to grow. A lot. By 5000 years ago, humans had established permanent cities, many of which would be destroyed and abandoned, and only rarely rebuilt, as a result of conflict and natural disasters, including climate fluctuations that were relatively mild compared to what is now bearing down on us. The human population continued to grow and expand geographically, building new cities and acquiring more and more new diseases. By 250 years ago, we had created industrial cities and the foundations of modern technological society.

In the past 50 years an exploding human population, rapid transit, and the globalization of economies have produced the real-time crisis that is now part of our daily existence. And the onset of accelerating warming and environmental disruption has intensified that situation. It seems that at least some of the climate change we are experiencing signals the end of the extended run of climate stability that marked the Holocene. But that is not the full extent of the problem. Today human beings drive considerable environmental perturbation and uncertainty. Anthropogenic input, including the injection of carbon dioxide and other greenhouse gases into the atmosphere, has a tremendous influence on the state of the climate, global atmospheric and sea temperatures, patterns of precipitation and drought, and the severity and frequency of the storm systems that sweep the planet.

Temperature and water ultimately mediate the structure of the biosphere and the distribution of diversity over time.[15] Atmospheric carbon dioxide has fluctuated throughout the course of Earth history, directly influencing global temperature and the availability and distribution of water. Relatively small changes in atmospheric carbon can have substantial cascading effects, as is demonstrated by the outcomes of atmospheric forcing from industrialization during the past 250 years. Ironically, the technological component of humanity arose during a period of relative stability in the global climate system, following the termination of the last major glacial advance of the Pleistocene just 12,000 years ago. The inception of industrial development in the 1750s perturbed that stability, accelerating warming and increasing climate fluctuations globally. The trajectory was fairly consistent for more than a century. Unusual temperature increases began in the late 1800s, spurring Arrhenius to make his calculations. The rate increased in the early 1900s, forming the base of the curve in what is called the "hockey stick." There has been considerable and spirited discussion about this hockey-stick model, but it appears to be an inconvenient truth that has been maligned but never disproven. That rate increase produced "anomalous" temperatures throughout the twentieth century,[16] but Arrhenius's calculations and observations of global temperature changes still hold true. Around 1970 there appears to have been another uptick in the rate of warming that was especially pronounced in the northern latitudes.

It's about Complexity

Climate, biology, and human society are all examples of complex systems. One hallmark of complex systems is that they are capable of surprise. This is because they are made up of interacting parts leading to different outcomes, depending both on the proportions of the different parts and on the context in which their interactions take place. Alicia Juarrero laid the conceptual groundwork for understanding complex systems in her groundbreaking book published just before the new millennium.[17] Following her lead, we will remind you at critical junctures in this book that when dealing with climate change and emerging disease:

It is never just one thing, and it always depends.

Some aspects of life can be explained but cannot be made simple. Humans interact in complicated ways with one another, with other members of the biosphere, and with the climate. Better food production has led to more people demanding more food, causing global poverty levels to rise. Improved public health measures allow more people to live longer, leading to more and older people demanding more public health services. Decreasing birth rates hold out hope for reduced population pressure but also diminish revenues coming into public health systems from the young workers who support the older, longer-living members of society. Multiple socioeconomic demands on land use lead to conflicts. Climate change leads to conflicts. An increasing variety of waste products deposited in the very places upon which humans are making increasing demands marks the end of the human production line. These activities are expensive, and soon the expense will be greater than we can bear.

Two manifestations of these complex interactions are critical. The first is *tipping points*, instances in which changes occur that impact ourselves and the planet irreversibly. Think of them as decisions made through action or inaction. Humans have failed to recognize tipping points that are occurring due to a burgeoning human population; agricultural advances; the ways in which landscapes are occupied, used, and modified; and a deepening reliance on technology driven by non-renewable power and energy resources. Combined, these effects have transformed humanity's world from the large and slow one of our ancestors to today's small and rapid one. Generation by generation, the transformations have been incremental and thus difficult to discern and identify. They are however, consistent with a world history of dynamic change, and increasing scientific understanding of that history has made it possible to understand that our current conditions of existence are increasingly problematic.[18] Within that world, our massive ignorance about potential pathogens and our indifference to that ignorance has brought us to the brink of a tipping point, beyond which our ability to respond to a crisis will be overwhelmed. Our capacity for responding to disease outbreaks after the fact is already eroding and will continue to decline if we do not find a way to buy time and mitigate the costs of outbreaks.

The second is *threat multipliers*. Commonly attributed to the U.S. military, threat multipliers are phenomena that exacerbate existing problems. Climate change, population growth and density, broad erosion of the public-health infrastructure, and globalization are threat multipliers for emerging disease.[19] Emerging disease, in turn, exacerbates poverty, famine, drought, conflict, and migration and is therefore a threat multiplier for those phenomena. This is why the world's militaries consider emerging diseases a national security issue. And if it is a national security issue for every country on this planet, it is truly a global threat. If we can avoid war by anticipating it, the effort was worthwhile; the public will only be unhappy if there is a conflict and the military never saw it as a possibility. Why should we not apply the same reasoning to anticipating future emerging-disease scenarios? If our predictions of a difficult future for humanity with respect to climate change and disease do not come true, we will be pleased to have people deride us as being overly pessimistic. But if our predictions come true, and they have not been anticipated, we do not want government officials to claim that there was no way to prepare for the disasters.

It's about Fear

Humanity is facing an existential threat. What we have been doing is not working, but something is holding us back from trying novel approaches. What is stopping us? In a word—fear.

Nothing is so much to be feared as fear.[20]

Although it is a popular trope among scientists that the public and their elected officials do not understand the seriousness of climate change, we think this is not so. On the contrary, we believe that the scientific community has done a good job of making people so afraid that many members of the public, and many of their officials, are in denial. Sadly, those who have tried to frighten humanity into responsible action have not applied what we know about primate behavior. Anyone who has seen our primate relatives in nature understands what we mean. Faced with a potential crisis, our cousins freeze in place, and when a potential crisis becomes a real crisis, they scatter, each one hoping that someone else is slower. We authors have lived our professional lives in an environment in which there were always funds available to have a workshop or a conference or a symposium, but rarely any—and never enough—money to put into action any of the recommendations from those activities. Facing a potential crisis, scientists and policy makers tasked with helping humanity to survive these crises hold meetings and make pronouncements; then they scatter to the winds and hope the leopard eats someone else.

It is in our nature as primates to freeze in the face of potential danger, scatter and then later make a plan—humans are shortsighted as well as fearful by nature. One might think that our acute sense of self-awareness would have brought with it an acute sense of the necessity to plan for future contingencies. But, in a twist of evolutionary irony, it appears that self-awareness has been—some say had to be—coupled with a pronounced capacity for psychological denial. The argument is something like this—the only way to take advantage of the evolutionary potential allowed by self-awareness and consciousness is to have some mechanism whereby you can keep from being paralyzed by the insight that the world is full of things that can kill you. Something that allows you to think, "I know the saber-tooth ate my cousin yesterday, but he won't eat me." Denial allows you to go out of the cave and down to the watering hole even after your cousin was killed; and if, in the process, you discover that the saber-tooth only appears every five days and then only early in the morning or late in the afternoon, you will have solved a problem that could not have been solved by staying in your cave.

Denial is one way we cope with legitimate fears, but at this critical juncture in human history, denial is our greatest enemy, because it can leave us paralyzed with fear at a time when we need to act. Humanity needs awareness and hope, not fear and despair.

Why Scientific Actions Are Falling Short

Science is an essential source of solutions to the emerging disease crisis, but it has also been an impediment to progress.[21] We tell children sitting in first-year university science classes that scientists are inherently cooperative people, pure seekers of truth for its own sake. This is not strictly true; humans do have a great capacity to cooperate, but scientists are encouraged to emphasize that which is unique over that which links group efforts, and to fight (literally) for priority. We are trained in this behavior from the moment we enter graduate school. And there are many opportunistic academics eager to tell us—despite manifest evidence to the contrary—that crass selfishness is our inescapable biological legacy, and that this a sustainable way to live. And yet, not all of the rhetoric about the cooperative nature of science is empty. Scientific progress occurs when scientists share the insight that cooperation not only can be a good thing for the advancement of human knowledge, but is essential for achieving their own personal goals. And when that kind of cooperation occurs in the context of improving the quality of life for all humanity, it is a beautiful thing indeed. Scientific objectivity is the result of cooperative standardization of the subjective insights of individual scientists.

Scientists Are Ambivalent about Cooperation

More than 25 years ago the eminent biologist Stephen Stearns spoke at an evolutionary biology conference in Montpellier, France. He boldly stated that the majority of European scientists—many of whom were in the audience—were treating the biodiversity and climate change crises as funding opportunities rather than existential challenges. He suggested that for the majority of biologists, it was business as usual, with a slightly altered title on their grant applications. Stearns felt this was the sluggish reaction of people who live in a professional social context characterized by "the normal tangle of jealousies and rivalries, and inclinations to trip and shove."[22] Within that milieu are scientists with diverse beliefs about climate change and disease, one impediment to cooperation. Some think that humanity is winning the war against disease, so there is no EID *crisis*. Some are skeptical about any link between climate change and disease. Some are skeptical that new diseases are even emerging. They point to decreasing birth rates, increasing life spans, increasing food production, improving health care, and an absence of mass mortalities throughout the world. Many of them have an almost religious faith in technological advances to save us from ourselves. Other scientists believe that the magnitude of the problem—both the emerging disease crisis and the global climate change within which that is embedded—is so great we cannot solve it.

Scientists love to receive credit for their discoveries, be it the configuration of the HIV genome or a beautiful, previously unknown metallic silver boa constrictor on a tiny island in the Bahamas. Specialization makes the pursuit of personal recognition easier, but this also impedes cooperation. Specialization for its own sake seems like a good thing—so long as you dream of efficiency and of your own career. If your area of specialization is small enough, nothing will ever change unless you make it change. Increasing specialization seems like a great career move for scientists in today's world, in which job security and research funding depend on making, and then promoting, novel discoveries.

And yet, the world of contemporary science is so complex that no area can be controlled by a single scientist or a single close-knit research group. Professional survival and success increasingly require a means of generalizing knowledge obtained through specialization, extending it to new problems or stretching it into new useful configurations. Otherwise, all the benefits are lost as a result of different groups having to invent the wheel repeatedly. Institutionalized failure to recognize this reality is called "stovepiping." And we have long known that stovepiping can be perilous: "If we do not hang together, we shall assuredly hang separately."[23]

Cooperation is often a good choice. If different researchers try to gain ownership or control of the same topic, all but one of them will fail and no one will gain, including the one "winner," who is left holding on to a piece of knowledge that no one else is allowed to use. If, on the other hand, those groups all cooperate, many people will get at least some credit. If we look at this as a class of risk/reward decisions, the greatest odds are that if you try for exclusivity and do not cooperate, you will likely not get credit, and you will be forgotten. This is because science is fundamentally a social system. And no matter how determined you are to be "the one," you are ultimately the person with the least control in determining that outcome. So let's just say that building bridges with your competitors is usually better than sitting alone on your island. It certainly leads to better outcomes for humanity. And we believe we live in a time when humanity is facing an existential threat.

The past decade has seen an explosion of interest in emerging diseases, and particularly in the interface between climate change and disease. Recent research is being done by many bright young scientists more eager to cooperate than their predecessors.[24] Their enthusiasm cannot compensate for the lack of a comprehensive conceptual framework that explains what we are experiencing, or for a unified set of protocols derived from that framework for coping with the problem. This book presents hopeful news for them.

Scientists Are Overmanaged

Bureaucracy destroys initiative. There is little that bureaucrats hate more than innovation, especially innovation that produces better results than the old routines. Improvements always make those at the top of the heap look inept. Who enjoys appearing inept?[25]

Science is an evolutionary system and must be based on principles of evolutionary *diversification* rather than static notions of efficiency.[26] The necessary attributes of evolutionary diversification are (1) the ability to specialize and generalize, (2) cooperation, and (3) division of labor.[27] Scientists can be extremely good at these tasks if they are allowed to be flexible and creative,[28] but rigid management systems rarely reward such behavior, although it is often essential to finding solutions to common problems.

Even when scientists recognize the need and have a framework for achieving success through cooperation, they must be allowed to act on that knowledge. Modern management practices have become exercises in denial and passive-aggressive behavior. *If I do something, the situation could get worse and then I will be blamed. If I wait long enough and do nothing, the crisis will most certainly pass. And if it does not pass, I can go into heroic crisis-management mode.* In the contact zone between short-term trivial priorities and long-term existential risk, therefore, modern management always opts for what is called "best practices." Too often, "best practices" seems based on the desire to give each manager a way to avoid culpability for making the wrong decision in the rare case that he makes any decision at all.

Once we begin to fear we cannot solve problems, we begin to feed the part of our mentality that avoids risk, and nothing is riskier than innovation. We begin to talk about managing our problems, based on a simple principle: "It worked in the past, so we will keep doing it, our only changes being more and faster and cheaper means of doing the same thing that worked before." There, don't you feel better? Think again. This approach to life only works if today and tomorrow are like yesterday, and yesterday was calm and safe. But our world is one in which today is manifestly not like yesterday and tomorrow will be different again. This is why scientists respond badly to being overmanaged—it engenders a sense of fatalism in those being managed that risks veering into nihilism.

Some blame "society" for not giving them enough money to get the job done. Trust is broken on all sides precisely when we need it most. The scientific community must not accept the way in which it has been infantilized by politicians and some media. We must show that we are adult enough to know when to cooperate in the face of a common crisis. It is unacceptable to be content with the status quo. Others have quietly given up and retired to the sidelines. The *Cassandra Collective* is a darkly humorous term referring to senior scientists specializing in all aspects of climate change research who have given up in despair and quietly abandoned ship, their Facebook pages now full of pictures of grandchildren and scenic vacations rather than the newest report or research article. This is an irreplaceable loss of expertise and experience that we cannot afford.

"Business as Usual" Is the Wrong Answer

You *rationalize* . . . You *defend*. You reject unpalatable truths, and if you can't reject them outright you trivialize them . . . evidence is never enough for you. . . . Temperatures rise, glaciers melt—species die—and you blame sunspots and volcanoes. . . . You turn incomprehension into mathematics . . .

It served me well enough.[29]

"Business as usual" is humanity's prevailing response to the interlinked threats posed by climate change, including emerging disease. It is not working.

What We Are Doing Is Costly

The past decade has demonstrated that EIDs directly affect our health, the integrity of ecosystems, and the sustainability of the primary food and water resources on which we depend. Even limited exploration of the current literature shows that the direct impacts of pathogens and the treatment of infectious disease are costly. We cannot be comprehensive or precise, however, about direct impacts such as mortality and morbidity, or the varied responses to infectious disease, because of the ways in which the data are presented: in scattered and often obscure sources, using disparate measures of socioeconomic "cost," reported by governmental organizations that have a vested interest in minimizing the reported costs. Table 1.1 contains some snapshots of the socioeconomic costs of pathogens and disease across managed and natural systems encompassing terrestrial, aquatic, and marine habitats to give the reader some sense of how costly disease is today. Direct costs are an imperfect indicator of the impacts of EID but serve to show the scope and depth of the historical challenge, and the degree to which accelerating climate and environmental change can influence new patterns for the distribution of pathogens and diseases. As disease emergence continues and expands, costs already approaching a trillion dollars annually—*an amount exceeding the GDP of all but 15 of the world's nations*—will rise. This, in turn, raises the specter that the economics of treatment costs and productivity losses owing to infectious disease could soon become unsustainable. Either the amount of money available for treating diseases will increase more slowly than new disease emergences, or the amount of money that must be transferred from other sources will create difficulties in the area of concern for the reallocated resources. As an analogy, consider the reallocation of water from agriculture to residential sources in Cape Town in late February 2018. That decision provided additional water for the residents of the city, but at the expense of likely increases in food prices and stress on agricultural lands. Beyond the direct economic health implications of EIDs are those related to variations in the efficacy of control measures (resistance and adaptation to chemical or pharmaceutical interventions), changing climate and environments, and the facilitation of vector development and resilience, enhancing new interfaces for exposure, transmission and emergence.

It seems clear that the annual cost of responding to emerging diseases is enormous and growing. The scientific literature concerning the costs of coping with EIDs that we have encountered since undertaking this project supports three general conclusions. First, it's easy to find articles talking about *potential* costs of EID. Secondly, it's far more difficult to find sources that give *actual* costs. And finally, we have not found even one article that provides unequivocal evidence that what we are doing with respect to EIDs is economically sustainable.[30] We believe that if there were such data out there, someone would have made

Box 1.1 Selected Snapshots of Socioeconomic Costs Due to Pathogens

General

Rosenzweig et al. (2001)

· Plant fungal diseases, nematodes, global: losses of $77 billion between 1988 and 1990

Pimentel et al. (2005)

· Invasive alien species including pathogens: overall impact of $120 billion annually in United States
· Plant pathogens: $33 billion; forest products $7 billion
· Animal pathogens: $14 billion
· Viral disease in humans: $6.5 billion (influenza); does not include emergent pathogens WNV, Zika, YFV, dengue

World Health Organization (2005); Conteh et al. (2010)

· Neglected tropical diseases (NTDs), global, Africa, Latin America, Asia: 543,000 deaths worldwide, 57 million disability-adjusted life-years (DALYs); include helminth and protozoan diseases (Chagas disease, human African trypanosomiasis, leishmaniasis, lymphatic filariasis, ascariasis, trichuriasis, onchocerciasis, schistosomiasis); equivalent to global burden of malaria, $3 billion annually)

Gebreyes et al. (2014)

· Emerging infectious diseases: zoonoses (source of 75% of EIDs), global: Over the past 15 years there have been more than 15 deadly zoonotic or vector-borne global outbreaks: (1) viral (hanta, Ebola, highly pathogenic avian influenza [H5N1 and recently H7N9], West Nile, Rift Valley fever, norovirus, severe acute respiratory syndrome [SARS], Marburg, influenza A [H1N1]) and (2) bacterial (*Escherichia coli* O157:H7, *Yersinia pestis*, and *Bacillus anthracis* [hemolytic uremic syndrome, plague, and anthrax, respectively])
· Since 1980 more than 87 new zoonotic and/or vector-borne EIDs have been discovered.
· The global economic burden due to zoonotic diseases is substantial.
· The World Bank estimate of the economic burden due to six zoonotic outbreaks in specific countries between 1997 and 2009 is $80 billion.
· Potential losses from a pandemic influenza outbreak could be $3 trillion.

Trtanj et al. (2016)

· Water-borne pathogens, North America: These are estimated to cause between 8.5% and 12% of acute gastrointestinal illness cases annually in the United States, affecting between 12 million and 19 million people

> Eight pathogens, all influenced to some degree by climate, account for approximately 97% of all suspected waterborne illnesses in the United States: the enteric viruses norovirus, rotavirus, and adenovirus; the bacteria *Campylobacter jejuni*, *E. coli* O157:H7, and *Salmonella enterica*; and the protozoa *Cryptosporidium* and *Giardia*. Expansion of *Leptospira* and *Vibrio* bacterial species and gastrointestinal maladies is anticipated with climate warming.

Bower et al. (2017)

- Chytrid fungus (*Batrachochytrium dendrobatidis*), global pandemic: amphibians; includes AU$15 million over five years to conserve 29 threatened frog species.

Mariculture and Fisheries

Shinn et al. (2014)

- Microparasites and macroparasites, global: fishes in aquaculture; losses range from hundreds of millions to more than $1 billion

French (2016)

- *Tetracapsuloides bryosalmonae* (proliferative kidney disease; myxozoans), Montana: salmonid fishes; fisheries closure, Yellowstone River; losses of $360,000–$524,000

Human and Animal Viral Pathogens

Meltzer et al. (1999)

- Influenza (human), United States: $71.3 billion–$166.5 billion

Guzman and Kouri (2003)

- Dengue, global: 50–100 million cases; costs unknown

Gilbert et al. (2008)

- Highly pathogenic avian influenza, Asia, 2005–2006: $10 billion

Fofana et al. (2009)

- Bluetongue virus, Europe: outbreak in France, 2007: $1.4 billion; Netherlands, $85 million

Kedmi et al. (2010)

- Epizootic hemorrhagic disease, Israel: $1.6 million–$3.4 million

Gubler (2012)

- Dengue, Puerto Rico: $46.5 million ($418 million over nine years)

Stevens et al. (2015)
- Epizootic hemorrhagic disease, midwestern United States: 42% mortality in domestic herds

Wappes (2015)
- H5N2 Avian Influenza, Iowa, 2015: $1.2 billion

Samy and Peterson (2016)
- Bluetongue Virus, global, United States, Europe: United States, $125 million annually; Belgium outbreak, $250 million

Panzer et al. (2016)
- Zika virus, Caribbean and Latin America: $3.5 billion short-term costs in 2016

Shepard et al. (2016)
- Dengue (human), global: 58.4 million cases; $8.9 billion

United Nations Development Programme (2017)
- Zika virus, Caribbean, Latin America: losses of $7 billion–$18 billion during 2015–17

Bloom et al. (2017)
- Influenza (human) pandemic, global: The World Bank estimates the anticipated gross domestic product (GDP) cost of a moderately severe global flu pandemic at about 5% of world income (roughly $3.5 trillion—a sum greater than Germany's GDP). The full cost of such an epidemic, including the lost value of health and longevity, is nearly an order of magnitude higher.

Centers for Disease Control (2017)
- Ebola virus, West Africa, 2014–2016 outbreak: $5.96 billion in response, $2.2 billion in lost productivity. The most severe impact of the Ebola epidemic, which began in Guinea in December 2013 and quickly spread to Liberia and Sierra Leone, has been in lost human lives and suffering.

Lee et al. (2017)
- Zika virus, six U.S. states: $183 million–$1.2 billion over 230 days; worst-case scenario: >$2 billion.

Aly et al. (2017); World Health Organization (2018)
- MERS coronavirus, global: emergent human zoonotic pathogen with near 40% mortality, discovered 2012

Bacterial Pathogens
Frenzen et al. (2005)

· *Escherichia coli* (human), United States: $405 million

Kirigia et al. (2009)

· Cholera (*Vibrio*), Africa, 2007: The real total economic loss attributable to cholera is estimated to have been $91,863,606 assuming a minimum regional life expectancy of 40 years; $128,136,952 assuming a regional average life expectancy of 53 years; and $155,993,261 assuming a maximum regional life expectancy of 73 years.

US Biologic (2018)

· Lyme disease (*Borellia* spp.), United States: 329,000 infections, >$3.5 billion. Substantial influence on tick vector distribution under climate forcing.

Protozoan Pathogens
Egbendewe-Mondzozo et al. (2011)

· Malaria, Africa: $12 billion productivity loss

Lee et al. (2013)

· Chagas disease, Global: $627.46 billion health-care burden; $7.18 billion annual productivity loss; $188 billion lifetime productivity loss

Plant Fungal Pathogens
Schumann and Leonard (2000); Voegele et al. (2009);
Singh et al. (2011)

· *Puccinia graminis* (stem rust or black rust), global: >$5 billion in losses annually; Substantial threat through emergence of new races (90% of wheat varieties grown worldwide are susceptible to Ug99)

Hovmøller et al. (2011); Pardey et al. (2014)

· *Puccinia striiformis* (yellow rust or strip rust), global, wheat: "The present large-scale epidemics of yellow rust, caused by aggressive *P. striiformis* strains that are tolerant to warm temperatures, may pose a severe threat to the world's wheat supply. Presence of two particular strains in high frequencies at epidemic sites at five surveyed continents may represent the most rapid and expansive spread ever of an important crop pathogen."
· Annual loss 1961–1984, $157 million; annual loss 2000–2012, $848 million. Changes reflect climate drivers of rapid global expansion and pathogen adaptation to control.

Human and Animal Helminth Pathogens

Wright (1972)

- Schistosome blood flukes, humans, global: across 71 countries with ca. 125 million infected. Annual estimates of losses attributable to reduced productivity: $445,866,945 in Africa, $755,480 in Mauritius, $16,527,275 in Southwest Asia, $118,143,675 in Southeast Asia, and $60,496,755 for the Americas. The total estimated annual loss worldwide amounts to $641,790,130, not including costs of public health, medical care, or compensation for illness.

Budke et al. (2006)

- *Echinococcus granulosus* (cystic echinococcosus), human, global: burden of disease $764 million
- *Echinococcus granulosus* (cystic echinococcosus), livestock, global: livestock losses $2.2 billion

Gasbarre et al. (2009)

- Helminths, cattle, United States: $2 billion.

Roeber et al. (2013); "Global Warming—A Rising Cost" (2017)

- Helminths, domestic ruminants (affecting food security and food safety), Australia: $1 billion
- Helminths, ruminants, global: >$10 billion

Grisi et al. (2014)

- Helminths, domestic ruminants (affecting food security and food safety), Brazil: $14 billion

Adenowo et al. (2015)

- Schistosome blood flukes, humans, global: In 2008, 17.5 million people were treated globally for schistosomiasis, including 11.7 million from sub-Saharan Africa; globally >730 million persons are at risk. More than 200 million individuals from Africa, Asia, and South America are infected. Schistosome infections and geohelminths account for more than 40% of the world tropical disease burden, with the exclusion of malaria. These pathogens are strongly responsive to climate change. Treatment in sub-Saharan Africa: $100 million over 5 years (cost of 1.2 billion tablets of praziquantel for 400 million individuals).

a very large noise about it. So we doubt that continuing to respond to each EID after it has made itself known makes economic sense.

We are cautiously sanguine that public and agricultural health management officials know this. We do *not* believe that they are part of any nefarious conspiracy. By virtue of their training and the bureaucratic structure created to implement that training with maximal efficiency, however, members of the medical-industrial and agricultural-industrial complexes have left themselves unable to cope with the reality of disease in a changing climate. We believe their sluggish response, which is being noted more often now,[31] indicates that they simply do not have a Plan B and are in denial. This book offers them a Plan B.

We Are Insensitive to the Threat

The 1918 influenza pandemic was a *low-probability/high-impact* disease emergence. In today's world of increased human population and population density, a repeat of that pandemic could wreak havoc on a global scale. And as *Time* magazine's May 4, 2017, cover noted, humanity is unprepared for such an event. Such possibilities are "black elephants"—a conceptual hybrid between a "black swan" (an unlikely, unexpected event with enormous ramifications) and "the elephant in the room" (a problem everyone knows about but pretends is not there).[32] Black elephants are problems that everyone knows about but no one wants to deal with, so they pretend they do not exist. When such a problem occurs, everyone pretends to be surprised, as if it were a black swan, something so rare we could not have prepared for it.

We are more worried about the steady accumulation of high-probability/low-impact pathogens in humans, livestock and crops throughout the world, a threat we believe is under-appreciated. No matter how mild, localized, and ephemeral, each high-probability/low-impact pathogen exacts a socioeconomic cost, especially at the time of its emergence. And even after resistance evolves, outbreaks become chronic and sporadic rather than acute and immediate, the pathogens persisting as *pathogen pollution*.[33] West Nile virus is no longer an acute problem in North America, but it will always be a chronic problem that requires some human and material resources to be expended in preparing for the possibility of recurrent outbreaks. Every year there are norovirus outbreaks in schools and on cruise ships around the globe. Few people die, so we do not call them pandemics. These outbreaks are most costly in terms of lost productivity. Whenever a child contracts a norovirus (or arenavirus, or rhinovirus—the list is long and growing) and must stay home from school, educational time is lost, the public health care system is engaged (at a cost), and at least one parent may have to stay at home, with an attendant loss of working productivity and family income. The socioeconomic damage caused by high-probability/low-impact pathogens is akin to death by a thousand cuts.

And yet, beyond a reference to antibiotic resistance, neither the World Economic Forum 2018 threat assessment nor the 2018 "horizon scan" by *Trends in Ecology and Evolution* mentions disease—either low probability/high impact or high probability/low impact—as a global threat.[34] This intellectual blind spot is the most important black elephant associated with climate change and disease.

Summary

The danger is great.
The time is short.
We are unprepared.
But we can change that.

Information about the diversity and distribution of known and potential pathogens is critical for limiting their socioeconomic impacts. Yet, our knowledge of the identities, geographic locations, and threat potential for the world's pathogens can be called at best fragmentary. At most, 10 percent of the world's pathogens have been documented; the rest remain utterly unknown, despite heroic efforts by a few determined individuals. This massive ignorance alone is reason to be concerned about our preparedness to handle the crisis. It's impossible to prepare for a threat whose very existence is unknown. We act as if EIDs were rare phenomena and go into in crisis-response mode, as if we expected each new outbreak to be the last one. The potential for EIDs is actually quite large, and climate change leads to pathogens' being more accessible throughout the world.[35] This makes the planet an evolutionary minefield into which millions of people, not to mention their crops, livestock, and pets, wander daily.[36]

What happens when the means of dealing with the conditions of life we have relied upon no longer work? Jared Diamond believes that this is the flash point at which previous human societies have collapsed catastrophically. Eörs Szathmáry believes that technological societies may have short life spans, and that species on other planets may have reached tipping points similar to the ones facing us and failed. In either case, we should be especially concerned about safeguarding human civilization.

We have been brutally shortsighted. How can we say something that harsh, when science has given us global travel, the Internet, and vaccines? All change is progress, and all progress is good, right? Right now, when we have never been more adept scientifically and technologically, our magic is not working as it should, particularly with respect to emerging disease: short-term successes in restricted places are more than offset by large-scale, long-term failures in places where climate change and associated human activities have brought pathogens to new localities and new environmental and socioeconomic contexts.

One of the fundamental goals of the Enlightenment was a desire to provide a sense of security based on simplicity, certainty, and control with respect to the interactions between humans and nature. Throughout this book we will encounter episodes in which those desires have led us astray, because the biosphere is an evolutionary system, and such systems are not simple, certain, or totally controllable. Accepting that reality is the key to pushing beyond denial and theoretical expectations contradicted by daily experience toward a more anticipatory and proactive approach for coping with emerging diseases.

There will be no single answer, no quick fix, to the crisis. Many people are working diligently to manage it, but catching up to and getting ahead of it requires a fresh perspective from scientific sources outside the usual suspects. We must take better advantage of the one process—evolution—that has been responsible for the recovery of the biosphere in all previous major climate-change events. We believe that the key to adopting policies allowing us to anticipate and mitigate emerging diseases lies in understanding the evolutionary context of ecological changes associated with climate-change events. We also believe part of the reason we are failing with respect to emerging diseases during this episode of climate change lies in a faulty understanding of how easily pathogens can encounter opportunities to colonize new hosts and move to new places. Our perspective suggests that while some of the answer is certainly anthropogenic, most of what we are seeing is a manifestation of an ancient underlying evolutionary dynamic. The impacts of an expanding human population and increased connectivity due to globalization reinforce the effects of climate change, leading to an accelerating rate of change that outstrips anything humanity has experienced before.

We plan for wars and weather-related disasters, but not for diseases.

Why is that? It is because we have been following a flawed coevolutionary paradigm based on the assumption that a pathogen can colonize a new host (emerge) only if it experiences genetic mutations that allow it to adopt the new host. And because genetic mutations leading to particular outcomes are not predictable, disease emergence cannot be predictable, and there is no reason to connect their occurrence with climate change. The EID crisis casts doubt on that paradigm. We therefore need an alternative explanatory framework in which what we experience is expected on theoretical grounds. From our perspective, the EID crisis is a health issue in only a superficial sense. It is more fundamentally an evolutionary and ecological issue, a predictable consequence of formerly separated species brought into close contact as a result of new opportunities provided by a changing global climate. Human population growth and rapid globalized commercial activity accelerate the rate of introductions, so outbreaks occur more frequently and over a wider geographic range than ever before. Science has advanced rapidly, but not as rapidly as EIDs and human needs, slowing the effective application of new scientific information. A powerful and highly risk-averse management class that has recently emerged, directing lines of research and mediating the exchange of scientific information, further exacerbates the situation.

Human beings, owing to their natural caution, have opted for biodiversity policies based on stasis and simplicity rather than change and complexity. We are too often paralyzed by a natural fear of complexity and an unknown future. We make a bad situation worse when we suggest that evolutionary processes are so prone to accumulating contingencies through time that it is impossible to predict their behavior, and thus to control or manage them. The potential for EIDs is a built-in feature of evolution. This explains why pathogens are not only good at finding us, they are also really good at surviving. Those species that best survive climate change will be the primary sources of EIDs now and in the future. There are many, not a few, evolutionary accidents waiting to happen out there, requiring only the catalyst of ecological alteration resulting from climate change, species introductions, and the intrusion of humans into areas they have never inhabited before.

All of these are happening *right now*. We are therefore writing this book *right now* because we believe humanity, at least in its highly technological mode, faces a threat to its existence *right now*. These are provocative statements. But if we cannot convince you that something is wrong with the way we deal with emerging disease, we cannot convince you that we need a different approach. What if we were to tell you that we have found a strongly repeating feature of evolutionary history that explains how the biosphere reacts to major external perturbations? What if we told you that this allows us to anticipate the future enough to mitigate harmful impacts on humanity while also keeping us from making the mistake of closing off evolutionary potential, which is the only known mechanism for renewing the biosphere following extinction events caused by such perturbations? And what if we told you this means we can actually do something to buy time and mitigate the impacts of EIDs? We can make a difference, but we have to work hard, we have to work fast, and we have to work smart.

What health professionals have been doing for more than a century is heroic, but safeguarding public and agricultural health has become a losing proposition. No wonder humanity is worried and afraid. But we cannot be prisoners of our fears.

So first of all let me assert my firm belief that the only thing we have to fear is fear itself—nameless, unreasoning, unjustified terror which paralyzes the needed efforts to convert retreat into advance.[37]

TWO

How Did We Get into This Mess?

The farther back you can look, the farther forward you are likely to see.
—ATTRIBUTED TO WINSTON CHURCHILL

The emerging disease crisis has caught us flat-footed, with devastating consequences and the promise of worse to come. We are seeing too many pathogens, in too many hosts, in too many places. For the scientific community, this is embarrassing and vexing. Haven't really smart people been studying these things for a long time? The answer is an unqualified "yes": many smart scientists and clinicians have studied pathogens of various kinds for more than 150 years, and we have accumulated enormous amounts of information. So how can it be that we were so unprepared? The answer to this second question is more complicated. Clearly, we had unrealistic expectations. And for scientists, that means something is wrong with the explanatory framework we have been following. Figuring that out requires us to dig a bit into the history of how we have thought about pathogens and their hosts. Some may find the thought of two chapters of history irritating, but there is a reason for it. We are going to find that human beings have ignored many warning signs—about climate change in general, and about disease in particular—for more than a century. We are in a mess. Understanding how we got into it is the first step toward offering meaningful assistance to a broad-based professional agenda attempting to safeguard humanity.

We wish to discuss the ways in which thinking about pathogen-host associations came to influence the thinking and research of particular scientists who set the stage for our modern practices. This will help understand how we arrived at a set of theoretical expectations that are so wildly out of sync with what is happening now and with what has happened in the past. And we want to expose the roots of a problem facing humanity in real time, to explain why current efforts are failing to solve (or even contain) the problem, and to set the stage for our proposed solution. The history of ideas about the nature of the relationship between pathogens and their hosts and their environments is a large and fascinating topic, and we hope it will be treated exhaustively someday. But that will not happen in the next two chapters. Here we want only to describe how the best research into evolutionary biology and the best health practices of the past century and a quarter have run off the rails. If we—and other like-minded scientists—are successful, human society may survive well enough that sometime in the next century a team of intrepid historians will be able to write the definitive history. We hope they will chide us only gently for our scholarly shortcomings.

Humans have been aware of organisms living on and inside other organisms for a long time. Everyone who ever prepared meat, fowl, or fish for food saw fleas, ticks, and lice on the outside and lots of worms inside. We have also known for a long time that humans host many of the same kinds of organisms. But questions of what those other species were doing and how they came to be where humans encountered them were not answered until rather recently in human history. Moreover, even when they were found in conjunction with disease outbreaks, those organisms were often considered to be an effect, rather than a cause, of illness and death. The transition in thought from "sick people have worms" to "worms make people sick" has only happened in the last 150 or so years. And that covers only those pathogens we can see with the naked eye.

It was not until the second half of the seventeenth century that Antonie van Leeuwenhoek's remarkable lenses showed us that we share this planet with a rich broth of invisible life, which he called animalcules. Leeuwenhoek had a glimmer of an idea that some of those microorganisms were associated with disease; notably, he was the first person to see the parasitic flagellate now called *Giardia lamblia*, a common (and annoying) intestinal pathogen, and he documented substantial microbial activity associated with gum disease and tooth decay. More than a century later—in 1796—an English country doctor named Edward Jenner produced preemptive protection against smallpox, using the precursor of the procedure we now call vaccination. This was an astonishing insight, because Jenner had no idea the disease was caused by a virus, something too small to be seen with the microscopes of his day. Indeed, it was not until more than half a century after Jenner's breakthrough that we were able to make a general link between van Leeuwenhoek's tiny denizens—and many more that even he could not see—and human suffering. We begin in the middle of the nineteenth century, with two scientific luminaries who seem not to have noticed each other to any great degree during their lifetimes.

Setting the Stage

Between 1859 and 1872 Charles Darwin provided the unifying theory of biology, publishing six editions of *On the Origin of Species* as well as related groundbreaking contributions, such as the foundation for all modern studies of sexual selection. And he was still able to work on barnacle taxonomy and spend time with his family. Darwin was a devoted husband and father who lost his 10-year old daughter, Annie, in 1851, likely from virulent tuberculosis possibly contracted as a result of her having being weakened by a previous scarlet fever infection. Some might say that modern medicine could have saved her easily and that this sort of thing could never happen today, but that is untrue. Tuberculosis and scarlet fever are among the diseases that are resurging globally and in resistant forms that we understand, ironically, as a result of Darwin's work. Darwin was aware of nonmicrobial parasites (at least enough to call them "disgusting"),[1] but he seems to have considered parasites to be a kind of predator and disease to be a name we give to situations in which the predator and prey are engaged in an evolutionary struggle beyond the normal efforts to survive.

When a species, owing to highly favourable circumstances, increases inordinately in a small tract, epidemics—at least, this seems generally to occur with our game animals—often ensue; and here we have a limiting check independent of the struggle for life. But even some of these so-called epidemics appear to be due to parasitic worms, which have from some cause, possibly in part through facility of diffusion amongst the crowded animals, have been disproportionately favoured; and here comes in a sort of struggle between the parasite and its prey.[2]

To contemporary readers, this passage sounds remarkably insightful, offering a glimpse of the notion that there is a causal relationship between pathogens and disease. In fact, Darwin seemed to have had an insight about the potential for what we would today call a coevolutionary struggle between host and parasite. Nonetheless, just a few pages later, Darwin retreated into more familiar territory

The dependency of one organic being on another, as of a parasite on its prey, lies generally between beings remote in the scale of nature. . . . But the struggle will almost invariably be most severe between the individuals of the same species.[3]

Darwin distinguished *mutual dependencies between distantly related species* from *independent struggles among individuals of the same species*. He saw host evolution as an outcome of host organisms in conflict with each other and parasite evolution as an outcome of parasites in conflict with each other. By not following up on his insight from a few pages before, Darwin failed to see that parasites and hosts might be part of each other's struggle for survival. The possibility of such mutual evolutionary modification would not be formalized until nearly a century after the *Origin* first appeared. By the mid-nineteenth century humans throughout the world had amassed an impressive array of medicines along with an equally interesting array of erroneous theories about what caused the diseases they treated with those medicines. Louis Pasteur provided the major breakthrough needed to begin the long process of scientific understanding of disease. Between 1858 and 1865, Pasteur produced definitive experimental evidence that there is no spontaneous generation of life, that fermentation is caused by the activity of microbes, and that what is now called pasteurization protects milk from spoilage by killing microbes. It would not take long for other scientists to connect Pasteur's insights with ideas that many agents of disease might also be microbial in nature.[4] Those scientists would subsequently accumulate a large amount of information about the ecology and behavior and the impacts of pathogens—that is, about where they live, how they are transmitted, and what they do. The critical answers about who they are in evolutionary terms would be unattainable until late in the twentieth century, but even so, at this time humans were not completely in the dark.

Long before anyone could explain illness, humans had knowledge of how to cope with many of its symptoms, mostly by drinking brewed extracts of certain plants.[5] A well-known example is the use of bark extracts from several trees in the genus *Cinchona* as a treatment for malaria. Having learned about it from indigenous people, in 1632 Europeans brought the medicine from South America to Spain, and then to Rome and other parts of Italy. Francesco Torti wrote a treatise in 1712 that established the efficacy of cinchona bark for treating "intermittent fevers," including those produced by malarial infection, but the active ingredient—quinine—was not isolated until 1820. (The search for relief from illness may also be how we encountered various drugs that became part of the religious-recreational aspect of most human cultures, but that is a different story.)

When Darwin published the *Origin*, medicating the ill was not the only approach to coping with disease. *Variolation* refers to a practice in which small amounts of material from lesions produced by a disease in one host are inoculated into another. Variolation of humans against sheep pox was mentioned by explorers in Africa as long ago as the sixteenth century but may have been attempted in China or India even before then. In many cases, variolation either produced no protection or actually induced the life-threatening disease it was supposed to protect against. Nonetheless, variolation was still being used in the mid-nineteenth century. Several attempts at immunization by variolation were made for sheep pox, which is closely related to smallpox in humans, and bovine contagious pleuropneumonia.

Edward Jenner noticed that cowpox, a disease of—not surprisingly—cattle, produced symptoms similar to but milder than those of smallpox in humans. He then used variolation techniques to introduce what he interpreted as a milder form of smallpox into humans and reported that those patients did not develop smallpox. Not knowing what caused smallpox, however, left people without an explanation for why vaccination provided protection against the disease, so there was no reason to think the approach could be applied more generally. Indeed, many did not believe Jenner's account of successes with his procedure.

This period was also a watershed for research in crop disease. Known from the Fertile Crescent in 1300 BCE, wheat stem rust was a serious problem in ancient Greece and Rome, recognized as early as the time of Aristotle (384–322 BCE). Each spring the Romans sacrificed red animals such as dogs, foxes, and cows to the rust god, Robigo or Robigus, during the festival called the Robigalia in hopes that the wheat crop would be spared from the ravages of the rust. This festival was incorporated into the early Christian calendar as St. Mark's Day, or Rogation, on April 25. Historical weather records suggest that a series of rainy years, in which rust would have been more severe and wheat harvests reduced, may have contributed to the fall of the Roman Empire.

By the seventeenth century European farmers recognized that barberry was somehow connected to stem rust epidemics in wheat. In France in 1660, the city of Rouen banned the planting of barberry near wheat fields. Two Italian scientists, Felice Fontana and Giovanni Targioni Tozzetti, independently provided the first detailed descriptions of the wheat stem rust fungus in 1767. Thirty years later Christiaan Hendrick Persoon, born in South Africa but destined for a life of science in the Netherlands, named that fungus *Puccinia graminis*. By 1854 the brothers Louis René and Charles Tulasne recognized that some rust fungi could produce as many as five spore stages. They were the first to link the red and black forms as different spores of the same organism.

Both wheat and barberry plants had been brought to North America by European colonists, who also recognized a connection between having barberry plants near their fields and stem rust epidemics in wheat. Laws banning barberry were enacted in several New England colonies in the mid-1700s. However, barberry continued to spread as pioneer farmers moved west. From farmyard plantings, barberry spread into fencerows and woodlots. Barberry bushes produce abundant berries that are attractive to birds and animals that feed on them and spread their seeds. Even if farmers stopped transporting barberry with them when they moved across North America, the plants always managed to find their crops. Using the farmers' belief that barberries were involved in some way with wheat rust, in 1865 Heinrich Anton de Bary, a prominent German mycologist, successfully inoculated barberries with spores from wheat plants and wheat with spores obtained from barberries. Once he established that the fungus had a complex life cycle requiring development on one host to produce spores that could infect the other, the search was on and many other known rust fungi were discovered to have similar life cycles, albeit with different hosts.[6] The laws enacted in seventeenth-century France and eighteenth-century New England banning barberries ushered in a third element in coping with disease: complementing medicating the ill and vaccinating those at risk, we added the notion of eradicating portions of biodiversity associated with transmitting a disease.

Ending the Nineteenth Century with a Flourish

The last quarter of the nineteenth century was tumultuous for evolutionary biology. Paradoxically, this was driven by a rapid acceptance of the fundamental notion that evolution was responsible for the world's biological diversity, coupled with the emergence of several research programs aimed at "fixing" or replacing Darwin's theory of how evolution occurred. By the end of the nineteenth century three distinct theoretical frameworks had emerged as the principal rivals in a race to replace Darwin. Each had a perspective about the significance and relevance of pathogen-host associations, and each has persisted into the early twenty-first century. Darwin initially considered geographic isolation an essential factor in producing new species—populations of an ancestral species living in isolation from each other might well be subjected to different forms or respond in different ways to the same form of natural selection. In later editions, he explored the notion that selection alone could produce new species.[7] Darwin's broad perspective produced two groups within the biological community: biologists who argued for sundering geographic isolation from the process of speciation by natural selection,[8] and biologists who believed that geographic isolation alone was responsible for producing new species.[9] This latter school of thought is exemplified by Hermann von Ihering, a German-born biologist who spent his career in South America. He observed strong similarities in the ecological associations between some temnocephalidean (flatworm) parasites inhabiting freshwater crayfish in New Zealand and those inhabiting freshwater crayfish in the mountains of Argentina.[10] Assuming that the temnocephalideans would never be found without the crayfish, he concluded that the contemporaneous species were derived from ancestors that had themselves been associated. Hence, he argued, South America and New Zealand must at one time have had freshwater connections. By the last decade of the nineteenth century, parasitologists had documented multiple examples of parasites associated with particular hosts in one place having relatives associated with relatives of those hosts living in other places. Von Ihering reasoned that this must mean that the pathogens and hosts experienced common episodes of speciation by geographic isolation, leading to simultaneous speciation events in both parasite and host lineages.[11]

Other ideas that emerged at the end of the nineteenth century represented fundamentally different views about the nature of evolution. Because we are focusing on ideas about pathogen-host systems, we will consider only two of these notions. Each was founded on the nineteenth century's fascination with the apparent fit of organisms to their surroundings, and on an intuitive sympathy for notions of progress, advancement, efficiency, even perfection.

Like the geographers' movement, the theory of *orthogenesis* developed as a response to what many scientists saw as an overemphasis on the role of natural selection in evolution.[12] Orthogeneticists proposed that change resulted solely from an internal evolutionary drive, the mechanism of which was never clearly delineated.[13] The Swiss botanist Carl von Nägeli saw evolution as a manifestation of an internal perfecting mechanism, evidenced by trends toward increasing specialization.[14] The version of orthogenesis that characterized twentieth-century research emerged near the end of the nineteenth century, promoted most notably by the German entomologist Theodor Eimer. Like von Nägeli, Eimer assumed that variation and further evolutionary options were constrained at the origin of life (he envisioned multiple origins) to such an extent that the pathway and outcome were predetermined. Eimer, however, rejected von Nägeli's idea of a positive perfecting internal drive. Rather, he insisted that there was an internal *adaptive* drive built into living systems. Eimer documented trends in which no one stage seemed particularly better adapted than another, trends that showed something like von Nägeli's perfecting tendencies, and trends in which the transformations became increasingly maladaptive. For example, paleontologists who espoused orthogenetic views believed that many fossil groups showed trends toward increasing size right up to the point at which they abruptly went extinct. As a result of his studies, Eimer believed that a full picture of evolution could only be achieved by linking all these different kinds of trends together into a unified narrative. And although the trends were most apparent in developmental sequences, it was in ecology that he found his key to linking the trends.[15]

Eimer believed that the orthogenetic process led to progressive ecological specialization, often indicated by increasing size or complexity of organisms and/or traits. These transformations would be seen initially as positively adaptive (in von Nägeli's sense). But that was only part of the story. At some point in the process progressive ecological specialization turned into dependence on other species, a loss of evolutionary independence, secondary simplification indicated by presumed reduction and loss of structures, and finally self-imposed extinction. It is with this latter part of Eimer's narrative that pathogens became seen as key examples of orthogenetic evolution. Their apparent extreme dependence on their hosts was taken as an indication that they had reached the penultimate stage in evolution—overdependence and overspecialization prior to self-imposed extinction. The propensity to create disease was, therefore, a by-product of evolution, indicating that pathogens had become maladaptive in their environments and would soon eliminate themselves. Eimer had given orthogeneticists a narrative, but at a cost. The perspective raised a paradox that would vex the orthogenetic movement: *If pathogens are in the final stages of evolution before self-imposed extinction, why do so many of them persist?*

Darwinism itself experienced an evolutionary transformation during the final quarter of the nineteenth century, emerging in response to a perception that the perfecting influence of natural selection was being underappreciated. The transition from Darwinism to neo-Darwinism was due mostly to the influence of the British sociologist and Lamarckian Herbert Spencer. One thing about Darwin's theory that irritated people like Spencer (not to mention Alfred, Lord Tennyson) was its lack of optimism. Unlike Lamarck's absolute adaptationism, Darwin's relative adaptationism included the possibility that species would sometimes be unable to cope with environmental change rapidly enough or appropriately enough to avoid extinction. Additionally, Darwin argued that many traits persist in environments in which they are viable but not excellent; in other words, they are "adequate." Such traits were the source of an organism's ability to cope with changing environments; if the environment changed in a manner that favored a previously marginal form, its increasing fitness in the new environment would characterize the evolutionary change. Though Darwin spoke of natural selection as being a "perfecting" influence overall, no particular change was guaranteed to make descendants "better" in any way; and in any event, the environments to which selection adapted organisms changed more rapidly than inheritance, so while evolution tracked environments, it always lagged behind. This was inconsistent with belief in a high degree of functional fit between organisms and their environments. Spencer, perhaps drawing on his early advocacy of Lamarckism, concluded that "selection" must refer to two complementary phenomena, one of which was traditional Darwinian selection while the other was a more general sense of selection that ensured that the traits available at any time would always include those that enhanced the fit of organisms to their environments.[16] Spencer's transformation of Darwinism from "survival of the adequate" to "survival of the fittest" marked the birth of *neo-Darwinism*.[17]

Orthogeneticists and neo-Darwinians seemed to have little in common. And yet, the neo-Darwinians ended up in the same paradoxical corner as the orthogeneticists: If natural selection is a perfecting mechanism, why are there so many pathogens? Unlike the orthogeneticists, the neo-Darwinians at least seemed to recognize there was a paradox.

But, ironically, their solution was to postulate a kind of evolutionary trend. While disavowing any kind of trends in character evolution as envisioned by the orthogeneticists, neo-Darwinians allowed themselves to postulate general ecological trends. One early neo-Darwinian solution to the paradox of why there are so many pathogens was to postulate that selection always reduced pathogenicity, producing more benign associations between pathogens and their hosts. This notion became so deeply embedded in scientific thinking that it is still a popular myth today.

When Darwin died in 1882, we still knew far more about what agents of infectious diseases did than who they were, but we were learning fast. Pasteur provided the first general theory of microbial disease, suggesting that van Leeuwenhoek's world of animalcules was full of agents of disease. But with that fearful recognition came a great sense of optimism—what we know can't hurt us.

Around 1850 arsenic was first used to treat nagana, a disease of cattle caused by flagellated protists called trypanosomes. This began a long period in which clinicians routinely used various compounds as medications for sick humans and livestock that were difficult to administer, often toxic, and usually ineffective. By the mid-nineteenth century humans throughout the world had amassed an impressive array of mostly (but not totally: remember quinine) useless medicines along with an equally interesting array of erroneous theories about what caused the diseases they treated with those medicines.

Pasteur used the term *vaccin* to designate all existing and future vaccines, not just the Jennerian vaccine. This became common usage in the international scientific community around 1880 and was shortly thereafter included in the French dictionary. Despite smallpox's peculiarities (e.g., it has no known reservoir host), the successful vaccine against it inspired the development of an energetic research program sometimes called *une maladie, un vaccin* (for each disease, a vaccine). From the beginning there was considerable overlap in development of animal and human vaccines: not only were similar diseases in humans and animals often caused by related microbes, but, in many cases, the same microbe caused disease in both humans and animals. Development of animal vaccines often led the way, presumably because the specter of experimenting on animals was less problematic to the public than experimenting on humans.

Pasteur's ideas had been quite controversial at first. When in short course, however, vaccines against fowl cholera (1876), anthrax (1878), rabies (1879–85), swine erysipelas (1881), cholera (1884), hog salmonella (1886), plague (1890), diphtheria (1894), typhoid (1896), bovine tuberculosis (1897), and foot-and-mouth disease (1897) were developed, both the germ theory of disease and the socioeconomic benefits of vaccination became obvious.

The same year the typhoid vaccine was developed, Robert Koch reported the first agar cultures of bacteria. His techniques were quickly refined by a member of his research group named Julius Richard Petri. Many of us have used a Petri dish in a high school or university biology class, but how many of us ever wondered who Petri was and what he did to warrant having a critical piece of laboratory glassware named after him? Think about that the next time you use a Petri dish as an ashtray. Within a decade many species of bacteria had been cultured following Koch and Petri's methods, revealing what at the time was considered to be an astonishing microbial diversity, but which we know today was barely scratching the surface. The microcosmos became even more complicated in 1892, when the first virus—the tobacco mosaic virus, a veritable giant among viruses, being almost as large as a bacterium—was isolated. This set the stage for a new century marked by the emergence of a global industry that remains a major source of hope. Whenever a pathogen was isolated and cultured, could a vaccine be far behind? *Une maladie, un vaccin*, indeed!

Natural history played a critical role in the emergence of a new approach to coping with disease. In 1881 scientists demonstrated, without knowing of the existence of viruses, that mosquitoes transmitted yellow fever, and in 1895 Ronald Ross and his team discovered that malaria—caused by a protist whose identity was known, but whose mode of transmission was not—was also transmitted by mosquitoes. Eradicating mosquitoes and their breeding grounds could now be seen as promising means of eradicating multiple major diseases. These findings amplified hope for disease control through widespread eradication of species responsible for their transmission. Jakob Ericksson's discovery in 1884, based on his studies in Sweden, that there were multiple distinct forms of rust fungi, each associated with a different plant host, should have given pause to the eradication notion but did not.[18]

The dawning of the microbial theory quickly extended to diseases caused by parasites that were neither viruses nor bacteria. By the last quarter of the nineteenth century parasitologists had amassed considerable information about the biology of various protists, worms, and arthropods infecting humans, crops, and livestock. This was a golden age for parasitology. Rudolf Leuckart, another member of the German scientific pantheon, was one of the most progressive parasitologists during the early days of acceptance of Pasteur's theory that organisms cause disease. He was the first to prove that the tapeworm *Taenia saginata* was transmitted to humans through cattle, while its relative *Taenia solium* was transmitted to humans through swine. He was part of the research group that first documented the life cycle of the parasite *Trichinella spiralis* in swine and humans, helping support Rudolf Virchow's campaign to create meat inspection laws in Germany. Evolution was not a critical part of Leuckart's world, but in this regard he was not exceptional. Health specialists then, as now, generally paid little attention to the evolutionary issues associated with pathogens. After all, for someone trying to control malaria, it matters little whether *Plasmodium* is on its way to self-imposed extinction or to a benign mutualistic association with humans at some point in the distant future. Right now it is causing human misery and death, and it needs to be thwarted, not explained. That attitude persists today: so long as what we are doing is working, there is no reason to change.

Texas cattle fever, a scourge of American livestock producers, was an early success story for the eradicators. It had arrived in North America in the 1600s, brought by cattle imported from the Spanish colonies in the West Indies and Mexico. It spread throughout the southern United States and began expanding northward as infected herds were driven across the plains to cattle markets. By 1877 northern ranchers were determined to stop southern herds from contaminating their own. With no organized control efforts, northern cattlemen sometimes sent armed parties to turn back herds coming from the south. Studies of Texas cattle fever, commissioned by the United States Department of Agriculture (USDA), were undertaken by John Gamgee, a British veterinarian, in 1868. The government's motivation may have been not only a desire to deal with a disease of great economic impact, but also a desire to restore goodwill between North and South following the Civil War. In any event, Gamgee made little progress. It was not until 1879, when the American veterinarian Daniel Elmer Salmon joined the USDA, that things finally began to move. Salmon found that healthy cattle that grazed on pastures where there had been infected animals contracted cattle fever. He also studied the geographic range of the disease, establishing a quarantine line between infected and disease-free areas from the Atlantic coast west through Virginia. Movement of livestock originating south of this boundary was regulated.

All along cattlemen had maintained a theory that ticks were responsible for cattle fever. Many scientists, including Koch himself, dismissed the laymen's theory. Salmon believed the idea had merit, because his own observations had shown a strong correlation between areas with high tick abundance and cattle fever. He put three scientists on his staff in charge of the investigation: Theobald Smith studied the microbial cause of Texas fever, Frederick L. Kilborne was responsible for testing the theory that ticks transmitted the disease, and Cooper Curtice turned his attention to the biology of the tick. In 1888 Smith discovered the protist that caused the disease. The organism, called *Piroplasma bigemina* (now known as *Babesia bigemina*), belonged to a group of blood parasites that cause babesiosis, an often-fatal disease of cattle, horses, and other animals. In 1890 Kilborne's experiments showed that ticks were necessary for the transmission of cattle fever. For his part, Curtice made several important discoveries about ticks themselves. He found that the tick that transmitted Texas fever spent its entire life cycle on the body of a single host and did not drop off the host after each molt like other ticks. After feeding, female ticks would fall to the ground and lay their eggs. After the eggs hatched, young ticks lived in the grass until they encountered a grazing animal, and the cycle was repeated. Control of Texas cattle fever thus was correlated more with eradicating tick populations than with developing a *Babesia* vaccine. It is difficult to overstate the significance of this finding: "For the first time in history it was proved conclusively that the essential factor of an infectious disease may be a microparasite that reaches its victims only through an intermediate host, and this revealed why a truly infectious disease may in no respect be contagious."[19]

At the same time, and on the other side of the Atlantic, rinderpest, the major plague of European cattle, was eliminated by the implementation of sanitary measures eradicating the ecological conditions that allowed the virus to flourish before the virus causing the disease had been identified and isolated. These measures did not extend outside Europe, and rinderpest was introduced into Africa, where it devastated cattle and buffalo populations, as well as those of other susceptible domestic ruminants and many wildlife species. Difficulties in applying the sanitary measures that were so successful in Europe leave rinderpest a potential problem today. By the end of the nineteenth century all of the elements of contemporary approaches to coping with disease were in place: *medicating* the ill, *vaccinating* those at risk, and *eradicating* species associated with transmitting the disease. This three-pronged approach to infectious disease was to serve humanity well for nearly a century.

That time also brought the first warning sign from the scientific community about climate change. Svante Arrhenius was a Swedish physicist, born the year Darwin published the first edition of the *Origin*. Arrhenius, often credited with establishing the discipline of physical chemistry, was the first scientist to calculate the impact of changing concentrations of atmospheric carbon dioxide on Earth's temperature. In 1896 he concluded that carbon dioxide output from human industry would be enough to create what he called the hothouse effect, now commonly referred to as the greenhouse effect but formally termed the Arrhenius effect in physical chemistry. He concluded that human industrial output would be enough to postpone the next Ice Age. Ironically, he felt this would be a good thing, because a warmer Earth would aid food production, which needed to increase in order to cope with the accelerating increase in human population.[20]

The First Half of the Twentieth Century

Vernon Kellogg spanned the end of the nineteenth and beginning of the twentieth centuries with considerable authority. In an influential book, Kellogg claimed that "everyone agreed" that traditional Darwinism was dead and that its successor would have to explain "how the right adaptation always shows up at the right time."[21] He dismissed Darwin's notion that traits permitting successful adaptation to new environmental conditions must have occurred in low frequency prior to the environmental changes in which they flourished.

Kellogg's ideas about the formation of new species were prescient, if underappreciated. He suggested that speciation was initiated by geographic isolation, and then completed by selection on the partitioned variation. Kellogg's views, however, were paradoxical when it came to pathogen-host systems: he wanted selection to play a role in evolution, but he also wanted host-specific parasites to be infallible indicators of geographic speciation.[22] His solution to this paradox was twofold: (1) parasites are highly attuned to their hosts (that is, they are highly host-specific and do not "straggle" much onto novel hosts), and (2) the host environment is so uniform that it buffers parasites from the effects of natural selection, complementing von Ihering's views.

This led Kellogg to recognize three categories of parasite-host relationships, based on geographic distribution and presumed dependence on particular hosts. The first consisted of closely related parasites occurring in association with single closely related hosts living in disjunct geographic areas. Following Darwin, Kellogg attributed such correlated parasite-host associations to genealogy: "I do believe that it is a commonness of the genealogy rather than a commonness of adaptation that is the chief explanation of this restriction of certain parasite groups to certain host groups."[23]

The second category comprised parasite species occurring in association with closely related hosts inhabiting distinct geographic regions. Kellogg believed these relationships reflected instances in which both the hosts and parasites had been geographically isolated but only the hosts had become different species. He hypothesized that if the environment was different enough in the geographically isolated regions to establish selection pressures leading to differentiation of the hosts, while the parasites' environment, consisting of parts of the host, did not change to the same extent, then the parasites would not speciate.[24] This was the basis for Kellogg's views that the presumed uniformity of the host environment buffered parasites from selection.

Kellogg's third category comprised parasites associated with two or more distantly related host species living within the same geographic region. Such "stragglers" posed an important problem for students of geographic isolation. Kellogg suggested that three major factors affected straggling: (1) physical contact between hosts (of the same or related species), (2) differential speciation between hosts and parasites, and (3) some inherent tendency for parasite species not to "straggle" (add a new host) even when given the chance.[25] After 1920 Kellogg and von Ihering no longer produced articles about pathogen-host evolution.[26] They left unfinished business behind.[27]

Maynard Metcalf succeeded Kellogg as the American champion of the geographic school of host and parasite speciation.[28] Working simultaneously with frogs and their opalinid parasites (large ciliated protists with multiple nuclei), Metcalf considered the Wegener hypothesis of continental drift to be supported by von Ihering and his successors. That being the case, Metcalf believed his research program was unified by what he first called the "parasite-host" method and later the von Ihering method.[29] While primarily interested in the effect of geographic factors on speciation, the geographers wove the threads of restrictions in host range throughout their writing.

Orthogeneticists considered parasites to be highly specialized beings that were no longer masters of their own destiny. Their evolutionary fate was sealed; they were totally dependent upon and molded by their hosts. From this belief sprang the two related concepts which formed the cornerstones of this research program: (1) *host specificity*, because only the study of strongly host-specific parasites would reveal the causal mechanisms underlying parasite-host relationships, and (2) *"true"* or *obligate parasites*, because only among the "true" parasites could one be certain that the environment to which the parasite responds evolutionarily is provided exclusively by the host.

Heinrich Fahrenholz,[30] an entomologist interested in lice, asked whether the presence of the same or related parasite species on different host species would allow researchers to draw conclusions about the relationships among the hosts, regardless of any geographical isolation or contact. He argued that parasites are as dependent on their hosts as free-living organisms are on their habitats. One would therefore expect members of the same host species to be inhabited by the same species of parasites, and members of different host species to be inhabited by parasites whose phylogenetic relationships would be as divergent from each other as were the phylogenetic relationships of their hosts. In other words, if parasites produce new species in response to their environments, and if the host is the environment, then it follows that parasites will produce new species in response to differences among hosts. This meant that host and parasite phylogenies should parallel each other. In a study of lice inhabiting catarrhine primates (Old World monkeys, such as baboons) and great apes, Fahrenholz wrote that the different species of lice were related to each other in the same way that their hosts were related.

Wolfdietrich Eichler, a German entomologist studying Mallophaga,[31] set the tone for his career with an early attack on the suggestion by von Kéler that at least some bird lice exhibited reduced host specificity—Kellogg's stragglers.[32] He argued that reports of the same parasite on different hosts meant one of two things: (1) the parasite had inhabited the common ancestor of the current host species (à la Fahrenholz) or, more likely, (2) there had been a mistake in identifying the parasite. Claiming that many of the examples discussed by von Kéler fell into the second category, Eichler concluded that one must base taxonomy on an understanding of the parasite-host relationship rather than on characteristics of the parasites themselves. This entrenched the perspective that hosts were the primary determinants of parasite evolution.[33]

There was abundant interest in orthogenesis in North America and Europe,[34] but the ideas remained deeply flawed. By 1950 orthogeneticists had created a circular argument—parasites are highly host-specific, so they coevolve with their hosts, and because they coevolve with their hosts, they are highly host-specific. Any conflicting or inconsistent observations, therefore, were erroneous or irrelevant because they failed to conform to the higher theory, which was taken as truth. Furthermore, there was no orthogenetic theory of inheritance that was consistent with experimental findings. And finally, like most members of the geographic school, orthogeneticists believed that host and parasite phylogenies should parallel each other. They thus joined those colleagues in being unable to test that assumption empirically because they lacked a method for reconstructing phylogenies.

Amazingly, none of these inconvenient truths slowed the orthogeneticists. Eichler, in particular, was unshakably convinced that the drive toward increasing host specificity was a universal trend among parasites. When faced with the suggestion that some parasites did in fact demonstrate widespread host preferences, he either dismissed the parasites as young and thus just beginning along the orthogenetic pathway to specialization or claimed that the parasites were members of different species. He supported his claim with information from population genetics, arguing that parasites inhabiting many hosts might appear identical but parasite populations inhabiting a particular host were really distinct species recognizable at the genetic, if not at the morphological, level.[35] And even if genetic evidence was lacking, orthogeneticists had an ingenious fallback to explain the anomalous situation in which a given parasite species seemed to be associated with more than one host species. They called upon something invisible but not imaginary—instincts and behavior. In 1899 the American zoologist Charles Otis Whitman wrote that "instinct and structure are to be studied from the common viewpoint of phyletic descent"; in other words, behavioral characteristics could be used as indicators of common ancestry. The orthogeneticists interpreted Whitman's statement in light of the presumed truth of their theory. They "knew" that each species of host had its own species of parasite, so the appearance of the same species of parasite in two or more different hosts was evidence that the parasites in each host were different species: they were clearly displaying changes in instinct (in this case meaning "host preference"). These changes occurred before changes in structure (morphology of the parasite) emerged.

Neo-Darwinian researchers at this time were primarily interested in the ecological relationship between particular pathogens and hosts, not in any larger-scale phylogenetic outcomes. This was the culmination of Kellogg's vision of neo-Darwinism, in which everything is narrowly adapted to a particular environment. The researcher's job was to figure out the adaptations on a case-by-case basis and explain how they emerged at the right time. This marked the beginnings of recognizing that the pathogens as well as their hosts might be important for under-standing the evolution of pathogen-host systems. This novel approach to understanding the evolution of interspecific associations found its most energetic advocates among those studying the interactions be-tween plants and phytophagous insects. Those associations often showed pronounced specificity with respect to hosts on a species-by-species basis yet showed no clear phylogenetic component with respect to relatedness among host species or geographic distributions. This triggered the insight that the host species was not the focus of evolu-tion; rather, *traits that the hosts possessed in common was the focus.*

Just three years after Kellogg published his book, and at the same time Fahrenholz published his studies, the Dutch biologist Émile Ver-schaffelt documented the chemical basis of plant odor and implicated it in insects' choice of plants for eating and egg-laying.[36] A decade later Charles Thomas Brues, an American entomologist with a back-ground in both tropical disease and agricultural research, summarized a growing body of literature indicating that something quite specific and conservative, yet not directly phylogenetic, was going on with phytophagous insects. Though technology remained limited, Brues suggested that the evidence supported Verschaffelt's findings that the host species itself was not what attracted preferential attention from particular insects. Rather, there was something—some trait—possessed by certain host plants that was the focus of attention.[37] This began the change from "the host species *is* the evolutionary resource" to "the host species *has* the evolutionary resource." About 15 years later, an American researcher, Vincent Dethier, began amplifying this line of research.[38] Dethier believed that the work of Verschaffelt and Brues showed that insects chose host plants based on scent and/or taste rather than nutritional value.[39] His first book, *Chemical Insect Attrac-tants and Repellents,*[40] summarized his work to that point by saying that these chemical features of plants repelled some insects and attracted others. Furthermore, he asserted that while all attractants are chemi-cal in nature, some repellents can be physical as well. And finally, he noted that the chemicals acting as attractants may be very specific in nature and yet widespread among plant species. Once technological developments allowed sensitive measurements of what insects were re-sponding to in an individual plant, this research tradition became crit-ical to research into the nature of ecological specialization. Ironically, neo-Darwinians accepted the orthogenetic notion of a progressive drive toward specialization and co-opted their argument that changes in behavior set the stage for new selection that led to the emergence of new traits. Both schools thus believed in form following function and in the centrality of explaining how the right adaptation showed up at the right time.

The heads of those readers who are still with us are now throbbing. Here in a nutshell is "the problem of ecological specialization": it was (1) invoked by some geographers as part of explanations for geographical speciation without selection; (2) used by others as evidence that selection was an essential part of geographical speciation; (3) treated by orthogeneticists as de facto evidence that neither geography nor selection had anything to do with pathogen evolution; and (4) treated by neo-Darwinists as evidence that neither geography nor orthogenetic trends have anything to do with parasite evolution.

This happens often in science: people agree about the observations, the basic data, but do not agree about what those data mean. The absence of an objective scientific method for reconstructing evolutionary history in enough detail to provide evidence bearing on the relative merits of their explanatory frameworks hindered all these research traditions. There was simply no way to test the hypothesis that pathogen and host phylogenies paralleled each other or why.

The first half of the twentieth century was also a boom time for disease and health, and not only for advances in treatment of headaches. By this time the "medicate, vaccinate, eradicate" approach had become well established, and intensive activities were under way in all three areas. And those activities were sorely needed—a burgeoning human population, increasingly crowded into cities in industrialized countries, might be largely free of the scourge of smallpox but still experiencing the devastating effects of measles, mumps, polio, tuberculosis, and influenza. In tropical and rural areas smallpox still existed, while yellow fever and malaria dominated the discussion, and a host of protistans and helminths bedeviled humans and their livestock worldwide.

One of the most amazing events of this time was the Scottish scientist Alexander Fleming's 1928 discovery of penicillin, an antibiotic produced by molds. Soon the search was on for naturally occurring microbes that produced substances useful in combating disease caused by other microbes. The emergence of chemistry as an industrial-grade enterprise led the way to identification and refinements of those newly found antibiotics. By 1934, for example, the antimalarial drug chloroquine, a synthetic version of quinine, had appeared.

Some vaccine researchers began to apply concepts from the new field of population genetics to health concerns at this time. Once again, veterinarians led the way. By the 1920s livestock disease scientists had noticed that diseases caused by pathogens seemed often to have a life of their own. The occurrence of a pathogen in the environment of susceptible hosts did not necessarily mean there would be an immediate disease outbreak. (The presence of a pathogen in an individual host did not mean that host would be diseased; consider the infamous case of "Typhoid Mary," as journalists nicknamed her—Mary Mallon, a cook for wealthy families in New York City who infected as many as twenty-two, and possibly more, people between 1900 and 1915. She was placed in forced quarantine twice after she refused to give up her work in food preparation.) So whenever there was a disease outbreak associated with a particular pathogen, not everyone got sick, not everyone got equally sick, and the disease outbreak did not continue indefinitely. This suggested that if there were immune individuals within a host population, there might be a threshold proportion of them such that when that proportion is reached, the pathogen should no longer be capable of sustaining itself.

The phenomenon was named "herd immunity" in 1923 by livestock health researchers and recognized within a decade as widely occurring.[41] Herd immunity brings good news and bad. The good news is that we expect that the more rapidly deadly a disease outbreak is, the more rapidly it will burn itself out. The bad news is that if the threshold level of immunity is not maintained, the disease returns periodically, and with a vengeance. Natural immunity in a population does not remain at a high level in the absence of regular epidemics, which kill hosts that are not genetically immune. Therefore, if we wish to maintain high levels of immunity in a "herd," it is essential that natural immunity be augmented by vaccinated immunity. It was ironic that the recognition of the value of mass vaccinations occurred at this time, when the cost, in time and resources, of producing new vaccines was beginning to skyrocket. For example, attempts to develop a bovine tuberculosis vaccine began in 1897 but were not successful until 1921.

Protecting herds against the consequences of foot-and-mouth disease (FMD) has been a concern for cattle breeders since antiquity. From World War I to the mid-1950s FMD was observed throughout most of the European continent in the form of widespread outbreaks occurring at intervals of approximately six years. In 1926 the first vaccine against FMD appeared. Once the difficult process of turning the virulent FMD viruses into safe antigens was mastered, the second problem was obtaining enough virus material for vaccine production. That technological hurdle was overcome shortly after the end of World War II. Thanks to mass vaccination, FMD is not currently an issue in Europe, but it is still prevalent in some parts of Asia and Africa and thus remains a threat throughout the world.

Beginning in 1912, efforts were made to develop a vaccine against yellow fever, but it would take 25 years for this quest to succeed. In 1927 the yellow fever virus was isolated, and efforts were made to derive a vaccine from attenuated samples of that culture. A decade later Max Theiler and Hugh Smith discovered a chance mutation in one subculture of the attenuated virus that allowed them to produce an effective vaccine. The strain with the mutation was called "17D," and 100 percent of vaccines administered today are made using this mutant. One might think that alarm bells would have gone off when it took such a fortuitous accident to develop a vaccine. Any such soul-searching, however, was quickly quashed when Theiler was awarded a Nobel Prize (not to mention the even greater honor of having a genus of parasites related to malaria named after him). So, time and chance notwithstanding, the yellow fever success story shaped our belief that we would be able to make vaccines for any pathogen, and that we would be able to vaccinate everyone at risk. This became the overall story of the development of vaccine technology and mass production that characterizes the present era.

Technical shortcomings still limited studies of microbe evolution, although much was being learned about their ecology, especially their transmission dynamics. This information was useful for mitigating disease through interrupting transmission. Little thought was given to the unintended consequences of such mitigations, or the evolutionary significance of the data being collected, although one brave soul studying wheat rust fungi followed this line of thinking and suggested that primitive host plants did not necessarily host primitive rust fungi.[42] The need for better diagnosis was never linked explicitly to taxonomy; commonalities in transmission modes among related species of pathogens was rarely linked to evolutionary conservatism.

Research on Texas cattle fever showed that ticks transmitted the pathogen responsible for the disease. This knowledge opened the way to better understanding of other diseases with insect vectors, such as malaria and yellow fever, both transmitted by mosquitoes. It also pointed toward a solution to Texas cattle fever: in order to control the spread of the disease, we would have to eradicate ticks. In 1905 the USDA began a cooperative effort with state governments to accomplish that task. The Cattle Fever Tick Eradication Program, begun in 1906, led to the eradication of the disease in the United States. Veterinary use of vaccines against bovine tuberculosis declined after World War II, when it became known that mass vaccination was more expensive and no more effective that attempting to eradicate the disease through the systematic slaughter of tuberculous cattle.[43]

Shortly after the end of World War II, however, there were indications that health goals were going to be more difficult to achieve than thought. In 1947, less than 20 years after the discovery of penicillin, the first documented appearance of microbial resistance to penicillin occurred. Not having the technology to fully understand microbial genetics, the implications of this new phenomenon were not fully appreciated. Two years later a potential turning point came—and went—when the eminent British evolutionary biologist J. B. S. Haldane wrote about the evolution of resistance to pathogens. That topic had been raised as early as 1905 in an article discussing Mendel's studies in association with the discovery of strains of wheat that were resistant to rust fungi. But, like the chance mutation in a laboratory strain that led to the yellow fever vaccine, encounters with genetic resistance to disease still seemed like fortuitous, and thus unusual, events. Haldane suggested that resistance should be studied as an evolutionary phenomenon, but that message went largely unheeded in the quarters where it could have made a difference, especially at a time when new understanding of climate emerged.[44]

We have a Serbian geophysicist and planetary climatologist, Milutin Milanković, working in Belgrade at the beginning of World War I, and later Budapest in the 1920s and 1930s, to thank for the foundations of our current understanding of the dynamics of Earth's climate systems. Remarkably, the insight into what is now the prevailing paradigm for exploring the past and future of global climate occurred in an Austro-Hungarian prison during 1914. Understanding sunlight, or *solar insolation*, and the energy arriving on Earth during the summer in the Northern Hemisphere was the critical component of that insight. Milanković correctly linked variation in insolation to three distinct planetary mechanisms to explain episodic events during the Ice Age on geological timescales of the Quaternary. The relationship of the Earth to the sun as it goes around it is one of *eccentricity*, which refers to the degree to which the orbit is circular or elliptical; this ellipticality produces 100,000-year climate cycles. Shifts in the tilt of the Earth during passage around the sun are called *obliquity*; increasing tilt produces greater seasonal extremes, in 41,000-year cycles. Wobbling motions of the planetary axis are called *precession*, driving variation or extremes of insolation seasonally in the Southern or Northern Hemispheres in 23,000-year cycles.[45]

Milanković cycles emerge from variations in solar forcing, or the levels of solar energy received on Earth, which result from complex interactions among these three mechanisms. Eccentricity in particular leads to varying levels of summer insolation in the Northern Hemisphere and in part explains the altitudinal distribution of snow; when snow accumulates and is persistent, glaciations are not far behind. At the other extreme, precession has been linked to monsoonal climate shifts, for example, either amplifying or depressing rainfall in Africa, resulting in the expansion or collapse of connections among ecosystems on different continents.[46] Summer insolation (as measured at 65 degrees north latitude) should gradually increase for the next 25,000 years. The eccentricity of Earth's orbit is expected to decrease over the coming 100,000 years, however, so a change in insolation should be dominated by obliquity. According to Milanković's calculations, our current climate stability should have continued for a long time, as inception of the next glacial cycle was not predicted for another 50,000 years.[47] Sadly, his calculations did not include the potential influences of accelerating climate warming in the Anthropocene.[48] As a consequence, accelerating change has been especially evident at high latitudes of the Northern Hemisphere since the 1970s when anthropogenic warming driven by carbon dioxide and greenhouse gas emissions reversed the incremental cooling trend of the past 2000 years.[49] The two critical pieces for understanding Earth's climate and accelerated anthropogenic warming were known more than a century ago but were never connected. Arrhenius believed atmospheric warming would be a boon to agricultural production and Milanković did not consider it in his calculations. Humanity continued to slumber in a false sense of security.

Summary

In the century following Pasteur's amazing insights, a unified approach to coping with the pathogens that caused infectious disease emerged: *medicate* the ill, *vaccinate* those at risk, and *eradicate* the species involved in transmitting the disease. The treatment of disease became a very pragmatic business. In the roughly 75 years following Darwin's death, by contrast, evolutionary biology underwent an upheaval, producing three different approaches to studying evolution. Evolutionary studies of pathogens became a very esoteric business, mostly unconcerned about the pragmatic matters associated with battling diseases those pathogens caused. As we leave that first century of research into evolution, disease, and climate, enormous advances have been made, controversies have been established, and technology has advanced.

But storm clouds are massing on the horizon.

THREE

Dawning Awareness

When the past returns to us with all its glory and pain, we don't know whether to embrace it or flee.
—Brian Herbert and Kevin J. Anderson, *HUNTERS OF DUNE* (2012)

Modern thinking about disease and evolution began quietly and slowly. A year before the centenary of the publication of the *Origin*, the distinguished American mathematician Charles Mode was a young scholar studying the genetic interaction between wheat and rust fungus pathogens. That year he published the first mathematical model of mutual evolutionary modification resulting from mutual antagonism between parasites and hosts, introducing a concept called *coevolution*.[1] Mode received little recognition for this accomplishment until recently, although that publication marked a critical juncture in our understanding of pathogen and host evolution. For the first time, scientists had to take into account the impact not only of the pathogen on the host, but also of the host on the pathogen. Technological advances in molecular biology and genetics during this period allowed researchers to produce the new data required by the new theoretical framework. The esoteric and the pragmatic could join forces, as Haldane had urged, but it would not happen quickly.

Neither geographers nor orthogeneticists could escape the issue of host specificity and the specialized ecological elements of pathogen-host associations. Most critically, neither group had an objective method for reconstructing phylogenies to compare the degree to which pathogen and host phylogenies coincided. Circular reasoning ensued. For the geographers, parasites occurring in similar but geographically separate hosts were assumed to be the result of speciation by geographical isolation, so one could assume that the parasites were related to each other phylogenetically in the same way that their hosts were. Orthogeneticists also assumed that parasites occurring in closely related hosts were related to each other phylogenetically in the same way that their hosts were related to each other, but without regard to geographic distribution. Lacking an independent method for generating phylogenies, how could either group test its fundamental beliefs? And how could they distinguish their results? Not a good situation.

The geographers had an additional, unacknowledged problem. Any given pair of species—pathogen and host—was assumed to have arisen in geographic isolation from their closest relatives. This implies that the common ancestor of those species had a broader geographic range than either of its descendants. And yet, that species would also be assumed to have arisen through geographic isolation. So how do the isolated ancestral species become widespread enough to be affected by an episode of isolation? And how do we detect and distinguish isolations and expansions? Most unsatisfactory.

Parasites . . . furnish information about present-day habits and ecology of their individual hosts. These same parasites also hold promise of telling us something about host and geographical connections of long ago. They are simultaneously the product of an immediate environment and of a long ancestry reflecting associations of millions of years. The messages they carry are thus always bilingual and usually garbled. As our knowledge grows, studies based on adequate collections, correctly classified and correlated with knowledge of the hosts and life cycles involved should lead to a deciphering of the message now so obscure. Eventually there may be enough pieces to form a meaningful language which could be called *parascript*—the language of parasites which tells of themselves and their hosts both of today and yesteryear.[2]

By the centenary of the publication of Darwin's *Origin of Species*, productive elements of the frameworks espoused by the geographers and orthogeneticists were being subsumed by neo-Darwinism. Neo-Darwinians had come to accept that geographic isolation provided the initiating conditions in most cases of species-formation, though they insisted that completion of the process required natural selection. They had also accepted many of the evolutionary trends recognized by the orthogeneticists—including those associated with pathogen-host evolution—but had injected selection as an explanation, called orthoselection, in each instance in which the orthogeneticists had failed to document a mechanism. Despite calls to integrate with mainstream evolutionary biology,[3] the majority of parasitologists remained attached to their orthogenetic roots and were apparently blissfully unaware of these discussions; neither the *Journal of Parasitology* nor *Parasitology*, Britain's leading parasitology journal, published any articles dealing with evolutionary concepts, nor did they offer any editorial comment about the centenary.

Mainstream evolutionary biology no longer considered the geographers and the orthogeneticists as belonging to distinct research programs, much less as competitors for center stage. Mode's initial article was published in the prestigious journal *Evolution*. Just six years later, an article in the same journal captured the attention of a broader population of evolutionary biologists.[4] Paul Ehrlich and Peter Raven proposed that the evolutionary diversification of plants and insects had been fueled by complex interactions involving mutual modification—notably of plant chemicals and insect olfactory and gustatory preferences.[5] They suggested that such coevolutionary dynamics might have a general phylogenetic context, but the fine details need not parallel the evolutionary history of the specific taxa involved. The distribution of insects among plants followed the evolution of host resources and the evolution of insects' abilities to utilize those resources, rather than the evolution of host species themselves. Within evolutionary biology, Ehrlich and Raven's vision of coevolution catalyzed research involving "two major groups of organisms with close and evident ecological interactions" that were not just associated with each other, but actually modified each other evolutionarily. Ehrlich and Raven did not cite Mode's articles on the subject, delaying the integration of basic evolutionary biology with the study of pathogen-host systems.

The Darwin centenary created little notice in the health world. The British medical journal *Lancet* carried limited discussion, reports of meetings, and a book review of Darwin's autobiography released in 1958. Two editorials and three short articles published between 1958 and 1960 included discussions of Darwin and Freud and of Darwin and Wallace, along with speculations on the cause of Darwin's poor health.[6] In the United States the *Journal of the American Medical Association* and *Journal of Tropical Medicine and Hygiene* carried neither editorials nor articles about the centenary. The *New England Journal of Medicine*, like the *Lancet*, published a short commentary taken from a lecture on the centenary focused on the centenary celebrations rather than any potential implications of evolutionary theory for medicine, as well as a short article about Down House, Darwin's residence, which served as a vehicle for discussing the transformation of religion, law, society, and science in the latter half of the nineteenth century. There were also two short perspectives published that floated various theories on the reasons for Darwin's many and various symptoms.[7]

The focus of disease specialists continued to be empirical and technological. Truly sophisticated culturing techniques and methods for nucleotide extraction and analysis emerged at this time, laying the foundation for modern studies of pathogen evolution. Disease researchers had always seen pathogen-host relationships in terms of a one-way conflict, pathogens attacking hosts, hosts requiring human intervention unless they happened to be resistant. Mode's articles showed clearly that a focus solely on host resistance to disease was inadequate; we now had to contend with the issue of pathogens overcoming host resistance as well.[8] It would seem inevitable, then, that when evolutionary concepts were placed center stage in understanding the nature of pathogen-host associations, the disease and the evolution people would finally get together. And indeed, such studies began to gain traction. They would not, however, become mainstream for another generation; "medicate, vaccinate, eradicate" seemed to be working well, mostly, and mostly was good enough.

Pathogen resistance to penicillin was first documented in 1947, and by 1975 microbial resistance to antibiotics was a consistent finding in clinical and field settings. This created an unanticipated drain on time and resources. From then on, producing medications in an attempt to stay ahead of the evolutionary capabilities of pathogens would be a large-scale and expensive preoccupation of the pharmaceutical industry, eventually leading some corporations to ask whether investing in antibiotics was worth their efforts.

With smallpox, typhoid, and yellow fever vaccines in place, no viral disease caused greater fear in the first half of the twentieth century than polio. Once the full attention of vaccine researchers turned to polio, there was no doubt there would be a polio vaccine, it was simply a matter of who would win the Nobel Prize. But polio turned out to be more of a challenge than most expected. The public was becoming a bit restive until, in the span of six years, not one but two polio vaccines were produced, by Jonas Salk in 1955 and by Albert Sabin in 1961. All three of the authors remember when our parents lined us up as youngsters to receive both vaccines.

After the success of the polio vaccine, work continued on developing vaccines for a number of pathogens. In 1962 the successful isolation of the rinderpest virus in cell culture led to a highly effective vaccine that has been the primary reason for the success of the global campaigns to eradicate the disease. Optimism and public expectations remained high, but a harsh reality was making itself known. Caught up in the celebration of each increasingly rare success, we lost sight of the increasing difficulties experienced in producing vaccines. For example, British researchers were aware in 1948 that there was a link between rubella in pregnant women and congenital birth defects in the children they were carrying. Despite that knowledge, no major effort against rubella was undertaken until a major outbreak occurred in the United States in 1962, just a year after the Sabin polio vaccine became available. That rubella outbreak led to a spike in children born with birth defects. The public outcry was tremendous, and significant amounts of funds and researchers were marshaled to the effort. Despite that, a rubella vaccine was not produced until 1969, seven years after the major outbreak in the United States began, and a full generation after British physicians first reported a link between birth defects and rubella in pregnant women.

By the 1960s and into the 1970s, herd immunity was at the core of discussions about the costs and benefits of mass vaccination programs, and at this point the term *herd immunity* became more widely applied. The immediate response was a resoundingly positive one: we can now control disease outbreaks through mass vaccination programs. Reality quickly set in. Mass vaccination programs require a manufacturing base for mass production. That became possible only in the decade following the end of World War II. They also need an infrastructure capable of delivering mass vaccinations. For pathogens affecting livestock and flocks of domestic birds in wealthy countries, all you need is an energetic veterinarian and assistance from the ranchers and farmers. For those affecting humans in wealthy countries, all you need is a unified medical profession urging the public to line up for vaccinations. But what about delivering vaccines on a level necessary to provide herd immunity in poor countries? Who pays for that? Most felt that as our technology became more sophisticated, vaccines would become easier to make and cheaper to produce and deliver. This turned out not to be the case.

This period of time also witnessed a shift in eradication efforts from coping with preexisting pathogens to dealing with emerging ones. Developing global commerce made it easier for pathogens to be transported to new areas. Surprisingly, in many cases we discovered that pathogens were perfectly happy to be moved from their place of evolutionary origin and might well take up with novel hosts whose evolutionary association with the original host might not be easily understood in terms of coevolution.[9] Even worse, once established they were not so easy to eradicate.

Myxobolus cerebralis is a member of a group of organisms called myxosporeans whose closest relatives are the group that includes jellyfishes and sea anemones. *Myxobolus cerebralis* lives in freshwater habitats, where it uses oligochaete worms (aquatic relatives of earthworms) as hosts for one stage of its life cycle and salmonid fish as hosts for the other stage. The pathogen has been associated historically with brown trout (*Salmo trutta*) in central Europe, where it provokes minimal pathology and disease despite living in the brain. Were it not for the importance of trout for food and sport fishing, *M. cerebralis* would undoubtedly still enjoy relative obscurity where it originated. In 1903, however, young nonnative rainbow trout (*Onchorhynchus mykiss*) introduced to Germany began to exhibit what became known as whirling disease. The parasite attacks the central nervous system and cartilage of young fingerlings, leading to skeletal malformations that cause the characteristic whirling symptoms as the fish try to swim; mortality in hatchery situations is nearly 90 percent in hosts other than central European brown trout. In 1956, *M. cerebralis* was introduced into the United States by infected fish imported from Europe. The parasite is now known in 26 states and provinces in the United States and Canada as well as twenty-five other countries globally. Introductions have been mediated by the movement of infected fishes, often rainbow trout, for aquaculture and hatchery rearing linked to sport fisheries. Local movements can be facilitated by persistence of the resistant spore stage released from decomposing fish, passage of spores in predatory fish-eating birds, and developing stages in oligochaetes. The pathogen impacts aquaculture and sport-fishing hatcheries and recently has begun to cause disease in wild populations outside central Europe—for example, seriously threatening the viability of some trout populations in the most important streams and rivers across Colorado and Montana.[10]

Increased appreciation for the socioeconomic impacts of disease led to the largest mobilization of humans and technology ever aimed at eliminating habitat suitable for the transmission of pathogens or their vectors. In the 1940s DDT was used to control insects—especially mosquitoes—that transmitted disease. It was so effective that by the 1960s parasitologists had begun to talk seriously about eradicating malaria from the globe. Two books published at the end of this period required that agenda to be reset.

The books were *The Ecology of Invasions by Animals and Plants*, by the British ecologist Charles Sutherland Elton, published in 1958, and *Silent Spring*, by the American marine biologist Rachel Carson, published in 1962.[11] Taken together, they presaged critical understanding of the emerging disease crisis. First, climate change was happening and would have widespread influence, particularly in terms of the global movements of plants, animals, and humans; and second, human technology for controlling disease incurred costs for the integrity of the biosphere. By 1972 the ecological damage done by indiscriminate use of DDT to control mosquitoes led to the chemical being banned. This was the first indication that dealing with disease could bring us into conflict with other agendas associated with climate change. The transmission dynamics of pathogens was key, being an indication both of the cohesive trophic structure necessary for the persistence of pathogens and of the places where targeted biodiversity destruction would be effective. In a world in which maintaining ecosystem integrity was becoming a pressing issue, conflicts of interest were inevitable. Public and agricultural health issues had entered the wildlands.

We now begin to emerge from the mists of history into contemporary research. After the Darwinian centenary, evolutionary biology began to move away from the search for "rules" as research continuously showed that the rules that were not flat-out wrong were more like guidelines. The coevolutionary framework provided by Mode and by Ehrlich and Raven fit this zeitgeist perfectly, catalyzing an explosion of coevolutionary studies that continues today. It also coincided with growing awareness of the reality of climate change and its likely impact on biodiversity and disease emergence. Critical to the development of scientific understanding in these connected areas was a methodological advancement called *phylogenetic analysis*, or *phylogenetics*.[12]

Ending the Twentieth Century

All evolutionary approaches to understanding pathogen-host systems up to the last quarter of the twentieth century floundered at critical points because they lacked a means of generating explicit phylogenies— histories of ancestry and descent—derived from objective evidence rather than from each researcher's a priori assumptions. A leading figure in the study of microbial evolution has bluntly stated that no proper studies of microbial evolution were possible until phylogenies of microbes became available.[13] Technological advances in molecular biology provided the raw data for generating microbial phylogenies, but data do not speak for themselves. We needed a way to evaluate those data with respect to what they might be able to tell us about phylogeny.

The elements of such an approach had long existed in the oral traditions of many taxonomists, including Darwin, and parts of it had even been formalized to some extent.[14] But it was Willi Hennig, a German entomologist, who provided the first explicit description of an objective methodology for reconstructing phylogenies. Based upon analyzing observable attributes of the organisms, the method was independent of theories about how they arose and therefore a powerful tool for studying evolution.[15] Phylogenetic trees have few moving parts, being made of branches, nodes, and traits, and yet they have proven to be extremely versatile in a wide range of evolutionary studies (fig. 3.1).[16]

The Geographers and the Orthogeneticists Unite: Cospeciation

In 1979 Brooks discussed two elements of parasite evolution that could be examined phylogenetically.[17] One of these was that common ground between geographers and orthogeneticists: *cospeciation*, instances in which hosts and parasites speciated together. When an entire group of parasites is the result of cospeciation, there will be complete congru-

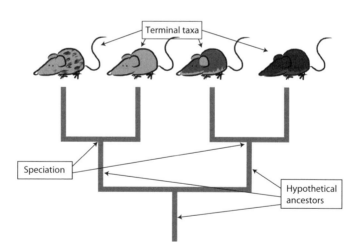

FIGURE 3.1   A diagrammatic representation of a phylogenetic tree showing terminal branches (known species), internal branches (hypothesized ancestors from which known species, or other ancestors, descended), and nodes connecting branches (hypothesized speciation events splitting an ancestral species into descendants).

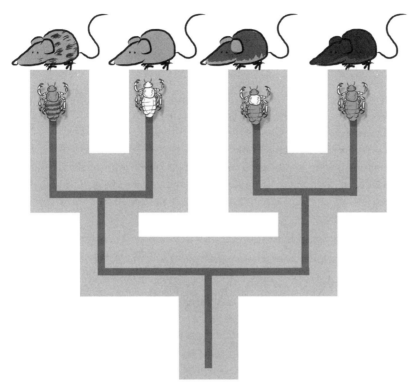

FIGURE 3.2 Phylogenetic tree of a hypothetical group of lice infecting our hypothetical group of rodents (from fig. 3.1). If we superimposed the two phylogenies, all the branches would align, indicating the hosts and parasites cospeciated.

ence between pathogen and host phylogenies (they can be superimposed on each other without a mismatch; see fig. 3.2).

Such phylogenetic association is not de facto evidence of any particular kind of coevolutionary process; in fact, cospeciation patterns may simply be the result of concomitant speciation by two or more ancestral species that happened to be living in the same area and experienced the same physical isolation. Support for this hypothesis offers relatively weak explanatory power.[18]

The *maximum cospeciation* perspective emerged from these initial discussions, based on the old assumption that pathogens are so highly specialized on specific hosts that they are effectively inherited with their hosts and thus track host phylogeny.[19] Such pathogen-host associations would be reinforced if pathogens and hosts experienced concomitant episodes of geographic speciation, so this perspective seemed to be an ideal way to unite the geographers and the orthogeneticists. Finding patterns of cospeciation was more important than any explanation for the mechanisms responsible for them. Secretly, the geographers would think cospeciation was due to geographic isolation, and the orthogeneticists would think it was due to the inherent orthogenetic drive. So long as comparisons of hosts and geography and parasites were all congruent, everyone could be happy.

As we mentioned above, Brooks's discussion of coevolution had two elements. The second reflected cases in which new pathogen-host associations were established when a pathogen colonized a distantly related host (Kellogg's "stragglers"). Colonization events could be detected by areas of disagreement between pathogen and host phylogenies (fig. 3.3).

The major problem for the maximum-cospeciation perspective was

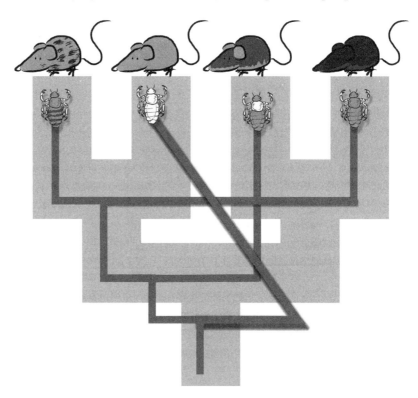

FIGURE 3.3 A different phylogenetic tree for our hypothetical group of lice infecting our hypothetical group of rodents (from fig. 3.1). If we superimposed the two phylogenies, the branches would not align, indicating that the hosts and parasites did not cospeciate. Rather, the parasites speciated in a context that involved moving to a host that was not their original one.

simple: when scientists began to compare pathogen and host phylogenies, they found little evidence of cospeciation.[20] Comparisons of parasite and host phylogenies undercut the assumption of the orthogeneticists about the universality of cospeciation; parasite and host phylogenies rarely show 100 percent agreement, and often far less than 50 percent. The news for the geographers was only slightly better—the geographic distributions of parasites and hosts still sometimes showed common speciation events caused by geographic isolation affecting both hosts and parasites, even when the host and parasite phylogenies did not match. However, the geographic school also suffered a serious blow: geographic isolation was not the only factor involved in the distributions of related species. In fact, there was evidence that isolation was episodic and that between episodes there was usually substantial movement; furthermore, that movement was often associated with the evolution of new species that did not conform to a simple geographic cospeciation scenario.[21]

The complement of cospeciation, called *coaccommodation*, concerns the evolution of the various traits associated with the nature of a given pathogen-host association (the phylogenetic element of the Vershaffelt-Dethier tradition).[22] In this type of study researchers seek to identify evolutionary episodes in which pathogens achieve novel host associations and then attempt to understand how they were achieved. This requires that we link two things. First, we need to distinguish episodes of association by descent from those of association by colonization.[23] Secondly, we need to ask whether each of the episodes of association by colonization of new hosts is linked with the emergence of novel traits. You might wonder how we could figure all that out without a time machine allowing us to visit the past. The answer is that genealogical inheritance is so conservative that every organism is a kind of time machine, carrying with it ancient inherited information. By the time of the phylogenetics revolution, two different models of coaccommodation had appeared. The more general of the two is called *resource tracking*, or *sequential colonization*. One of four things can happen to a pathogen when it attempts to colonize a new host. First, it may not succeed. Secondly, it may colonize the new host species and maintain its identity because the new host possesses the same resource as the ancestral host.[24] As far as the colonizer is concerned, the new host is simply more of the same habitat—we humans see that the pathogen has added a host, but from the pathogen's perspective it is occupying more of the same habitat. Thirdly, the pathogen may initially colonize the new host species because it possesses the same resource as the ancestral host but may then diverge evolutionarily anyway (perhaps owing to geographic isolation).[25] Finally, and least commonly, the pathogen may colonize a host species representing a novel resource. If the colonizer already possesses a trait that allows it to access the novel resource, it may establish itself on the new species. If the colonizer does not speciate, it has simply expanded its resource base. If the colonizer does speciate, we have the emergence of a new pathogen species using a new resource. A special case of this scenario is called *phylogenetic tracking*.[26] In that scenario, the evolutionary sequence of events is as follows: a new host species evolves, characterized in part by an evolutionarily modified form of the required resource. This new species of host is then colonized by individuals from the pathogen associated with the ancestral host. Some of the pathogen individuals adapt to the new form of the resource, eventually producing, in their turn, a new species.[27] And so the cycle continues. In order for this to occur, the ancestral host species must persist after the speciation event that produces descendants bearing the modified resource; otherwise, the ancestral associate species would have no resource base to support it through the colonization phase. This dynamic would give rise to cases of apparent cospeciation in which the pathogens were nonetheless younger than the hosts (this would later be rebranded as *delayed cospeciation*). The second class of models encompasses what are now called *coevolutionary arms races*. These are refinements of the coevolution models by Mode and by Ehrlich and Raven. The primary assumption in the arms-race models is that coevolving associations are maintained by mutual adaptive responses (mutual association with mutual modification). Variations on this theme have similar beginning assumptions: (1) pathogens reduce the fitness of their hosts; (2) hosts that, by chance, evolve traits that make them resistant to the pathogens (defense mechanisms) will increase their fitness relative to their undefended brethren, and the new defense mechanism will spread throughout the species; and (3) some pathogens will, in turn, evolve a counterdefense allowing them to survive in association with the previously protected hosts. If this trait confers a fitness advantage on those individuals (e.g., through access to an unused resource base), the counterdefense mechanism will spread throughout the pathogen species. The pathogen species will now enjoy unhampered dining on the previously protected members of the host species, until the cycle begins anew. We have documented many cases in which insects that feed on specific plants possess the ability to detoxify specific secondary chemical compounds sequestered in plant tissues.[28] But in most cases it is uncertain whether either of the traits evolved in the context of the association. As Dethier pointed out in the 1940s and 1950s, insects may be attracted to plants having compounds that the insect can already detoxify. Also, because insects are often able to "unfriend" a host that turns out to be unpalatable, how could there be any evolutionary mechanism forcing the insect to associate with an unpalatable host in the hope that the insect could produce offspring that could detoxify the host or forcing the insect to stay in association with a host if it evolved a toxic secondary compound?

Given that cospeciation is far from common, we must assume that many pathogen-host associations are established when pathogens add new hosts or change hosts. But the great majority of pathogens have small host ranges, often only a single host. So there is little evidence of cospeciation, yet most pathogens are associated with a small number of host species. How can a pathogen be so restricted in terms of host range and yet evolve by changing hosts? The traits that enable pathogens and hosts to maintain specialized associations must be connected in some way with how they change host allegiances. If that is true, however, it would seem that all of the traits associated with the specialized relationship must be unique to each evolutionary episode. That is, each host switch must be accompanied by "the right adaptation showing up at the right time." The way to investigate that inference is to ask whether any of the specializations persist through changes in associations—that is, whether they are phylogenetically conservative. This would seem to be the case—after all, if we say "malaria" most readers will immediately think "mosquito" (transmission dynamics) and "blood cells" [and possibly "liver"] (microhabitat preference), even if the precise species of malaria, the precise host, and the country of origin are not specified. Clinicians of all kinds have accumulated such knowledge anecdotally for centuries under the rubrics "diagnosis" and "control," taking their successes for granted and rarely thinking that evolution was the source of their insights.

In 1980 Dan Janzen wrote the first of a series of insightful articles asking "when is it coevolution," in which he noted that simple association between pathogens and hosts (herbivorous insect in Janzen's case), no matter how predictable and exclusive, was not, by itself, evidence of mutual evolutionary modification worthy of being called true coevolution.[29] Five years later he coined the term *ecological fitting* to explain cases in which a given pathogen occurred in association with one host in one locality and a different one in a second locality. Janzen felt that this meant that while most pathogens might be ecologically specialized, they still possessed flexibility in host use.[30] In between those articles, Janzen and Paul Martin published an article discussing what they called *evolutionary anachronisms*. To them, an evolutionary anachronism was a persistent ancestral trait that may have functioned in a particular way in the past but no longer served that function owing to extinction. Their exemplar was large thorns on some neotropical trees, which they postulated had served as a defense against being eaten by Gomphotheres, members of an ancient and honorable, but extinct, group of large herbivores related to elephants.[31] The lack of communication between ecology and systematics at that time meant that Janzen never saw that ecological fitting in ecology and coaccommodation in systematics might be pointing at the same thing; "anachronisms" indicated phylogenetic conservatism in specialized ecological traits, which could be a source of ecological fitting.

In Europe, mistletoe (parasitic plants of the order Santanales) seem to love trees standing in or near flooded ditches. An arms-race scenario might assume initially that waterlogged trees must be at higher risk of infection, their defenses having been weakened. That initial weakness would somehow provoke a coevolutionary response in the host that would then lessen the cost of being infected. But what if heavily watered trees are healthier than their water-stressed relatives, and thus more attractive to mistletoe, and also more able to afford to be infected? If they can pay the freight, so to speak, it's not a disadvantage. If this is the case, it may be coevolution, there may even be mutual modification, but there is no arms race. Alternatively, what if mistletoe infection is a lottery, infected trees simply being ones that happen to be visited by birds that disseminate mistletoe seeds, who choose trees living near water because the birds need water too? If so, then the fact that many trees are uninhabited while some are heavily infected may not mean anything about the details of the interaction between the pathogen and the host species; rather, it may be more a matter concerning the details of the transmission dynamics of mistletoe itself. And if the number of infected trees killed by heavy mistletoe infection is sufficiently small, mistletoe-induced mortality might occur, but not often enough for selection to play a role in mitigating that cost.

This range of possible explanations for a seemingly simple plant-plant association takes us back to the Verschaffelt-Dethier school of thought, which recognized that the chemicals produced by—and stored in—plants may be attractants to some and repellents to other insects. Insects that are attracted to the scent of secondary compounds they can metabolize and repelled by those they cannot face a far easier time making decisions about food and oviposition sites than those attracted to plants containing noxious secondary compounds, requiring the evolution of detoxification mechanisms. From our perspective, we are uncertain how pathogens and hosts can afford the continual conflict implied by coevolutionary arms races when simply avoiding hosts that are, or have become, noxious is a cheaper option. What links these observations to coaccommodation is the sense of evolution's taking the line of least resistance, leading to resolution of conflicts of interest rather than intensification and prolongation of conflict. If evolution is a process of selective diversification, it must be fundamentally a process of conflict resolution in which both of the parties in conflict manage to persist.[32]

In order to make some progress in this area, we need to be more analytical about the evolution of ecological specialization. By the early 1980s a research program emerged using phylogenies to link ecological specialization with phylogenetic conservatism.[33] Through such research we discovered that phylogenetic conservatism in functional traits of all kinds, including those associated with transmission dynamics and microhabitat preference for pathogens, is by far the norm and not the exception.[34] Parenthetically, this supports the traditional Darwinian view that "the right adaptations at the right time" are traits that were already in existence before the "need" arose. The candidate for integrating these emerging problems in coevolutionary studies was the geographic mosaic theory of coevolution (GMTC).[35] The GMTC is based on some standard ideas: that the world is composed of specialists and generalists, and various evolutionary phenomena produce different proportions of each; that only specialists participate in coevolutionary interactions; and that coevolutionary interactions act to limit host ranges either by progressive resource specialization (resource tracking) or mutual modification (coevolutionary arms races). The GMTC improves upon previous proposals, however, by suggesting that not only can specialists evolve from generalists, but they can, with the help of geography, evolve from other specialists. Coevolution can occur when a generalist parasite is distributed among multiple hosts—each representing a different resource—in different geographic locations, where differential selection produces new specialist parasites associated with each different host. Or it can occur when a pathogen is restricted to a single host, but that host is isolated geographically into populations that are subjected to different selection pressures. In that case each host species is in reality a mosaic of coevolving populations. The GMTC allows but does not require reciprocal evolution of defense and counter-defense traits, and it specifies an element of geographic isolation leading to diversification. But that diversification does not need to be consistent with any cospeciation scenario.[36] It is still, for our purposes, incomplete.

The phylogenetics revolution did not produce a new theory of pathogen-host evolution, but it cast doubt on all contemporary alternatives, leaving a conundrum: how can associations be so specialized as to require the pathogens and hosts to be together geographically long enough for concomitant geographical speciation, and yet still be able to switch hosts and expand geographically, as virtually all studies of pathogen-host systems have shown? All major explanatory frameworks of the twentieth century agreed on one thing: host-range changes (increases or alterations) should be difficult to achieve. We had theories suggesting that ecological specialization might allow parasites to move around geographically but would rarely allow them to switch hosts; we had theories that said ecological specialization would not allow parasites to move around much geographically and would rarely allow them to switch hosts. But phylogenetic studies indicate that most associations are established by some form of host colonization, and that the key traits allowing pathogen-host associations to persist and evolve are both highly specialized and phylogenetically conservative.

How wonderful that we have met with a paradox. Now we have some hope of making progress.[37]

This is the parasite paradox: parasites are ecological specialists and yet (a) specialists and generalists do not occur equally, specialists predominating, even though specialists evolve from generalists; (b) there are many emerging diseases occurring more rapidly than seems reasonable if they require chance evolution of novel host-use capabilities; and (c) phylogenetic comparisons routinely show high levels of host-range changes in the evolutionary diversification of pathogen-host associations.[38]

The parasite paradox is particularly germane for understanding emerging diseases. Pathogens ought not to be doing what they are doing, but they do. This affects humanity. Negatively. A lot. Resolving the paradox is critical for studies of, and sustainable responses to, the emerging disease crisis. We have expected pathogens to be extreme ecological specialists, locked in coevolutionary arms races with their hosts that should progressively decrease the chances that they will be able to switch to new hosts readily. In other words, the coevolutionary process itself should be our best firewall against emerging disease. And yet, emerging diseases are occurring in real time and more often than can be accounted for by random mutation of genetic information that fortuitously allows parasites to infect novel hosts. Phylogenetic studies told us that solving the conundrum was going to be complicated, and would require integration among different viewpoints, each of which wanted to be self-sufficient as an explanatory framework but clearly was not.

Meanwhile, on the health front, technology was advancing rapidly, theoretical ideas were coalescing, money was being spent in ever-increasing amounts on "disease," and yet it was all beginning to fall apart. A well-known adage in competitive tennis is "Never change a winning game, always change a losing game." By now it should be clear that we are playing a losing game with respect to dealing with disease. So we have to admit there is something lacking in our ideas about how the world ought to be, figure out what that lack is, and bring our reason and our experience together to forge a new paradigm and new means of coping with emerging disease. Otherwise we are lost.

A New Framework

By the end of the twentieth century we knew more and understood less than ever about pathogen-host associations and how to cope with them. We knew the shortcomings of the twentieth-century paradigms but did not have a unified alternative paradigm. We had documented parallel lines of research in evolutionary biology, characterized by a belief that each group had a self-contained and complete explanatory framework. More often than not, however, the groups simply appeared to be unaware of each other. In some cases, the research groups reacted to each other antagonistically, to everyone's detriment. In modern parlance, the study of pathogen-host systems was heavily "stovepiped"—highly specialized and without any mechanism for generalizing results across research groups or any sense that different research groups could benefit from collaboration and division of labor.[39] Against all expectations, we have achieved a hopeful breakthrough, which we call the Stockholm Paradigm, integrating a number of existing research programs: historical ecology; ecological fitting and sloppy fitness space; the geographic mosaic theory of coevolution; the oscillation hypothesis; and the taxon pulse hypothesis.[40] In the next four chapters we will introduce and explain the conceptual glue that binds these research programs into the Stockholm Paradigm and show how they contribute to understanding the nature of the emerging disease crisis and, more important, insights into what we can do about it.

Our joy at being part of the group that has achieved a synthesis that eluded scientists for more than a century is tempered by the implications of that synthesis for a world caught in the throes of a major climate-change event. While we were busy trying to sort out these tangled conceptual issues, health professionals were finding that the end of the twentieth and beginning of the twenty-first centuries were far more challenging than they might have imagined.

Technological advances in microbiology made possible detailed examination of the evolutionary and ecological context of microbial pathogen-host associations and comparison with what we knew about the evolution of larger pathogens, such as ticks and tapeworms. The first phylogenetic studies of rust fungi, for example, supported earlier claims that plants and their rust fungi do not cospeciate.[41] The resulting explosion of studies of microbial evolution, the beginnings of genetic engineering, and the focus provided by the coevolutionary arms-race framework provided substantial optimism to those engaged in the fight against disease. The emerging field of parasite immunology found a comfortable fit in the 1960s with the arms-race idea. Evolutionists might see the arms race as a means of long-term conflict resolution and diversification, but epidemiologists and clinicians saw it more literally. For them, we are in a war against some elements of biodiversity that are attacking us, our children, and the plants and animals upon which we depend for life and livelihood. The desired end point, therefore, is their elimination. "Medicate, vaccinate, eradicate" continued to seem a productive way to achieve that goal. But the warning signs that had emerged in the second half of the century were becoming more significant. Little did anyone know that the fight would soon hang in the balance and we would be facing possible defeat.

Technical difficulties in producing effective new medications that emerged early in the twentieth century continued into the early twenty-first. Melarsoprol B, still the only effective drug against African sleeping sickness, was introduced by the end of World War II. Other antiparasitic agents such as suramin and pentavalent antimonials became available in the 1920s, and pentamidine isothionate and sodium stibogluconate appeared in the 1940s. Since then, the development of new antiparasitic treatments has stagnated, not only because of the complexity of protozoan and helminth pathogens, but also because of a low economic incentive for diseases that mainly affect people in developing countries. Increased global travel and exchanges, as well as severe parasitic infections in immune-compromised patients, have led to renewed interest in antiparasitic therapy. Parasitic diseases remain an affliction of hundreds of millions globally.

Many drugs were initially developed in veterinary medicine after around 1980.[42] Antiparasitics target the parasitic agents of the infections by destroying them or inhibiting their growth. Broad-spectrum antiparasitics are drugs that are efficacious in treating a wide range of parasitic infections. At the present time antiparasitics remain a cornerstone of control and potential eradication programs. They are the rationale for the development of programs for mass drug administration. Initial approaches to development of antiparasitics led to the near-eradication of the Guinea worm, *Dracunculus medinensis*, and the agent of river blindness, *Onchocerca volvulus*.

The aim of preventive chemotherapy is to avert the widespread morbidity that invariably accompanies helminth and other infections—and sometimes leads to death.[43] Early and regular administration of the anthelminthic drugs recommended by the World Health Organization (WHO) reduces the occurrence, extent, severity, and long-term consequences of morbidity, and in certain epidemiological conditions it contributes to a sustained reduction in transmission. By the beginning of the twenty-first century, pharmaceutical researchers understood that there was an evolutionary basis to the issue of increasing resistance to antibiotics. If antibiotic resistance is an example of natural selection in action, what is called the Red Queen hypothesis suggests that attempts to produce technology to outpace the emergence of resistance will be futile.[44] The more and stronger the antibiotics with which we challenge microbes, the stronger the selection for resistance, and the faster the resistance occurs. Our experience bears this out.

Major challenges are seen in the continued development of resistance and loss of efficacy for many of these compounds (such as widespread benzimidazole resistance seen in ruminant and equine nematodes, and expanding resistance to ivermectin, once considered the magic bullet for control of ectoparasitic arthropods and nematodes). Veterinary drugs following the trend of five to ten years between drug introduction and the emergence of resistance, established with penicillin more than half a century ago, include the phenothiazines (1940), benzimidazoles (1961), avermectins (1980), and macrocyclic lactones (1991).[45] Furthermore, only 1 percent of new drugs developed between 1975 and 1999 were antiparasitics, which raised concerns that insufficient incentives existed to drive the development of new treatments for diseases that disproportionately affect low-income countries. Commercial sources are producing fewer medications, and resistance to the ones we have continues to evolve.

Phylogenetics has been recruited to help speed the search for new versions of medications whose efficacy was being curtailed as a result of evolutionary processes. For example, one of the most significant recent contributions to cancer therapy was the discovery of a substance called taxol, produced by a tree species called the American yew. When the therapeutic properties of taxol became known, botanists using phylogenetics suggested that a close relative of the American yew, the European yew, should be examined, and in short order the related anticancer agent Taxotene was discovered. A phylogenetic analysis of coneflowers and their relatives was used to suggest which species are most closely related to, and thus most likely to possess therapeutic properties similar to, *Echinacea*.[46] Such research continues to prove fruitful; a recent study showed that the closest relatives of those species of *Cinchona* already known to have high concentrations of quinine in their bark also have high concentrations.[47] But none of this was enough. Not sustainable.

Despite attempts to medicate our way out of parasitic disease, parasites continue to exert a substantial impact on human well-being and food security among, for example, domestic ruminants and fishes in aquaculture.[48] Not sustainable.

Recognizing this continuing challenge led to new public sector and public-private partnerships, including investment by the Bill and Melinda Gates Foundation. From 2000 to 2005 twenty new antiparasitic agents were developed or in development. Despite those efforts, a new antimalarial cost approximately $300 million to develop in 2005, with a 50 percent failure rate.[49] Not sustainable.

More people and more crops and livestock are medicated than ever before. But in proportion to the risk, are we moving ahead, holding our ground, or slipping backward?

Vaccination efforts in the last quarter of the twentieth century also lagged, for three reasons: (1) the easy vaccines had been developed, and even the significant technological advances to that point were not sufficient to speed up the process; (2) the "who is going to pay for this/who is going to receive the benefits of this" socioeconomic syndrome plagued the development of new vaccines as well as new medications; and (3) too many new pathogens were emerging, each one inducing a demand for a vaccine.

Like medicating, vaccinating continued the twentieth-century trend, becoming more time-consuming and more expensive to produce owing to unanticipated genetic variation, diversity, and complexity among pathogens. Despite technological advances,[50] *une maladie, un vaccin* was overrun by too many new pathogens emerging at the same time, making it too costly and time-consuming to vaccinate our way out of the problem. In addition, most of the risk groups lived in countries too poor to pay for mass vaccination programs, and countries that could afford them saw no reason to do so.

Targeted destruction of elements of biodiversity essential to disease transmission has had some successes. Rinderpest was declared to be globally eradicated in 2011. Foot-and-mouth disease eradication protocols in Europe have worked so well since the 1960s that mandatory vaccinations were banned in 1992, raising hopes that this pathogen might, like smallpox and rinderpest, be eradicable. If so, it will join a very select list indeed. Overall, the approach has become problematic owing to conflicts with agendas related to climate change and biodiversity protection. Targeted biodiversity destruction for food production enhances health and reduces susceptibility to disease but has become problematic in the same way, reminding us that the various challenges posed by climate change are interconnected and synergistic. And yet, as we began losing ground with respect to medication and vaccination, eradication seemed more and more attractive; the benefits of eradication are seen by some to be worth the risk of public condemnation. In September 2006 the WHO declared its support for the indoor use of DDT in African countries, where malaria remains a major health problem, claiming that benefits of the pesticide outweigh the health and environmental risks.

Now we come to modern awareness of global climate change, within which the crisis of emerging diseases is inextricably embedded. This includes concerns about human population and food production; about habitat alteration affecting human habitations, agriculture, and biodiversity; and about industrialization and pollution. Farsighted people saw all these factors coming together in a very unpleasant way and recognized that it all had to do with "growth," particularly the notion that the solutions to all our problems was "more growth." In 1972 the Club of Rome commissioned a group of mathematical modelers to project humanity's future if we continued along those lines—what they called the business-as-usual scenario. They published their results in a book called *Limits to Growth*, which issued some grave warnings about the future.[51]

Summary

The twenty-first century linked all the socioeconomic problems addressed by *Limits to Growth* with those involved in coping with known diseases and the new phenomenon of emerging diseases. We believe it was no coincidence that the emerging disease crisis imposed itself on our awareness as the twentieth century drew to a close and the twenty-first began, when the long-predicted global climate-change impacts began to manifest themselves.[52] At a time when we thought we would be celebrating our dominion over the Earth, climate change poses the greatest challenge our species has ever faced. The scope of the problem has been underrecognized owing mostly to professional specialization and bureaucratic territoriality. The situation in which we find ourselves, characterized by a lack of communication and collaboration among research programs with common interests, has its roots in academic sins of omission, namely preoccupations about research specialization and personal agendas. Current and widely held concepts of specificity or narrow host range imply that pathogen-host relationships should not change readily. But they do. Our ability to anticipate and mitigate the impacts of disease emergence catalyzed by climate change depends on achieving a better understanding of complexity in the biosphere, in particular its fundamentally evolutionary nature.

Let's get started.

FOUR

Back to the Future

We're just recycled history machines.—Jimmy Buffett, *DON'T CHU-KNOW*

Kellogg assigned Darwin's original formulation to the historical dustbin of ideas superseded by modern thought.[1] Much of the telling of evolution since then has viewed Darwin's *On the Origin of Species* as the foundation for, even the foreshadowing of, the predominant twentieth-century evolutionary framework, neo-Darwinism.[2] That framework is an approach of elegant simplicity, in which evolution is reduced to a trinity of "variation, selection, adaptation." As a result, discussions often begin with the final sentence of the sixth chapter of *Origin of Species*: "Hence in fact the law of the Conditions of Existence is the higher law; as it includes, through the inheritance of former adaptations, that of Unity of Type," suggesting that Darwin believed natural selection was the law of the Conditions of Existence.

Paradoxically, many of the same discussions criticized Darwin's writing style, wondering why it took him 195 pages to get to the point. Had Darwin been smart enough, or had he had a good enough editor, the point should have been the climax of his fourth chapter, the one specifically on natural selection, rather than this conclusion two chapters later. That would have avoided the misleading impression that while natural selection was a significant part of the process, it was not *the* process. There is a different perspective on this harsh assessment of Darwin's writing abilities. Effective scientists write according to a fairly standard formula, described to us as, "Tell them what you're going to tell them; tell them; tell them what you told them."

If Darwin was following that formula, the statement above occurs at a point where the author is summarizing what he has told his readers. His "Tell them what you're going to tell them" should be near the beginning of the book, prior to all that empirical evidence, perhaps in the second paragraph of the main text:

There are two factors: namely, the nature of the organism and the nature of the conditions. The former seems to be much more the important; for nearly similar variations sometimes arise under, as far as we can judge, dissimilar conditions; and, on the other hand, dissimilar variations arise under conditions which appear to be nearly uniform.[3]

Most historians consider this passage little more than a general repudiation of Lamarckism. However, we believe it is fundamentally important. From the beginning, Darwin proposed that evolution was an outcome—in today's terminology, an *emergent property*—of asymmetrical interactions between two causal agents, each of which has its own properties relatively independent of the other, producing outcomes that are not readily predictable from knowledge of the properties of either one alone.

Darwin's conception of the nature of the organism was explicit: it is in the nature of the organism to produce offspring; to produce offspring that are similar but not identical to each other; and to be able to act in their own behalf. The most important aspect of all this is that these aspects of the nature of the organism obtained regardless of the nature of the conditions (the environment). Without a high degree of autonomy from the nature of the conditions, there would not be more organisms born than there are resources available for them to survive ("reproductive overrun") and thus no struggle for survival. This creates what is known as "Darwin's necessary misfit" between organisms and their environments,[4] what Darwin envisioned to be the usual conditions of existence.

While nineteenth-century philosophers were fascinated with ways in which organisms seemed perfectly to "fit" their surroundings, Darwin was too good a naturalist to be fooled by such wishful thinking. He recognized that the key to understanding evolution stemmed from the ways in which organisms were able to persist despite *misfits* between their surroundings and themselves. Finding such misfits to be common and universal, Darwin then recognized that not all viable members of all species were able to cope equally well with the nature of the conditions in which they found themselves. Nonetheless, generation after generation, all species produced offspring exhibiting varying degrees of misfit with their surroundings, and in far higher numbers than could be sustained by environmental resources. This, Darwin reasoned, must lead to a struggle for survival on the part of those organisms proportional to their degree of misfit. When the inherent overproduction produced variety in traits critical for survival, organisms possessing traits that were functionally superior in that particular environmental context would survive best. Whenever an environment changed, those organisms that already had the functions necessary to survive in the new environment would do so, whereas those who lacked them would not; what is good today might not be good tomorrow.

Natural selection was the *result* of the conflict created, because the conditions of existence included the autonomy of the nature of the organism as well as the nature of the conditions. And the outcome of natural selection was the partial or complete elimination of those variants that were so "misfitted" that they could not survive. Selection is a local, immediate, and intimate phenomenon, acting on existing conditions, not previous ones. And thus, "natural selection is the indirect action of changed conditions."

The issue is not "how do organisms survive in the current environments?" but rather, "how do organisms survive in changing environments?" This means that natural selection was not the higher law, it was a *consequence* of that law. The final paragraph of the sixth chapter of *Origin of Species*, and not just the final sentence, clearly shows Darwin's perspective:

It is generally acknowledged that all organic beings have been formed on two great laws—unity of type and the conditions of existence. . . . On my theory, unity of type is explained by unity of descent. The expression of conditions of existence . . . is fully embraced by the principle of natural selection. . . . Hence in fact the law of the Conditions of Existence is the higher law; as it includes, through the inheritance of former adaptations, that of Unity of Type.[5]

Darwin's "higher law" governs the total conditions of existence, which he defined as the sum of interactions between the nature of the organism and the nature of the conditions.

Darwin proposed two rich visual metaphors to help readers understand his bold vision—the tree of life and the tangled bank. From our perspective, these metaphors encapsulate the fundamental complexity of biological evolution, reinforcing Darwin's framework of great scope and generality:

As buds give rise by growth to fresh buds, and these, if vigorous, branch out and overtop on all sides many a feebler branch, so by generation I believe it has been with the great Tree of Life, which fills with its dead and broken branches the crust of the earth, and covers the surface with its ever-branching and beautiful ramifications.[6]

The tree-of-life metaphor is more than an accounting scheme; it is a symbol of a major part of the evolutionary process. Living systems are not only capable of acting in their own behalf; they regularly take the initiative, using what they have inherited. Metaphorically, the present is the state in which biological systems create their own futures based on their own pasts. Organisms carry so much of their history with them that most explanations for their appearance and function stem from their past.

In Europe a sycamore is a maple (*Acer pseudoplatanus*), while in North America a sycamore (*Platanus occidentalis*) is what Europeans call a plane tree (*Platanus orientalis*). Darwin's metaphor of natural classification as a phylogeny enables us to understand why North American sycamores and European plane trees resemble each other so closely, why their ecological preferences are so similar, and why they are able to hybridize so readily, despite having such dissimilar common names. Specific points of origin in space and time play integral roles in explaining the properties of species and the organisms that make them up, most important how they interact with their surroundings, including other species.[7] Darwin was explicit about the significance of a phylogenetic tree:

As it is difficult to show the blood relationship between the numerous kindred of any ancient and noble family even by the aid of genealogical trees, and almost impossible to do so without this aid, we can understand the extraordinary difficulty which naturalists have experienced in describing, without the aid of a diagram, the various affinities which they perceive between the living and extinct members of the same great natural class.[8]

Darwin's phylogenetic tree metaphor contrasted with a progressive view of diversity embodied in the *scala naturae*, in which "lower forms" were replaced by "higher forms." Thus, the only illustrated metaphor Darwin ever provided in any edition of *Origin of Species* specifically underscored the notion of selective accumulation, rather than replacement, of diversity. Darwin was well aware of what that accumulation of diversity looked like:

It is interesting to contemplate a tangled bank, clothed with many plants of many kinds, with birds singing on the bushes, with various insects flitting about, and with worms crawling through the damp earth, and to reflect that these elaborately constructed forms, so different from each other, and dependent upon each other in so complex a manner, have all been produced by laws acting around us. These laws, taken in the largest sense, being Growth with reproduction; Inheritance which is almost implied by reproduction; Variability from the indirect and direct action of the conditions of life, and from use and disuse; a Ratio of Increase so high as to lead to a Struggle for Life, and as a consequence to Natural Selection, entailing Divergence of Character and the Extinction of less improved forms. Thus, from the war of nature, from famine and death, the most exalted object which we are capable of conceiving, namely, the production of the higher animals, directly follows. There is grandeur in this view of life, with its several powers, having been originally breathed by the Creator into a few forms or into one; and that, whilst this planet has gone circling on according to the fixed law of gravity, from so simple a beginning endless forms most beautiful and most wonderful have been, and are being evolved.[9]

This metaphorical statement evokes visions of complex ecosystems produced by selective accumulation of diversity. It also explicitly underscores Darwin's view that natural selection is an emergent property, reinforcing his conception of the law of the conditions of existence.

So, at the end of the *Origin*, Darwin "told us what he told us," reflecting back to the second paragraph of the book, where he "told us what he was going to tell us," and in the 425 or so intervening pages, he "told us." The tree of life, embodying the conservative nature of inheritance in the nature of the organism, explains why we find species doing similar things living in dissimilar conditions. The tangled bank, embodying the fact that biodiversity occurs in complex ecosystems comprising many species, explains why we find species doing dissimilar things while living together in similar conditions.

Darwinism was highly unusual for a nineteenth-century theory. Darwin suggested that evolution was not so much a process as an outcome of the interaction of two classes of phenomena (the nature of the organism and the nature of the conditions), each following its own rules but nonetheless spatially and temporally entwined. Furthermore, the two classes of phenomena were not coequal—one was more important than the other. And yet, the "greater" phenomenon inevitably creates conflict, while the "lesser" leads to conflict resolution.[10] We think this is the reason so many scientists of his day readily accepted his fundamental propositions but almost immediately began to try to "fix" the theory. Darwin did not postulate a predictable set of outcomes for a given set of conditions under the external control of a general and powerful natural law, as expected for acceptable scientific theories in the mid-nineteenth century. In modern parlance, we would say that Darwinism was an early (perhaps the earliest) example of a complex systems theory.

Breaking the myth of a perfectly adapted biosphere was one of Darwin's priorities: what kind of well-oiled machine routinely produces so much waste? Natural selection was a perfecting mechanism, but perfection was never achieved, and that was a good thing for species coping with changing environments. Restoring the myth of perfect adaptation was the priority of twentieth-century evolutionary biology. Kellogg wrote that Lamarckians, orthogeneticists, and neo-Darwinians "all agreed" that the core issue for evolutionary theory was explaining how the right adaptation shows up at the right time.[11] As we noted before, Cope focused on a key issue: if evolution is *survival* of the fittest, how do you explain the *origin* of the fittest?[12] But by the dawn of the twentieth century, the Darwinian notion that what was necessary for survival must already be present—in low numbers, in the background or on the margins, but present—or extinction would result was not a sexy enough answer. Biologists craved something more direct, more powerful sounding, more heroic, mythic, even magical. This craving was not new, as Whitman suggested in 1899:

The pouting instinct is supposed to have arisen de novo, as an anomalous behavior, and with it a new race of pigeons. . . . The incubation instinct was supposed to have arisen after the birds had arrived and laid their eggs, which would have been left to rot had not some birds just blundered into "cuddling" over them and thus rescued the line from sudden extinction. How long this blunder-miracle had to be repeated before it happened all the time does not matter. Purely imaginary things can happen on demand.

Protests notwithstanding, twentieth-century biologists continued to build their heroic, mythic edifice. But none of their variations on a simple and seemingly all-encompassing framework anticipated the emerging disease crisis.

The Nature of Capacity

In a Darwinian system, the capacity for life—the nature of the organism—flows from inherited information. Transmitting that information from the present into the future gives living systems a distinctly historical nature. The capacity to reproduce, develop, mature, and reproduce again produces both variable and conservative outcomes; we are all unique and yet we all have family resemblances. And some of those family resemblances are ancient indeed—the fact that all the cells in our bodies have a nucleus surrounded by a membrane is a family resemblance so old that it embraces a billion years of family history. This is the reason Darwin asserted that similarities among species are overwhelmingly due to common ancestry, not to living in common environments.

Another capacity of inherited information is the maintenance of life. Organisms use inherited information to impose themselves on their surroundings, extracting matter and energy necessary to persist long enough to reproduce. The inherited information determines the kinds of matter and energy needed for survival (a rosebush and a horse have different needs even if they are living in the same environment) and how they are extracted from the environment. Inherited information determines the capacity to act in one's own behalf, regardless of the conditions—in other words, the capacity to make the best of a bad situation. Some organisms survive despite the fact that they live in conditions in which they do not thrive. And that leads to an important insight about the nature of life.

Inherited information allows organisms to be both *explorers* and *exploiters*.[13] All organisms exploit the usable parts of their environments as much as possible and in so doing explore those environments. Through reproduction and inheritance, each organism also explores its genetic possibilities. Depending on its inheritance and the situation in which it finds itself, every organism is capable of oscillating between exploiting and exploring its surroundings. And no matter how much life at any one point in time is biased toward either exploitation or exploration, organisms that mostly do the one never lose the ability to do the other. It seems to be in the nature of the organism to engage in the maximum amount of exploitation with minimal loss of capacity to explore, to the extent allowed by inherited possibilities and limitations.

Functional Capacity: Ecological Fitting

Evolutionary and ecological explanations for traits associated with specific functions tend to stress increased efficiency, especially in exploiting resources necessary for survival. As Darwin noted, such explanations come with a caveat: if you are too specific in your needs, preferences, or abilities, you could be at risk for extinction if the conditions of life change in such a way that those specific functions no longer allow survival. Evolvable life is able to cope with changes—small and large—in the environment. *Ecological fitting* is the active expression of this functional capacity.[14] It lies at the heart of the parasite paradox, being the key to understanding how pathogens can have very specific ecologies yet still be able to change their geographic distributions and hosts readily when the conditions of life change. Three classes of phenomena contribute to ecological fitting.[15]

Co-option: How New Kinds of Associations Arise

One of the major criticisms aimed at Darwin was the perceived absence of transitional stages in either the fossil record or in living species. If natural selection operates through a series of small, gradual changes, where are the transitional stages? One of Darwin's answers involved changing the function of an existing structure.[16] In some cases this occurred when two organs performed the same function, thus setting the stage for selection to increase the efficiency of one, then modify the second, and now superfluous, organ for a different function. Darwin believed that this type of dynamic was "an extremely important means of transition" in evolution. In other cases, an organ that served a major and a minor function was modified to serve the latter at the expense of the former. In either such process, the starting conditions might be obscured and the transitional stage missing.

Darwin also discussed traits that served no apparent function because they arose as the by-product of evolutionary processes other than natural selection (e.g., the "complex laws of growth" or the "mysterious laws of the correlation of parts"). He believed that such traits were an important part of organismal evolution because they might eventually acquire a function that was necessary for survival in a new environment: "But structures thus indirectly gained, although at first of no advantage to a species, may subsequently have been taken advantage of by its modified descendants, under new conditions of life and newly acquired habits." In each case imagined by Darwin, important evolutionary change was accomplished without creating new structures.

In the first half of the twentieth century, the French biologist Lucien Cuénot championed the term *preadaptation* for traits of little or no importance at their point of origin that nonetheless played a critical role down the evolutionary road by allowing organisms to take advantage of new conditions. Unfortunately, his mechanism for how shifts in functionality could occur was a somewhat confusing amalgam of Darwinian and Lamarckian themes: "One could say . . . that the need and the organ create function; in the individual functioning changes the organ most efficiently through the effect of usage, and finally selection intervenes to eliminate descendants that cannot handle the new conditions."[17]

This confusion over mechanism clouded his central message: traits arising under one condition could allow the transition to a new environment or way of life without the need to evolve new structures. Cuénot spent 30 years expanding the definition of preadaptation to include the co-option of useful traits to serve a new function. He was also a member of the first generation of modern geneticists, inspired by Gregor Mendel, who attempted to delineate how the units of heredity worked. Many of these early researchers focused on the seemingly random process of mutation, arguing that adaptation was an incidental or accidental by-product. From the preadaptation perspective, this created a paradox: if the production of preadaptations was a random process, why did adaptation appear to be so organized with respect to the environment? Cuénot grappled with the paradox but never managed to answer it in a satisfactory manner.

Neo-Darwinians of the day disagreed about preadaptation. Dobzhansky dismissed it, writing that "'preadaptation' is a meaningless notion if it is made different from 'adaptation.'"[18] This ignored Cuénot's point that preadaptations might, after their function had been changed (co-opted), appear to be adaptations to the current environment, when in fact they were not; their origin and raison d'être predated the current role they fortuitously found themselves playing. Richard Goldschmidt reduced preadaptations to random micro-mutations, arguing that only organisms with "chance hereditary mutant combinations for life under changed conditions" could survive in a fluctuating environment.[19] But he echoed Cuénot's proposal that individuals preadapted in this way might enter an empty niche and there survive and propagate, eventually producing new species.

George Gaylord Simpson elaborated this latter theme, returning to the Darwinian notion that organisms carried within them a storehouse filled with deleterious and neutral mutations accumulated over time, any one or combination of which, under changing conditions, might allow them to move from one adaptive zone (e.g., living in the water) to another (living on land). Once in the new zone, the population would rapidly adapt to it. He thus believed that the preadapted trait is not adaptive with respect to the environment in which it originated, but that it eventually becomes adaptive in the new zone. He concluded that such traits were "of tremendous importance because they afford an explanation of quick, radical shifts in adaptive types."[20]

A decade later Simpson wrote that biologists were using the term *preadaptation* in nine different ways, eight of which assumed that the process was adaptive at all stages, and one of which supposed a transition from nonadaptive to adaptive.[21] Simpson's categories were foreshadowed by Darwin's and Cuénot's initial recognition that preadaptations originated in two different ways: (1) a neutral or deleterious trait acquiring an adaptive function in a new environment, and (2) a useful (adaptive) trait changing function. The confusion, then, surrounds the origin of the trait: was it initially adaptive or not? Regardless, the outcome of a preadaptive trait was always the same: evolutionary change happened rapidly because the material for change already existed, for whatever reason, in the organism. There was general agreement about the importance of preadaptation, coupled with dissent about which one of the two possible pathways was more important to the production of a preadapted trait in the first place.

The glaring problem with the word *preadaptation* is that it implies there is a direction or purpose to evolution (to make the trait adaptive), when in fact it is impossible to predict with any degree of accuracy the future of a biological character based on its current state. Natural selection is an intimate affair, a process of the here and now with ramifications for the future that we can recognize only in hindsight. Stephen J. Gould and Elisabeth Vrba eliminated the specter of evolutionary foresightedness by replacing *preadaptation* with the concept of *exaptation*.[22] The process by which the trait switches function is called *co-option* (Gould and Vrba used the correct form, *co-optation*, but it never caught on), a general term encompassing all possible processes. Co-option is critical to Darwinism because without it, it is often impossible to explain dramatic movements into new environments or the sudden appearance of "new" traits without unintentionally invoking a Lamarckian "need" for the trait that led to its origin. As Darwin and Cuénot proposed, the prior existence of a trait permitted the transition from one environment to another.

Persistent traits co-opted to perform novel functions is common in interspecific associations.[23] For example, co-option explains how dipterans, hymenopterans, and coleopterans have become parasitoids,[24] and co-options featured prominently in the evolution of plant-herbivore and plant-pollinator interactions among *Dalechampia* (Euphorbiaceae) species.[25] Deciduousness in oaks originated as a mechanism for coping with seasonally dry habitats but turned out to function equally well in seasonally cold habitats;[26] in fact, many of the adaptations to living in xeric conditions among plants bear close resemblance to the adaptations to living in cold conditions.

Co-option is key to understanding the origins of pathogen-host systems. Without co-option, we are left with a variety of almost magical scenarios, most of which involve waiting for a fortuitous mutation to arise. If preexisting traits are capable of performing new functions that allow new conditions to be utilized, we have a relatively cheap and nonmystical mechanism for achieving the evolutionary transition. Because pathogens require hosts, it seems reasonable that hosts existed before pathogens arose. It also seems reasonable that the ancestors from which each group of pathogens arose were themselves not pathogens. Did those ancestors evolve an entire suite of traits specific to being pathogens, or did preexisting traits allow them to become pathogens through co-option?

The life-history traits of parasitic flatworms, for example, do not differ from the life-history traits of their free-living relatives, indicating that these species do not have a parasitic mode of life, but rather a platyhelminth mode of life, which was co-opted to function in a host-parasite context.[27] It is easy to see that the basic conditions of life with which a free-living flatworm had to contend, feeding on decomposing material in low-oxygen conditions in the mud flats of a tidal marsh, might not be that different from those it encounters living in the intestine of a vertebrate. If that free-living ancestor evolved a novel kind of thick skin that helped it fend off other decomposition-feeding organisms in its surroundings, members of that ancestral species might well have been protected when eaten by a vertebrate, allowing it to survive the new challenges posed by life in the vertebrate's intestine. Such a seemingly simple episode of co-option probably set the stage for the diverse group of trematodes and tapeworms that taxonomists now call the Neodermata (literally, "new skin").

Phylogenetic Conservatism: How Associations
Persist and Proliferate

When Darwin wrote that similarities between species should generally
be assumed to be the result of common ancestry, he underscored the
fundamentally conservative nature of inheritance. You expect a litter
of kittens to be kittens, not puppies; mammals, not birds; vertebrates,
not invertebrates; animals, not plants; and so on. From this it follows
that if the capacity for ecological fitting stems from the flow of inherited information, most of it will be due to historical legacy.

Just how conservative are the functional traits that provide species with the capacity to survive and persist? The noted entomologist
Herbert Ross proposed that approximately one in every 30 speciation
events in a variety of insect groups was correlated with geographic dispersal from the primitive climatic zone to a derived one, or with shifts
from the ancestral condition to any subsequent state in reproductive
behavior, ecology, or host preferences.[28] He concluded that such shifts
were consistent with, but much less frequent than, phylogenetic diversification. Furthermore, he believed there were no predictable patterns
explaining the shifts that did occur and suggested that such transitions
constituted a biological "uncertainty principle." By the beginning of
the twenty-first century phylogenetic studies had uncovered ample
evidence of conservative and specific resource use supporting Ross's
view.[29]

We have already seen that evolutionary conservatism plays a role
in co-option and the origin of associations such as pathogen-host systems. Now we will see that this conservatism plays a role in maintaining a diversity of such associations. Evolutionary conservatism may
lead to a situation in which two or more closely related pathogen species share common aspects of their biology inherited from a common
ancestor, though they may live in different places and inhabit different
host species.

Back to our "think malaria" comment. No matter what species is
involved and no matter where on this planet it occurs, all malaria is
transmitted by mosquitoes. Clinicians have long made use of this reality without understanding, or at least without acknowledging, that it
is fundamentally a matter of evolutionary conservatism. It allows us to
engage in unified efforts to combat malaria worldwide by eradicating
the elements of biodiversity (species) associated with mosquito production. But there is a cost to this element of historical conservatism that

is often unrecognized. Evolutionary conservatism produces ecological homologues, species associations that could conceivably replace each other or be interchanged in ecosystems without a fundamental change in trophic structure. Ask any six-year-old child to describe what the rabbits in their countries do—how they behave, what they eat, who eats them, where they make their homes—and you will get a very similar response. And it will be a correct response in each country—all rabbits share many aspects of their biology in common because they inherited those capacities from a single common ancestor. This may seem insignificant until you see rabbits from one country settling into another, competing with the native animals that have very similar preferences and requirements. Now replace "rabbits" with "malaria," and "country" with "host."

We can now see the significance of evolutionary anachronisms, traits that evolved in an environmental context that no longer exists. Rather than representing a wasteful evolutionary cost,[30] they have a particular value when it comes to ecological fitting. Because they are traits that functioned well in past environmental conditions, anachronisms allow species that retain them to survive if anything like those ancient environmental conditions recur. And the fundamentally conservative nature of biological inheritance guarantees that evolutionary changes are often slower than environmental changes.

Plasticity: Making the Best of Bad Situations

Not only are organisms able to act in their own behalf, they are flexible in the way they go about doing it. Most organisms are capable of doing different things in different situations. This is the built-in flexibility of organisms postulated by Darwin, more commonly referred to today as *plasticity*.[31] Plasticity becomes important in situations in which a pathogen—by chance or by choice—finds itself in a less than ideal situation.

Pathogens have specific host preferences and fine-tuned sensory responses to their environments when it comes to finding hosts. Hosts, however, are largely indifferent to the aspirations of pathogens. As a consequence, for most pathogens finding a preferred host is something like participating in a lottery with a time limit. Scientists who believe pathogens are "locked in" on one or a few optimal hosts to the exclusion of all others assume that most pathogens do not find one of the preferred hosts, die, and are thus eliminated from evolution. This reduces pathogen biology to a brute-strength approach: simply blast the environment with enormous numbers of offspring (buy *many* lottery tickets) and enough will find preferred hosts to produce the next generation.

Plasticity changes the rules of the find-a-host lottery in the pathogen's favor. As Darwin envisioned, selection is absolute for those who cannot find what they need to survive and reproduce. Survival is paramount in evolution, but surviving does not necessarily require thriving. Pathogens with the capacity to survive in suboptimal hosts are good examples of evolutionary plasticity. Plasticity thus allows individual pathogens to provide a positive answer to the question, what happens if your preferred host isn't available?

Ecological Fitting and Traits Specific to Pathogen-Host Systems

Despite differences in life-history traits among all pathogens, it seems there are no differences in evolutionary outcomes regarding their associations with hosts. That means higher order dynamics supersede life-history evolution—the various components of ecological fitting. Recognizing the significance of this for understanding pathogens begins 80 years after *Origin of Species* was published, with an article by the Russian Aleksandr Aleksandrovich Filipchenko, one of the first true parasite ecologists. Filipchenko analyzed different nineteenth- and early twentieth-century definitions of parasitism and disagreed with using a notion of "harming the host" to define the phenomenon. To his mind it was impossible to define one object using characters of another. In this case "harm" refers to the host, because the host is damaged by parasites. Thus, harm is not a character of the parasite itself; it is a characteristic of parasitized hosts. Filipchenko underlined that parasitism is an ecological phenomenon and defined parasites as organisms whose ambient medium was another organism (the host). Thus, parasites may be considered as an ecological group of organisms—for example, aquatic organisms, whose ambient medium is water. He suggested that Hegner first introduced this approach in parasitology,[32] but the ideas were developed more fully by Valentin Alexandrovitch Dogiel and his colleagues in what was then the Soviet Union.[33]

This perspective leads us to seek evidence for ecological fitting in traits associated with *transmission dynamics* (how they find a host) and with *microhabitat preferences* (in what part of the host they live). For traits associated with these aspects of pathogen biology to exhibit potential for ecological fitting, they should be both highly specific and evolutionarily conservative.

Transmission Dynamics

Biologists have examined two fundamental components of pathogen transmission dynamics. The first is ecological, encompassing (1) the sequence and number of hosts,[34] (2) the type of host in which each stage develops,[35] and (3) the site within the host in which each stage develops.[36] The second is developmental, concerning stages leading to reproductive amplification of infective forms that increase the likelihood of infection in the ecological setting in which a particular pathogen resides. There is no doubt that every pathogen species exhibits highly specific ecological and developmental traits associated with its transmission dynamics.[37] Phylogenetic analysis has consistently shown us that those traits are also evolutionarily conservative.[38]

Microhabitat Preferences

Now we examine the other critical element of evolutionary survival and persistence: microhabitat preferences. Scientists have asked why such preferences should be so specialized. After all, just finding a host in the first place is something of a lottery, so why not utilize the whole host once you get there? And indeed, in a number of microbial infections, bacteria, viruses, or fungi may be found widely dispersed throughout a host's body when the host is acutely diseased. But in most cases, the pathogen resides in a particular part of the host. There is no shortage of theories about why this should be: (1) pathogens are extreme resource specialists, and preferred or required resources may occur in only limited parts of the host (a subset of this argument is that pathogens find microhabitats where they can hide from potential host effects, such as immune responses); (2) every pathogen that finds a restricted place in the host reduces the chances of competition with other pathogens; and (3) every pathogen that lives in a predictable and restricted site has a better chance of finding opportunities for reproduction.[39] More than likely, all of these factors are in play in different proportions in particular cases,[40] because all have one thing in common—they help the pathogen cope with its conditions of existence and thus can be acted upon by selection. The more specific and restricted the nature of the conditions, the smaller the target selection has and the more effectively it can act.

Just as transmission dynamics are specific enough to serve as clinical diagnostic features, so are microhabitat preferences. Among the digeneans producing disease in humans, *Schistosoma* species occur in the circulatory system, *Fasciola hepatica* and *Clonorchis sinensis* occur in the liver and bile ducts, and *Paragonimus* species occur in the lungs.

Microhabitat preferences are also evolutionarily conservative. For example, the rivers of South America east of the Andes are home to more than 30 species of stingrays that make up the family Potamotrygonidae. Like all elasmobranchs (sharks, skates, rays, and their relatives), potamotrygonids have intestines that appear to be very short compared to those of other vertebrates. If you examine the interior of the intestines, however, you discover a spiraling architecture that creates multiple compartments and a lot of space for food processing and pathogen habitation. One of the groups of pathogens inhabiting those stingrays is several species of the tapeworm genus *Potamotrygonocestus*. Members of each species of this genus live only in the anteriormost chamber of the spiral intestine of whichever stingray they inhabit.

Finally, microhabitat preferences are not linked to cospeciation. Different species of paramyxoviruses (causal agents of measles, mumps, Newcastle disease, rinderpest, canine distemper, and human parainfluenza) often infect hosts that are not closely related, yet each one always resides in homologous tissues of the hosts they do infect. Likewise, host colonization by viruses may be facilitated by phylogenetically conserved receptors on cells, and by the capacities of viruses to use those receptors.[41]

The world of pathogen *capacity* is populated by species possessing traits that are specific, conservative, and flexible. Exploitation is thus maximized without restricting exploration capabilities. In short, pathogen capacity is associated with a wide range of ecological fitting options. We next examine what the world of pathogen *opportunity* looks like.

The Nature of Opportunity

The nature of capacity stems from the nature of the organism; so, does the *nature of opportunity* stem from the nature of the conditions? In large part it does. Darwin, however, used the term to refer to the context in which the opportunity for survival occurs for any given species, and that is an outcome of the interaction between organisms and their environments. Organismal capacities to exploit and explore their environments allow, but do not guarantee, survival. This is because neither capacities nor the environments in which they can function are unlimited. We call the arena of opportunity "function space."

The members of each species use a small but predictable portion of function space. Inherited capacities provide specific ways that organisms impose themselves on the environments where they happen to find themselves. This is because inheritance produces highly similar offspring regardless of the conditions. Each species produces more offspring in each generation than the function space available for their survival. For each species, the portion of function space in which the differential survival takes place is *fitness space*. Fitness space contains all organisms that survive the struggle and reproduce; in technical terms, they have positive Darwinian fitness, whereas all others have zero fitness and thus are not represented in fitness space.

Simply having the capacity to live in a given fitness space, however, does not lead to evolutionary persistence and diversification. Environments change, and fitness space changes with them. If natural selection filtered all offspring down to only those whose capacities closely and narrowly matched opportunity, the only outcome of changing conditions would be extinction. We might envision the *operative environment* of a pathogen species to encompass everything defining a host as a viable resource. But at any given time, the pathogen is unlikely to inhabit all possible hosts in all possible places; therefore, its *current operative environment* will be a subset of its overall operative environment. This means that no matter how specialized the pathogens, they will always, given the opportunity, have some ability to survive in novel operative environments, that is, susceptible hosts not inhabited in the current operative environment. Higher proportions of some kinds of offspring rather than others may survive and reproduce in different operative environments, but even if those produce viable offspring, not all of them are as fit as their parents.

Darwin felt it was significant that organisms were capable of surviving in less than optimal conditions.[42] This suggests that even though fitness space for each species is a subset of function space determined by the nature of the organism, it is expansive. Within fitness space, not everything that could happen does happen. Not everything that does happen occurs all the time, or everywhere it could, or in equal proportions. And finally, not everything that happens is optimal for the conditions of the time and place. Where organisms thrive is only a subset of where they can survive. And since it is in the nature of the organism to explore as well as to exploit, variation allows organisms to investigate novel operative environments presented by the circumstances in which they find themselves.

Opportunity can thus act as both an evolutionary filter and a facilitator if current operative environments are subsets of all possible operative environments. No matter how narrow selection is in any particular place in association with any particular host, evolution will occur within *sloppy fitness space*,[43] a by-product of direct selection under some other set of conditions (the ancestral environment). The sloppier the fitness space, the more room there will be for species to alternate between exploitation and exploration, so the greater the chances are that some of their offspring will survive in some form. During periods of environmental stability, we tend to see an accumulation of members of a species that are exceptionally good at exploiting those very predictable surroundings. But during periods of environmental change, the members of the population that are better at exploring will survive. So long as the conditions fluctuate between stable and changing, the frequency distribution of relative exploiters and relative explorers within a species population will fluctuate as well.

This reveals an interesting insight about evolution. Life itself creates opportunities for yet more life; in today's biosphere, the environment of every species—pathogen or not—is composed primarily of other species,[44] and therefore the nature of the conditions is determined, to a great extent, by the capacities of other species. If fitness space is sloppy, there is an overwhelming amount of function space available for pathogens, because every time a new host species evolves it creates even more possible fitness space for pathogens. This is why there are so many pathogen species on this planet.

Sloppy fitness space can be visualized thusly:[45] Consider a bivariate world whose dimensions define all possible combinations of two resources, X and Y (fig. 4.1). At a given point in space or time, the local or *realized* conditions are defined by the subset of X and Y that exists. For an evolutionary unit (genome, species, even a group), its *fundamental* fitness space is all combinations of X and Y for which positive fitness can be achieved. *Realized* fitness space is that portion of fundamental fitness space accessed to meet the realized conditions. The extent to which fitness space fits any particular set of realized conditions will be a function of time and selection pressures, but the sum of all conservative elements of inheritance, called *evolutionary lagload*,[46] ensures that the fitness space will always remain in large part a function of retained history. The imperfect fit between fundamental fitness space and realized fitness space at any given point in time or space is propor-

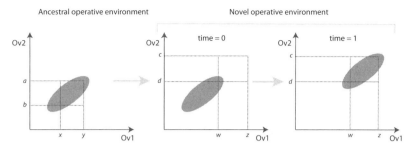

FIGURE 4.1 Graphical representation of a hypothetical fitness space and the relationship with changing environmental conditions represented by two variables (Ov1 and Ov2). From left to right, despite the change in environmental conditions (i.e., the change on the rectangle describing the variation of the environment), the species illustrated in this example does not go extinct locally, thanks to the overlap of its fitness space with the conditions of the new environment (or novel operative environment). Over time (time = 1) and as the species is subjected to new selection pressures, the fitness space of the species will accommodate the new local conditions.

tional to how sloppy the fitness space is. From environmental perturbations on a regional or global scale to small-scale perturbations such as the local extinction of a host species, it is this sloppiness that provides the essential ability of life to cope with rapidly changing conditions (fig. 4.1a) and to explore new options across the physical landscape (fig. 4.1b). Following disturbance or dispersal and depending on patterns of reproductive isolation, wandering through sloppy fitness space both facilitates evolutionary diversification by exposing organisms to novel conditions and promotes evolutionary stasis if there are many ecologically fit populations experiencing contradictory selection pressures and significant gene flow.

Fitness space is inherently sloppy. Even if you are specialized, your realized fitness space (RFS) is a restricted portion of your fundamental fitness space (FFS). If your RFS is specialized, then whatever selective forces are impinging on you, they will not affect other elements of your FFS, so FFS cannot be optimized tightly. If your RFS is generalized, it is a relatively large and more diverse portion of FFS. That means you will be responding to multiple selective forces, so again your FFS also cannot be tightly optimized. The good news is that if it is the nature of FFS to be sloppy, it should always be a source of evolutionary potential, complementing the indefinite variation produced by inheritance that is part of the nature of the organism. At the same time, fitness space will be sloppy in proportion to how much of FFS is RFS. All the factors we have discussed above suggest that RFS should always be smaller than FFS, and a recent study has confirmed that this is empirically true.[47]

Surprisingly, this Darwinian perspective makes it difficult to imagine an effective coevolutionary arms race on more than the local scales envisioned by the geographic mosaic theory of coevolution.[48] In other words, if there is an arms race locally, it will not affect the pathogen or the host globally, and any interactions that affect all associations cannot be a restricted case of mutual modification, so there cannot be an effective arms race beyond the local association. This is why strong coevolutionary interactions cannot limit the emergence of disease during climate change.

Opportunity Space for Pathogens

Studies of the evolution of pathogen-host systems tend to focus on hosts and geography. Yet to this point we have mentioned neither. This is because geographical location and hosts are not inherited capacities of pathogens. They comprise opportunity space for pathogens, elements of their fitness space. As the founders of pathogen ecology (Hegner and Filipchenko in particular) noted, pathogen fitness space is made up of connections in geographic space, host range, and specific ecology related to transmission dynamics. Fundamental fitness space is made sloppy by discontinuities in two factors. First of all, geography, hosts, and trophic structures have inherent discontinuities. Secondly, if, as we have noted before, the nature of the organism is paramount, the need to exploit the local surroundings for survival outweighs the need to explore, and this will tend to produce multiple localized occurrences of pathogens. Selection tends to favor exploitation, which will tend to restrict realized fitness space but increase fundamental fitness space. In biological terms, this means that the occurrence of the pathogens will be "patchy."

Geography is the most fundamental element of opportunity. It is the matrix that carries hosts and the ecology of transmission and is key to both connecting and isolating fitness space. We can estimate this by documenting as many features of the known geographic locations for the pathogen (part of its realized fitness space) as we can, then determining where else on the planet such conditions exist. Complementing this is the recognition that pathogens must navigate organized ecological diversity in order to transmit their offspring from one host to another. This is the bridge between geography and hosts—here we are asking where in the world we might find a suitable ecological structure in which a given pathogen could survive. If we know the transmission dynamics of the pathogen, we can determine where such an ecological structure occurs. Combining the geographical and trophic information using *ecological niche modeling*,[49] we can determine much about the fundamental fitness space of a given pathogen. The greater the mismatch between where the pathogen *does* exist and where it *could* exist, the sloppier will be its fitness space. So we can also use ecological niche modeling to assess just how sloppy fitness space is for any given pathogen. And the sloppier the fitness space, the greater the chances will be that a given pathogen will be able to survive changes in its surroundings.

For pathogens, the resources necessary to maintain life and produce the next generation come packaged in other living organisms. Hosts have geographic distributions and are embedded in organized ecological structures, so host-based pathogen fitness space is a subset of geographical-ecological fitness space. But because hosts are the entities that are "diseased" by pathogens, we focus more on them. In this case, we estimate the pathogen fitness space through phylogenetic comparisons. Hosts have the resources necessary for parasite survival. If we know the phylogenetic distribution of those resources, we can determine all the species that could host the pathogen. And once again, the mismatch between known hosts (realized fitness space) and potential hosts (fundamental fitness space) tells us how sloppy fitness space is. Studies performed in laboratory settings have demonstrated that at least some pathogens are capable of surviving in many different hosts, including many with which they never associate in nature.[50]

The phylogenetic distribution of host resources falls into four classes (fig. 4.2). For three of those classes, the resources have a common evolutionary origin, so their occurrence in any given organism or species is the result of inheritance from a common ancestor; they are *homologous*. In the most extreme case, the resources evolved in a single host species. Traits that are not ancestral are called *apomorphic*, and apomorphic traits restricted to a single species are called *autapomorphies*. So for this class we would say the host resources are autapomorphic (fig. 4.2a). In the second class, the host resources are shared by an inclusive group of related species; such groups are called *monophyletic* because they are the collective descendants of a single ancestral species. The traits that diagnose such groups are novel traits that arose in the common ancestor, so they are apomorphies; and they persist (sometimes in modified form) in its descendants, so they are called *synapomorphies*. The host

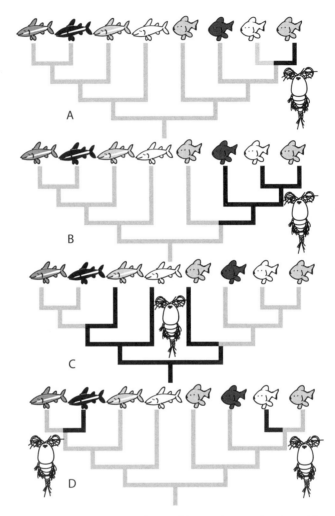

FIGURE 4.2 Parasites are resource trackers. In this example, a species of parasitic copepod explores distinct numbers of host species (fishes) that express those features considered resources to the parasites (identified in the phylogeny by the color black). (A) Among the host species, only one species presents the necessary resource to the parasite. (B) The resource required by this parasite evolved in an ancestral host species and is expressed in all its descendants. Hence the parasite species can use all of these host species. (C) The resource is plesiomorphic (i.e., is present ancestrally in the host lineage depicted in the figure) and subsequently lost in both branches. The parasite is found in basal species that retained the resource present in the ancestral fish species. (D) The resource is found in two distantly related host species as a result of independent evolution.

resources for this class are synapomorphic (fig. 4.2b). The third class of homologous distribution of resources encompasses groups of species sharing resources that are persistent ancestral traits, called *plesiomorphies*. If the plesiomorphic trait evolves within the host group, then the hosts that retain the ancestral trait do not, on their own, form a monophyletic group, even though they inherited their trait from a common ancestor (fig. 4.2c). Finally, resources that have evolved independently in more than one species are called *homoplasies*. In such cases, the host species do not form a monophyletic group, nor do they share the resource traits as a result of inheritance from a common ancestor (fig. 4.2d). Sadly, we rarely can depict the full phylogenetic distribution of the required host-based resources because we rarely know what they are. All is not lost, however.

If a 1:1 correspondence between pathogen and host phylogenies means cospeciation, disagreement between the pathogen and host phylogenies should indicate "not cospeciation." But the idea of cospeciation is intuitively appealing to many biologists and is a deeply embedded principle in certain research programs. It should come as no surprise, then, that methods have been developed to support interpretations that any evidence of disagreement between host and pathogen phylogenies represents a "cryptic" case of cospeciation, so long as the hosts involved are closely related (in other words, we have no evidence for the missing pathogen species that would give us a 1:1 correspondence, but we "know" they must be there). This is known as *cophylogeny analysis*.[51] Cophylogeny as cryptic cospeciation is controversial because the number of ad hoc assumptions needed to support such inferences can become increasingly large: how many "missing species" do you need to invoke to create a perfectly congruent phylogenetic tree for pathogens and their hosts? The Stockholm Paradigm, however, allows us to see cophylogeny analyses in a more productive light. What if episodes of increasing host range and apparent host switches are evolutionary manifestations of the potential for sloppy fitness space to facilitate pathogen survival when host opportunity changes? In such cases, cophylogeny studies might serve as analogs of ecological niche modeling, giving us an estimate of host-based fundamental fitness space by estimating the potential host ranges for a group of pathogen species.

Let's consider an actual case. The Intermountain West of North America is a tumultuous landscape of towering isolated mountain ranges and plateaus, with mosaics of forests, grasslands, and deserts. The area also teems with chipmunks, small and perpetually busy striped rodents belonging to the genus *Tamias*, subgenus *Neotamias*. Chipmunks are related to squirrels in an ancient family of rodents (Sciuridae) whose history links Eurasia and North America.[52] Chipmunks have been accompanied on their evolutionary journey by characteristic pathogens, among them nematodes called pinworms. Pinworms do not have a complex life cycle involving multiple hosts, but their transmission dynamics are specialized in another way: infection occurs among parents and offspring only in natal burrows. As adults, pinworms live in the cecum near the junction of the small and large intestines, although females may migrate along the large intestine to the rectum to lay eggs that will be dispersed into the environment. Human pinworms, *Enterobius vermicularis*, also live primarily in the vestigial cecum we call the appendix, with females migrating to the anus to disperse eggs. This is the reason there are occasional reports of appendicitis caused by large numbers of pinworms packed into the appendix, and the reason pediatricians and knowledgeable parents watch for children scratching in an indelicate manner as a sign of pinworm infection.

Rauschtineria eutamii is a chipmunk pinworm. It comprises six genetically distinct lineages infecting ten species of *Neotamias* (19 of the 23 species have been examined) distributed among members of three of the five recognized parasite subgroups.[53] Cophylogeny analysis based on 17 species of *Neotamias* (fig. 4.3) found no evidence of cospeciation, despite the fact that the parasite lineages are one another's closest relatives and all the hosts are members of one subgenus. The specialized transmission dynamics and microhabitat preferences attest to the specialized ecological nature of *R. eutamii*, so we cannot assume that it is a cryptically generalized species in any sense except its host range.

We conclude that the parasite is specialized on some aspect of western chipmunks involving traits that are persistent ancestral characteristics of the host group (see fig. 4.2c). The RFS of the parasite is the 10 documented hosts, whereas the FFS likely includes at least all 23 species in the group. As anticipated, the FFS is larger than the RFS,[54] suggesting that, given the opportunity, this parasite could readily colonize twice as many chipmunk species as it currently does.

Rauschtineri eutamii living in different hosts in the same area are more closely related to one another than they are to members of their own species living in the same hosts in different places. So the parasites are more localized geographically than their hosts, and the known host range is a function of hosts periodically expanding their geographic ranges and then becoming isolated. *Neotamias* has experienced a complicated history of recurring geographic expansion and isolation during the last 2.75 million years,[55] and the cophylogeny analysis

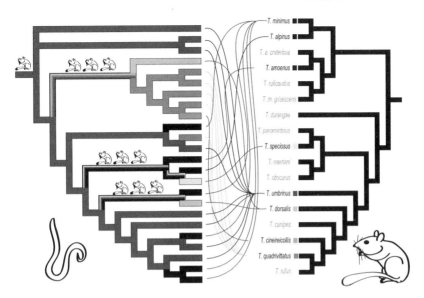

FIGURE 4.3 Cophylogeny of pinworms (*Rauschtineri eutamii*) and western chipmunks (species of *Eutamias* [*Neotamias*]). The host tree shown here contains 17 of the 23 species of *Neotamias* for which sufficient data were available, 10 of which host *R. eutamii*. Parasite data for eight of those hosts were analyzed (those for *T. rufus* and *T. canipes* were not included in the study). The diagram, called a *tanglegram*, depicts parasite associations with hosts connected between their respective phylogenies. Each of the subgroups of *R. eutamii* originated from one or more host range expansions, beginning with different species of chipmunks in each case. Parasites are distributed among three of the five species-groups of western chipmunks.

documented at least three episodes of host range expansion correlated with geographic range expansion.

At this point one might assume that the host range would be a subset nested within the ecological possibilities that would be a subset nested within the geography. But in the real world, things are not so simple. Parasites can't always live everywhere their hosts can live—fitness space for the hosts may exclude some parasites and include others; and on short time scales and small spatial scales, there may not be any parasites. This has led some to believe that the success of some introduced host species is due to their landing in an "enemy-free zone." The reverse is also true: many parasites have geographic ranges larger than the range of the preferred host(s), and there will always be "local" pathogens capable of engaging with any introduced host. Once again we see that sloppy fitness space provides means of both filtering and facilitating pathogen-host systems. As a consequence, the relationship between pathogens and geography is complex, as are the relationships among pathogens, hosts, and geography.

There are additional, more subtle elements of pathogen fitness space —things like salinity, temperature, humidity, seasonality, and changes in the abundance of hosts associated with those variables. Cryptic genetic variability (accumulated low-frequency alleles) and plasticity also add to the picture. Pathogen fitness space is indeed large and sloppy.

The very term *sloppy fitness space* invites divergent reactions. For some, it is reassuring to think that, no matter how specialized species become, there is always room to maneuver, room for innovation, room for escape. If it also implies some degree of disorderliness in the operation—well, that's life. Others, however, find the term disquieting. They find comfort in the idea that every species occupies a particular place in the order of things, and they wish to believe that evolution makes that place a very snug fit indeed. They thus use terms like *optimal* and *fittest* to refer to the state in which they wish species to exist in their surroundings. And yet, if the nature of the conditions changes, the species that were fittest in the previous conditions are doomed to extinction.

Most important, sloppy fitness space seems to resolve Darwin's paradox, because opportunity is both filter and facilitator. For this to be true, in the first place, capacity must always be greater than opportunity, and we know this is true because selection occurs. Unless capacity is greater than opportunity, there can be no selection, no matter how specific and constrained capacity appears to be—because history retained (plasticity, co-option, conservatism) is not always history expressed. We also know that existing capacity is substantial, that the ability to store and use previous capacity is large, and that the system from which new capacity emerges is one capable of producing indefinite variation. In the second place, opportunity must be changeable. And finally, capacity must be able to take advantage of changes in opportunity—the role of *ecological fitting*.

The astute reader is beginning to hear alarm bells: knowing what a species is exploiting at any given place and time is not enough to anticipate its future activities. We may talk about *removing* opportunity (e.g., eradicating the part of biodiversity associated with transmission) in disease control, while in reality we are only *changing* opportunity—and if we are lucky, we change opportunity in such a way that the pathogen cannot take advantage. But our track record so far suggests that the pathogens are cleverer than we are (of course, many of them have had more than a billion-year head start to experiment with trial and error), so when opportunity changes, pathogens emerge in unanticipated ways.

The Nature of Evolution: Ecological Fitting in Sloppy Fitness Space

The operational aspect of capacity is ecological fitting, which encompasses traits of the organism that are both specific and variable as a result of conservatism, co-option, and plasticity. This is what allows species to "awesome up" when the right opportunities present themselves. The operational part of opportunity is fundamental fitness space, defined as everywhere on the planet that the capacities of any given species would result in nonzero fitness. Fundamental fitness space is larger than realized fitness space (the fitness space that is actually occupied), the difference being proportional to how sloppy fitness space is. The interaction of ecological fitting in sloppy fitness space thus defines the arena of evolution in which each variable can serve as both filter and facilitator, depending on the situation. Once again, *it is never just one thing, and it always depends.*

Organisms that survive in sloppy fitness space are capable of exploring and exploiting, of generalizing and specializing to an extent made possible by inherited characteristics and depending on the contingencies of the conditions of life in which they find themselves. How they accomplish this at any given place and time depends not only on capacity but also on opportunity. In fact, Darwin suggested that while the nature of conditions did not cause evolutionary innovations, changes in the nature of conditions seemed to allow such innovations in a general way: "Indefinite variability is a much more common result of changed conditions than definite variability."[56]

And since each species is not subjected to the same selection pressures at different places in its range, or at different periods of time within a given place, only species that are capable of ecological fitting will survive. All organisms are oscillators, with the built-in ability to generalize and specialize in their sloppy fitness space according to the conditions. It's a built-in feature of life and creates a bias toward persistence that is its own reward, regardless of relative costs.

We tend to think that "specific and predictable" means "limited options." But Darwin realized that the dualistic interaction between the nature of the organism and the nature of the conditions could produce predictable and specific outcomes without sacrificing options. That is because, in our terms, capacity is always greater than opportunity. And that is a good news/bad news situation. The bad news is that this means there will always be reproductive overrun and conflicts of interest among the members of each generation. The good news is that information is not distributed randomly, optimally, or maximally in fitness space. It is distributed functionally, and function is part of the nature of the organism. Information space is thus structured but variable—a system of indefinite variation. And while it is variable, it is far more conservative. That means there will be conflicts even if fitness space is sloppy; but the offspring that survive such conflicts will be variable because fitness space is sloppy. And what a pathogen is able to do, given the opportunity, can be as important as, or even more important than, how well it performs in explaining reactions to climate change. Changes in the nature of the conditions are best understood by knowing the nature of the organism. *Survival in new environments allows, but does not require, new capacities specifically fit to that environment* (fig. 4.4).

Organisms as oscillators exploit and explore according to the intersection of ecological fitting and sloppy fitness space. They will tend to be biased toward exploitation, because survival always requires exploitation but—when conditions are highly stable and predictable—need not also require exploration.

Exploitation-biased activities will tend to restrict realized fitness space and will be facilitated by discontinuities in fitness space. For a pathogen species, the ultimate exploitation-biased existence would be a species occurring in a single place, with a single host, and transmitted in a single way. These are the conditions under which opportunity acts as a filter, and such things as strong coevolutionary interactions would be favored if they arose, especially if they occurred in geographically isolated situations.[57] A species living in such conditions would be highly specialized within realized fitness space. Because fitness space has been assumed to be highly optimized and limited, some believe that host changes could not occur without prior specific mutations permitting it. But given the enormous potential capacity provided by plasticity, co-option, and evolutionary conservatism, strong local coevolutionary interactions may restrict realized fitness space locally without affecting fundamental fitness space.

Exploration-biased activities, by contrast, occur when the nature of the conditions changes in such a way that connectedness among elements of fitness space increases, amplifying opportunity. In such circumstances, opportunity acts as a facilitator. Exploration can be as subtle as reproduction, but that is apt to be mostly unsuccessful if available

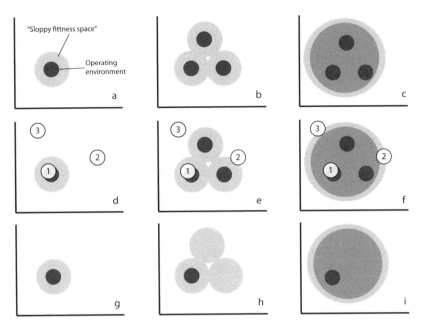

FIGURE. 4.4 A schematic representation showing two dimensions of the operative environment associated with hypothetical host resources. Sloppy fitness space can allow fitness in an area outside the operative environment to which the parasite is adapted (light gray circles). Panels a, d, and g illustrate a specialist, adapted to a single host resource; panels b, e, and h illustrate a polyspecialist, which has adapted independently to three different resources; and panels c, f, and i illustrate a true generalist, with a more general host recognition and tolerance system that allows it to utilize any resource that falls within the dark grey area. The open circles in panels d–f represent three novel resources. The specialist in panel d can colonize resource 1, which is more or less identical to the ancestral resource, but not resource 2 or 3. The polyspecialist in panel e can colonize resource 1, but also resource 2, which falls within its sloppy fitness space. And the generalist in panel f can colonize all three resources. In panels g–i only one host is available, so all three species of parasites will appear to be specialists, but their ecological and evolutionary potential will be very different. Redrawn and modified from Agosta et al. (2010).

fitness space is occupied and no new connections can be made. But if there is always reproductive overrun, there will always be offspring produced that are not the fittest. Add to that plasticity, co-option, and phylogenetic conservatism, and you have a system in which no matter how limited the realized fitness space is at any one place or time, capacity will always be greater than opportunity.

Successful exploration-biased activities allow the realized fitness space of the species to expand, so the pathogen becomes more generalized in fitness space. This can be manifested in many ways—the pathogen may remain with its customary hosts and use its customary transmission dynamics, but the geographic range of all species involved can increase. The range of hosts used may increase. The number of modes of transmission may increase. And, of course, there may be a combination of factors.

Ecological fitting is a key element in exploration. If specific environmental cues and resources are widespread, or if traits can have multiple functions, then the stage is set for the appearance of ecological specialization and close (co)evolutionary tracking, without loss of the ability to establish novel associations. Exploration also sets the stage for new capacities to evolve; it is not the result of earlier evolution of new capacities. Virtually every species of pathogen, for example, has some ability to perform and persist on hosts other than those with which they are customarily associated. In addition, some of the inherited flow of information allows members of pathogen species to persist at the margins until their time comes, and newly accessible fitness space gives them access to novel hosts with which they may establish a more successful relationship. We now discuss three examples to give you an idea of how this capability works in the real world.

The popular name of *Fascioloides magna* is the giant liver fluke. Immense by the standard of most digeneans, *F. magna* is native to North America, where it is primarily a pathogen of various species of ungulates (hoofed livestock, deer, and their relatives). Adults live in the liver, and their eggs pass out of the host in the feces. Some wind up in the water, whereupon a larval form—the miracidium—hatches from the egg and burrows into a snail. A complex series of developmental and reproductive events then occurs in the snail, producing large numbers of larval clones called cercariae that swarm from the snail into the aquatic environs. The cercariae alight on aquatic vegetation (as well as on the exoskeletons of crayfish and the shells of mollusks), where they encyst and hope that a free-ranging cervid or domestic ungulate will chow down on the vegetation and provide aid and comfort to the next generation of *F. magna*. This pathogen has been introduced into Europe, where it also lives as adults in the livers of herbivorous mammals and is transmitted by snails, which shed cercariae that encyst on the surface of aquatic vegetation (fig. 4.5). This species is specialized in its transmission dynamics and its microhabitat preferences, yet it is capable of finding suitable realized fitness space in geographical locations far from its origin because the specific ecological context necessary for survival of the pathogen is widespread.[58]

Like *Fascioloides magna*, *Haematoloechus floedae* is a North American

FIGURE 4.5 Diagrammatic representation of the life cycle of the deer and livestock liver fluke *Fascioloides magna*. Adults live in the liver, and their eggs pass out of the host in the feces. If feces wind up in the water, the miracidium hatches from the egg and burrows into the snail. A complex series of developmental and reproductive events then occurs in the snail, resulting in large numbers of larval clones of the miracidium—called cercariae—swarming from the snail into the aquatic environs. There, the cercariae alight on aquatic vegetation (as well as on the exoskeletons of crayfish and the shells of mollusks), where they encyst and hope to be eaten by an ungulate or cervid.

native digenean, described originally as living in the lungs of bullfrogs, *Lithobates catesbeiana*. This digenean's eggs pass out in the feces of the frog host into the aquatic environment, where, unlike *F. magna*, they are ingested by the snail host, hatch (releasing the miracidium), and pass through the same kind of complex developmental and reproductive cloning events that occur in *F. magna*. In the case of *H. floedae*, however, the cercariae that swarm out of infected snails penetrate and encyst in the aquatic dragonfly larvae called naiads. There they sit, encysted and waiting, until a bullfrog eats the infected naiad or the metamorphosed adult, whereupon the next generation begins (fig 4.6). Twenty years ago, during an inventory of parasites in the Área de Conservación Guanacaste, in northwestern Costa Rica, two local frog species, *Lithobates taylori* and *L. cf. forreri*, were found to be infected with *H. floedae*. The only consistent differences between the North American and Mesoamerican populations of *H. floedae* were three synonymous substitutions in sequences of the cytochrome oxidase I gene, indicating that the Mesoamerican lung flukes were recent introduc-

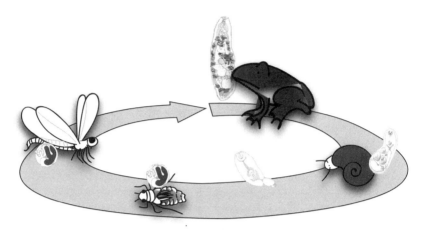

FIGURE 4.6 Diagrammatic representation of the life cycle of the frog lung trematode *Haematoloechus floedae*. Adults live in the lungs of frogs, and eggs pass out in the feces of the frog host into the aquatic environment. There the eggs are eaten by snails, whereupon they hatch; the miracidium penetrates the snail and undergoes the same kind of complex developmental and reproductive cloning events that occur in *F. magna*. In the case of *H. floedae*, however, the cercariae that swarm out of infected snails penetrate and encyst in dragonfly naiads. There they sit, encysted and waiting, until a bullfrog eats the infected naiad or the metamorphosed adult, and the next generation begins.

tions rather than ancient relict populations.[59] There are other reports of *H. floedae* outside its original geographic distribution, but always in areas where *L. catesbeiana* has been introduced and where it remains the primary—though not the only—frog host. The Costa Rican finding added intrigue. Some detective work involving collections of the national museum in Costa Rica revealed that, a generation before the discovery of *H. floedae* in northwestern Costa Rica, *L. catesbeiana* had been imported into Costa Rica in a failed attempt to grow them for commercial food purposes. No one examined the imported frogs for parasites, especially for parasites that have multihost life cycles, such as digeneans. The prevailing assumption was that such parasites could never become established in a place where only one of the tightly coevolved hosts occurred. The most likely point of entry for the parasite into Costa Rica is the capital, San Jose, which is in the mountains, hundreds of kilometers southeast of Guanacaste. How could *H. floedae* get from San Jose to Guanacaste? Thanks to what we know about the conservative nature of transmission dynamics, we can anticipate that *H. floedae* became established in local snails, odonate naiads, and aquatic frogs in the area of San Jose where *L. catesbeiana* was introduced. But how did it get to Guanacaste?

The short time frame involved seems to rule out gradual migration of the parasite and its hosts from San Jose to Guanacaste. This leaves an additional possibility: adult dragonflies are outstanding long-distance dispersers, so it is feasible that *H. floedae* was introduced to Guanacaste from San Jose by infected dragonflies. There are reports of dragonflies being infected with the cercariae of various species of *Haematoloechus* while they were naiads and still carrying infective parasites even after they metamorphosed into adults. In each case—Georgia, the southwestern United States, San Jose, Guanacaste—*H. floedae* was able to survive by exploring local fitness space through ecological fitting and encountering essential elements of fitness space: integrated aquatic ecology including snails, dragonflies, and frogs whose diet regularly includes dragonfly naiads.[60]

Studies of frog parasite communities gave us our first view of just how comprehensive ecological fitting in sloppy fitness space could be.[61] One might think the community structure of digeneans infecting frogs living in such diverse environments as temperate deciduous forests in Bohemia and North Carolina, temperate grasslands in Nebraska, and wet and dry tropical forests in both Mexico and Costa Rica would be structured quite differently. Yet, each community encompasses a great diversity of digeneans that show the specific and conservative patterns of transmission dynamics and microhabitat preference common to all pathogens: members of the related genera *Megalodiscus*, *Diplodiscus*, and *Catadiscus* all live in the rectum of the frog host and pass through a snail host, and their cercariae encyst on the surface of crayfish and snails that are eaten by frogs; all members of *Haematoloechus* live in frog lungs and have the same snail-odonate naiad (and in some cases other aquatic insect larvae)—frog cycle as we described for *H. floedae*; members of the related genera *Gorgodera* and *Gorgoderina* live in the urinary bladders of their frog hosts, use pelecypods (clams) as molluscan hosts, and may infect insect larvae or larval frogs, becoming adults either when the infected larval frog metamorphoses or when infected insect larvae or larval frogs are eaten by adult frogs; members of *Halipegus* live in the buccal-esophageal area (eustachian tubes, under the tongue, in the esophagus), utilize snails, and then encyst in arthropods that are eaten by the frogs; members of the Lecithodenriidae live in the small intestine or sometimes in the gall bladder and use snails and aquatic insects as hosts; members of *Cephalogonimus* live in the small intestine and use snails and tadpoles; and various other digeneans congregate in the small intestine, having gotten there by encysting in various aquatic invertebrates and anuran larvae and being eaten (in one group the cercariae actually encyst in the skin of the frog host, which infects itself when it eats its shedding skin).

There is—as expected—little evidence of cospeciation in the particular associations within and between these communities. There is, however, evidence of cophylogeny. That is, by far the majority of parasite species, and by far the major abundances, in all these communities occur in frogs belonging to the family Ranidae. Viewed from another perspective, in all cases there is a strong correlation between how much time the frog host spends in water and what diversity and abundance of trematodes it hosts. There is also a strong correlation between parasite abundance and the foraging habits of the frogs: active and opportunistic foragers hosted a high diversity and abundance of parasites. And finally, there is a correlation between body size and parasites: in all communities, some frog species are smaller than the intermediate hosts carrying parasite larvae. All three variables—habitat preference, feeding behavior, and body size of the frog host—are specific and conservative traits of the frogs themselves. That is, they are all part of the nature of the organism of the frog hosts, as well as part of the nature of the conditions of the parasites, whose transmission dynamics are also specific and conservative. This accounts for the cophylogenetic associations in the absence of cospeciation. And additional studies from the temperate grasslands, marshlands, and forests of Argentina, as well as from the Pantanal of Brazil, show the same pattern, with one exception. There the largest, most aquatic, and most opportunistic predators are members of the family Leptodactylidae, not Ranidae.[62]

What do these communities of diverse parasites, living in diverse hosts in diverse habitats, have in common? The startling answer is that they are fundamentally the same, structured by specific and conservative inherited features of the nature of the organism for hosts and parasites. These features are so powerful that community structure is largely independent of the nature of the conditions; the temperate forests and temperate grasslands of North or South America and the marshlands, tropical dry forests, and tropical wet forests of Central and South America all show the same pattern.

Summary

Scientists have spent more than 150 years trying to explain pathogens by referring to their hosts, because the socioeconomic importance of pathogens involves their impact on hosts. If we continue to study pathogens in that way, we will never learn why we face a daily barrage of news about pathogens moving into new hosts, including us and the species upon which we depend for our survival. More important, we will never know what we can do to mitigate their impact on us.

We must understand emerging disease from the perspective of the pathogen. This is because hosts are not inherited capacities of pathogens; they are part of the conditions of life for pathogens, and Darwinism is based on the notion that the nature of the organism is the most important element of understanding evolution. Like all life, pathogens are oscillators, capable of both exploiting and exploring their surroundings. Survival, however, means retaining the ability to explore even when there is nothing new to explore or no need to explore. The nature of the conditions always changes, but if you have no ability to explore the new world of those changes, to find new viable fitness space, you will eventually go extinct. The capacity to explore at any time—ecological fitting—is a built-in part of the nature of the organism. This allows pathogens to respond to environmental perturbations via evolutionary diversification.[63] In a Darwinian world, organisms try to survive, and it is important to remember that pathogens are fundamentally concerned with surviving, not with attacking.

The Stockholm Paradigm thus resolves the parasite paradox and provides a novel framework for studying the evolution of pathogen-host systems from the pathogen's perspective. Now we will begin to build on that to help us understand some implications of the emerging disease crisis.

FIVE

Resolving the Parasite Paradox I: Taking Advantage of Opportunities

Not all those who wander are lost.—J. R. R. Tolkien, *LORD OF THE RINGS*

When we examine pathogen evolution from the pathogen's perspective, we focus on their capacities for exploiting existing viable hosts, and for exploring fitness space where previously unexposed susceptible hosts live. It is therefore not enough to enumerate the specific ways in which particular pathogens are coping with their immediate surroundings. We need to understand how they came to have those capacities, what capacities they have that we might not see at any given place and time, and what those capacities may allow them to do when faced with new opportunities.

When opportunities are limited, ecological fitting in sloppy fitness space leads to exploitation-biased evolution, in part because exploring the boundaries of fitness space generally fails. When opportunities abound, by contrast, ecological fitting in sloppy fitness space leads to exploration-biased evolution, and exploring the boundaries of fitness space is more often successful. The proportion of realized fitness space to fundamental fitness space determines—in a relative sense—how generalized or specialized you are. And temporal trends determine whether a species is specializing (exploitation-biased) or generalizing (exploration-biased) in fitness space. In this view, the ability to generalize or specialize within fitness space is a built-in feature of living systems, so any species can oscillate between generalizing and specializing themselves in fitness space as their surroundings change, no matter how specialized their previous situations. The process of specializing refers to decreases in realized fitness space, whereas the process of generalizing refers to increases in realized fitness space. Each species, at any place and time, can be said to be relatively specialized or relatively generalized in fitness space, depending on our understanding of the full extent of fundamental fitness space for the species in question. Consider a species whose fundamental fitness space is known. Remember that fitness space is fundamentally sloppy—that is, realized fitness space is always a subset of fundamental fitness space. If the species occupies more than half its fundamental fitness space, we could say the species is relatively generalized in that fitness space. If it occupies less than half of its fundamental fitness space, we could likewise say it is relatively specialized. More important, if we know something about the species' history, we could determine whether the species has become (or is becoming) more specialized or generalized.

The Oscillation Hypothesis: Generalizing and Specializing and Back Again

The *oscillation hypothesis* postulates that species, as inheritance groups of organisms, should exhibit oscillating behavior that emerges from the collective oscillating tendencies of individual organisms.[1] On relatively short time scales and small spatial scales, oscillations appear as localized changes in host range associated with fluctuations in environmental conditions. Large-scale evolutionary diversification of pathogens involves an initial phase in which host range increases as opportunities allow the pathogen to generalize itself in fitness space, which in turn sets the stage for the emergence of new pathogen lineages, each specializing itself in fitness space.

There are four ways to initiate host-range oscillations: (1) altering existing trophic structure so that previously inaccessible but susceptible hosts become apparent to the pathogen; (2) bringing new susceptible hosts that have never encountered the pathogen into the ecosystem; (3) expanding the pathogen into new geographic locations where susceptible but previously unexposed hosts live; and (4) expanding pathogen capacity by the accumulation of evolutionary novelties through time.

None of these factors implies any particular relationship between host range changes and rates of diversification. They provide opportunities to establish new associations through episodes of generalizing in fitness space, which then set the stage for diversification through subsequent specializing in fitness space. The geographical and temporal extent and the biological magnitude of such oscillations depend on the circumstances causing increasing or decreasing connections in fitness space.

The oscillation hypothesis is a convenient explanation for evolutionary changes in host range that is logically consistent both within the Stockholm Paradigm and with empirical data. But is it likely to be the general pattern for changes in host range during evolution? The answer to this question involves a surprisingly complex and interlocking set of issues that is best examined using the powerful tools of mathematical modeling.

The Complex Problem of Complex Systems

Complex systems are those in which the whole is greater than the sum of its parts. A collection of individual things thus becomes a complex system depending on the kinds of interactions the individuals exhibit. In order to fully understand the properties of a complex system, therefore, it is fundamental to understand how individual members of the system interact with each other based on relatively simple rules that nonetheless produce properties that could not be deduced by delineating only the behavior of each individual. We cannot predict the general behavior of city traffic only by understanding how each person drives; we must take into account how multiple drivers, each following (more or less) the same general rules, interact with one another in a variety of situations. Generalities about a system as a whole that cannot be discovered simply by knowing the interactions among elements of the system are *emergent properties*. Host range oscillations of pathogen species are emergent properties that cannot be predicted from the oscillations of any individual pathogen. Disease associated with pathogens is an emergent property of associations between pathogens and their hosts and cannot be predicted from the properties of the pathogens alone. This strongly supports Filipchenko's assertion that host-pathogen systems should be studied from the point of view of pathogens' own ecology, in the context of their interactions with their hosts;[2] that is, host-pathogen associations should be studied as complex systems.

Life is inherently complex. Cells associate with one another to form an individual; individuals interact to form a population; populations interact to form a community. Each level of complexity presents unique emergent properties because of the interaction of actors at the lower level—properties of populations emerge from the interaction of individuals, and populations interacting with each other produce emerging properties of the communities. Darwinism is based on two elements of complexity. The first is that evolution itself is not a process per se, but rather an emergent property of the interaction between the nature of the organism and the nature of the conditions. The second is that the history of biological interactions is significant in explaining their properties—how and where they emerged is important in understanding what they are doing today. Indeed, it is possible that Darwinism was viewed with great skepticism because it was the first theory of the behavior of complex systems, articulated more than a century before physicists and chemists "discovered" complexity.[3]

Much of our understanding of complex systems has come from *network analysis*. The interactions among the elements of a system analyzed using network analysis can reveal properties of the system that could not have been predicted by understanding how one or a few elements of the system behave. If the system is complex, a relatively simple set of rules describing interactions among members of the system can produce many outcomes, such as multiple instances beginning at the same point but producing different outcomes.[4]

Network analysis of pathogen-host systems has provided two important insights relevant to the Stockholm Paradigm. First, networks connecting pathogens and hosts (sometimes called epidemiological networks) store great amounts of potential information in an easy-to-access structure, allowing great flexibility in the face of changing conditions. The second is that many pathogen-host associations fall into two general categories of complex behavior: those in which the pathogen becomes established with difficulty in a given host but spreads rapidly among members of a host population once that occurs; and those in which the pathogen becomes established readily in a given host but spreads slowly once established. The Stockholm Paradigm suggests that these categories are not fixed properties. Rather, the oscillatory capacities of pathogens allow any given species to behave in either manner, depending on the particular host species. Pathogens that easily infect and spread slowly in wild hosts may only rarely infect humans or livestock or crops, but when they do they may spread rapidly.

Computer simulations offer another way to understand complex systems. The technique involves constructing biological systems that are allowed to evolve following simple but realistic rules. The idea is to evaluate whether those rules can generate predicted emergent properties. Perhaps the most comprehensive class of models is called agent based models (ABM) or individual-based models (IBM). In these models, individual actors (automata) are simulated, using properties associated with them (e.g., reproduction, death, mobility, and behavior). These simulated individuals can interact with each over many generations. The results often reveal the nature of complex properties observed in natural conditions. Subsequently, we can compare these properties with patterns observed in nature and construct hypotheses about their corresponding relationships, including insights about patterns and processes that are otherwise not expected by researchers.

ABMs developed to test predictions about some of the elements of the Stockholm Paradigm revealed exciting insights into pathogen-host associations.[5] These models are based on the minimum number of assumptions required for realistic behavior, with pathogens modeled individually while potential host populations are represented as single entities. They assume a simple pathogen-host relationship influenced by the relative fitness of the pathogen with respect to the host. Each pathogen population is composed of individuals with the potential to mutate and reproduce, with offspring migrating to other hosts in the surroundings. The survival of each individual phenotype of the pathogen produced during the simulations depends on a survival curve defined for each host species (fig. 5.1). Thus, all pathogen automata are assigned specific rates of mutation, reproduction, and emigration (attempts to colonize new host species). The hosts are modeled as the value of the pathogen phenotype that presents 100 percent survival on that specific host species (the mean value of survival) and as the survival curve for the pathogen phenotypes that have more than 0 percent survival on the same host species, distributed normally around the mean. The normal curve used during the simulations represents, in many ways, the fit between the pathogen phenotypes and the host species. The host population is modeled as an invariable resource, so the curve is basically a function of the survival of the distinct pathogen phenotypes in that modeled host species. Each phenotype summarizes the collection of factors that influence the establishment and maintenance of the pathogen-host system (i.e., the relative status of the defense mechanisms of the host species, the ability of the pathogen to be transmitted in that host, the host resources

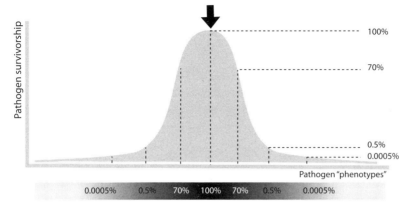

FIGURE 5.1 Assumptions of the mathematical models. The compatibility curve of the graphic defines the survivorship and fitness of each parasite phenotype on a specific host species during simulations. Both hosts and pathogens are modeled as a unidimensional value. The host species is represented as an unchanged resource in which the pathogen may achieve maximum survivorship, so the distance between the pathogen phenotype and the host "optimum" determines the survivorship of the pathogen phenotype.

required by the pathogen, and the virulence and pathogenicity of the pathogen).

More than one pathogen variant may present the same relative compatibility within the host population. A simplified example is the relative fitness of hypothetical pathogens that have, within their population, individuals that express two phenotypes. One phenotype may have enhanced ability to evade the host's immune response but low reproductive potential, while a second may be not as efficient in escaping the host's response but has a high reproductive potential. Fitness is not dependent on a single feature or gene but represents an integration of many phenotypic features of each pathogen. For simplicity, phenotypes with similar fitness in a host species are called *variants*.

Pathogen populations in the simulations evolve via selection by the host being exploited while encountering other host species. The compatibility curves of host species represent the pathogen's fundamental fitness space, and the portion of fundamental fitness space occupied by the pathogen is its realized fitness space. The fundamental fitness space of a certain pathogen is expected to be larger than the realized fitness space (fitness space is expected to be sloppy) because opportunity is never maximal—that is, not all host species that make up the fundamental fitness space can be reached or are available to the pathogen at any given time (fig. 5.2).

111

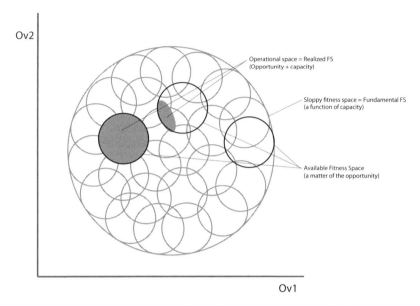

FIGURE 5.2 Graphic representation of sloppy fitness space for a species of pathogen (circles within the large circle), including potential (circles outlined in gray), available (circles outlined in black), and actual host species (partially or completely gray-filled circles).

Models, Theory, and Reality

The first simulations explored how ecological fitting allowed pathogens to incorporate new host species into the spectrum of resources used by the pathogen.[6] Opportunities to colonize a new host involved offering hosts randomly in every generation (fig. 5.3). If the pathogen established a new population associated with the new host, the pathogen population associated with the original host was no longer "followed" by the program—the data generated by the original population were no longer recorded.

The second simulations tested the oscillation hypothesis:[7] namely, whether the temporal fluctuation between specializing and generalizing pathogen lineages could represent an emergent property derived from the interaction between pathogens and hosts owing to variations in the opportunity to alter host range by ecological fitting. This model operates in opportunity space comprising a fixed community of host species, representing distinct resources, that are equally distributed in the resource space. Limits to fertility were arbitrarily defined; pathogens could not reproduce among themselves when their virtual genomes differed by more than 20 percent. Thus, the simulations allowed speciation associated with the dynamics of host range changes.

Jumping Hosts Is Easy Even When Capacity Varies

Both models supported the predictions of the Stockholm Paradigm and uncovered some striking similarities with the results of empirical studies. They showed that new host species can be added easily through ecological fitting; no new mutations are required. The successful establishment of new associations depends solely on the survival of the pathogen variants in the original host that emigrated to the new host. Following successful colonization, sexual reproduction and mutations generated new variants, as would be expected. This allowed the pathogen population to persist and become modified in the recently colonized fitness space according to the selective pressure imposed by each new host. Colonizing a new host thus did not require evolutionary innovations but simply set the stage for them to emerge after the new host was colonized, as anticipated by the Stockholm Paradigm.

In general, the more diverse the phenotypic variation of the pathogen species, the greater its capacity to successfully incorporate new host species into its realized fitness space (fig. 5.3). However, as expected for a complex system, the relationship between variability and the capacity to successfully change hosts is not linear. Even when pathogen phenotypic variability is null (i.e., the pathogen population is represented by a single variant), new hosts were successfully colonized more often than expected. This is an unanticipated outcome of the model that has some important consequences for understanding the emergence of infectious diseases.

In a complementary manner, continuous increase in the variability of the pathogen population does not necessarily result in corresponding continuous ability to colonize additional hosts; there is a limit (fig. 5.3). Capacity determines the ability of the pathogen to ecologically fit to new host species, and it also increases with time owing to the incorporation of evolutionary novelties and recombination. The modeled accumulation of variability in the pathogen was intended to simulate the accumulation of evolutionary novelties in real species through time. These novelties are usually termed spontaneous mutations because they are neither caused by, nor arise in response to, environmental changes (in this case, novel hosts).

Increasing variability expands as a larger spectrum of new potential

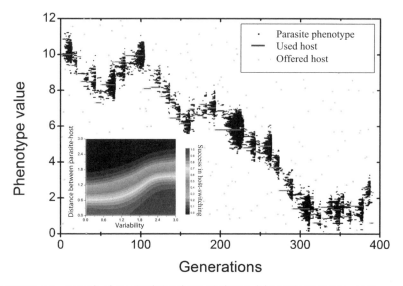

FIGURE 5.3   Example of one simulation (large graphic) and the synthesis of many similar simulations (insert) of the mathematical model. In the large graphic, the vertical axis denotes the value of the phenotype for the host species (gray dots and lines) and for the pathogen (black dots), and the horizontal axis the number of generations of the pathogen. The pathogen occupies a single host species but at every generation is offered different host species (gray dots), which it may or may not colonize. A colonized host is represented by a gray line, and the black dots associated with it represent the phenotypic composition of the pathogen population. The insert shows how the success in the colonization of new host (color coded) changes with different levels of pathogen variability (horizontal axis) according to the distance between the pathogen population and the new host species (vertical axis). Redrawn and modified from Araujo et al. (2015).

host species becomes available, also making the pathogen fitness space sloppier. There are gaps too large for pathogens to cross at a given point of time, in a single event or multiple episodes of host range expansion. That actual host range expansions will lag behind potential host range expansions gives us some comfort when thinking about emerging disease because it suggests that pathogen host ranges are not limited solely by opportunity—that is, pathogens are not inherently capable of infecting every host on the planet.

Successful host range expansion is easily achieved in many cases because it is not dependent on specific pre-evolved mutations. This reflects what our phylogenetic studies show has happened in the past, what we experience today, and what we believe we will continue to see for the foreseeable future. Given any opportunity, pathogens trade up to new hosts more easily than is convenient for our current health services. Emerging disease owing to pathogens colonizing us, such as the well-publicized agents of SARS, HIV-AIDS, Ebola hemorrhagic fever, Zika fever, Lyme disease, and mad cow disease, as well as others that are not so well-known, such as the agents of Hendra virus disease, Q fever, Marburg hemorrhagic fever, monkeypox, and herpes B virus can no longer be considered rare events unlikely to be repeated. In fact, they are likely only the tip of the iceberg.

The models suggest that pathogens continuously explore potential fitness space as much as possible. This includes the surprising insight that pathogens that colonize suboptimal hosts may nonetheless persist indefinitely. And *those associations are likely to be associated with disease as a result of the poor fit between the pathogen and host.* This leads us to reassess the significance of what has been reported as "accidental infections" or "accidental hosts" in pathogen inventories and then ignored. We, and the plants and animals upon which we depend, are exposed to continuous attempts of colonization by pathogens that may present a potential for the emergence of significantly dangerous diseases. For instance, how many times have you, or some member of your family, been informed, "You have a viral infection," without being told exactly which virus is the culprit? The ease of surviving in a new host species supported by the model simulations suggests to us that many of those indeterminate viral infections may represent attempts at colonization by new (and sometimes recolonization by old) pathogen species. Many of these attempts do not become successfully established thanks to an inherent incompatibility of the particular genetic variant of the viral species. In other cases, medical intervention offers them no opportunity to spread in the new host population. Regardless, pathogens probe host fitness space relentlessly, and when they encounter viable portions, they readily take up residence.

Demographics and Variation during Colonization

Every time a pathogen adds new host species to its repertoire (known by some as "spillover"), the new pathogen population exhibits significantly reduced variability when compared to the original population (compare figs. 5.2 and 5.4b and d). The reduction in variability of the newly established population is caused by genetic bottlenecks known as the "founder effect." This phenomenon is well-known in population biology as an important agent facilitating rapid change. A numerically and genetically limited population of colonizers will be responsible for

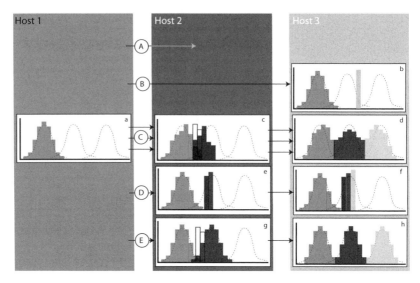

FIGURE 5.4 Selection and demography during the incorporation of new host. Population demography (histogram of frequency distribution of phenotypes) is represented on the putative compatibility curves for the three hosts represented in the figure. From the top down: (A) An attempt by pathogen population (a) in host 1 to colonize host 2 was unsuccessful, resulting in the death of the pathogens on that host. (B) The pathogen population (a) on host 1 was successful in colonizing a distant host species (host 3), and the resulting new population (b) has limited variability owing to the founder's effect. (C) Colonization of new host species (hosts 2 and 3) by (a), with continuous genetic exchange between pathogen populations on the three hosts, results in the accommodation of the frequency of phenotypes (or genotypes) of the pathogen population to each respective host species (c and d); the white bar in (c) represents the frequency of phenotypes of the new population that tend to accommodate to the new resource with time (d). Variants from the old pathogen populations are continuously exchanged among distinct host populations, reducing the depth of the "valleys" between frequency distributions (d), and, hence, diversification into distinct species is hindered. (D) Successful colonization of a new host species (host 2) associated with a reduction in variability by the founder's effect (e); however, this approximates the phenotypic variability of the pathogen to the compatibility curve of a third host species (host 3), which host can be incorporated into the pathogen host's range (f). (E) Same as C but without continuous genetic exchange between populations established on different host species; accommodation of the new populations depends on accumulated evolutionary novelties (e.g., by mutation). Accommodation increases the capacity of the population in host 2, which facilitates dispersion into host 3. The valleys between the frequency distribution of pathogen populations are well marked owing to selection at the margin of each respective host.

the growth and differentiation of the new population associated with the host species. The reduced variability of the colonizing pathogen population under these circumstances likely represents biased sampling of the original pathogen population either by chance (i.e., only a few variants have the opportunity to reach the new host species) or through immediate negative selection effects on the variants that arrive in the new host species (such that only some variants survive).

This process is well-known for several virus species during transition to a novel host species.[8] For instance, the genetic diversity of the SARS-CoV virus is considerably larger in bats (its putative original host) than in humans (a recipient host). The same is true for the Zika virus: the genetic variability of the virus collected from African localities includes the variability observed for Asian and South American populations.

Low Variability, High Risk

The initial reduction in variability associated with a founder effect event during the establishment of the new populations of pathogens may represent potential and unforeseen problems to the health community. The elements that compose the relative fit between pathogen variants and the host species are derived from phenotypes that make it possible for the pathogen population to become fully established in the new host. Within the parameters included in the definition of "fit" in the model is the value of R_0 of the pathogen, signifying the average number of secondary infections arising from one infected individual in a completely susceptible population. This widely used measure in epidemiology qualifies the transmissibility of a given species of pathogen in the host population. When R_0 is equal to or greater than 1, the pathogen can be transmitted successfully within the new host population and colonization is achieved; with $R_0 < 0$, the colonization is unsuccessful. The model suggests that if any fit exists, the variants of a specific pathogen can rapidly colonize the new host. The founder effect favors those pathogen phenotypes from the original host that are best able to cope with the new host or resource (figs. 5.3 and 5.4). Whenever opportunity presents itself, those few pathogen variants can rapidly colonize the new host population, which, for the pathogen, will appear as small variations of the original resource. Transmission is fast and effective within the new host population. This means that the more random and unanticipated the host colonization, the more likely it is we will experience an emergence. This is how the same pathogen can commonly infect some hosts and spread slowly within their populations, and rarely infect other hosts but spread rapidly within their populations when successful infections do occur.

Epidemiological data about the development of disease syndromes for many species of EIDs support the pattern suggested by the model.

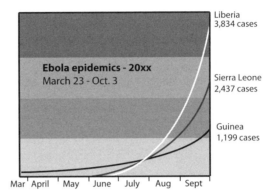

Liberia
3,834 cases

Ebola epidemics - 20xx
March 23 - Oct. 3

Sierra Leone
2,437 cases

Guinea
1,199 cases

Mar   April   May   June   July   Aug   Sept

FIGURE 5.5   Exponential growth of cases of Ebola. Limited genetic variability of the EBOV in the novel human host does not minimize the impact of the virus on the human population. On the contrary, variants that are effective in causing disease and spreading among humans may have been preselected in a similar host species. Despite the low variability of the virus lineages, spread of the disease is extremely rapid, and the illness is overwhelmingly fatal. Source: Centers for Disease Control and Prevention (CDC), https://www.cdc.gov/vhf/ebola/outbreaks/2014-west-africa/case-counts.html.

Abundant examples in the literature include reports of rapid spread of infectious pathogens including Ebola virus (EBOV; fig. 5.5), SARS-CoV, and influenza viruses in humans, and Q fever virus in goats and cattle. Agents of many EIDs, such as Hendra viruses, herpes B, and Q fever in humans appear to have limited ability to colonize populations of new hosts and effectively cause epidemics. This class of pathogen may produce disease in the population of the new host species, but with limited transmission. The model recognizes these events among the many unsuccessful colonization attempts (fig. 5.3, insert). While they could conceivably cause extensive epidemics, they do not represent a productive means of incorporating new host species and establishing a new pathogen-host association. Nonetheless, they always pose a threat of occasional recurrence with unpleasant effects.

Danger at the Margins: Edge Effects and Disease

Variants located in the tails of the distribution of compatibility (fit) of the pathogen-host association in the model are *marginal variants* (see fig. 5.1). Despite being a poor fit to the host, these variants are valuable members of the pathogen population in several situations, including those involving host range expansion in changing environments. The poorly fit portion of the original pathogen population, independent of the variability of the original population, represents the variants that hold the greatest ability to colonize the most distant host resources. While these variants may have low fitness in the original host association, they can be highly fit in another host species (fig. 5.3).

We need special focus on these marginal pathogen variants that are attempting to colonize the tails of the compatibility/fitness curve of new hosts (fig. 5.1) and thus have a poor fit with the new association. The colonizing variant may be in this marginal position because of characteristics directly associated with the pathogen (e.g., poor transmissibility) or with the host (e.g., efficient immune response to the pathogen). It is in this region of the fitness curve that we expect to see high pathogenicity and virulence. This scenario is especially significant for transient associations but may apply to associations that persist for long periods of time.

Before proceeding, we need to introduce two terms. *Pathogenicity* is the potential of the pathogen to cause disease in a single host specimen, while *virulence* is the actual measure of how much damage a pathogen causes to the host. Virulence is strongly correlated with the capacity of the pathogen to multiply within the host (either in one part of the host or throughout the host). From an evolutionary perspective, pathogenicity and virulence indicate that a pathogen is exploring the limits of its fitness space, not that it has mutated into an enemy attacking a new host. And the interaction of each phenomenon may produce different outcomes. Pathogenicity may be low, but if virulence is high, the damage caused in the host is magnified and leads to disease. Contrariwise, an association of low pathogenicity and low virulence may not even result in disease. Variants that, in association with certain hosts, are linked with pronounced virulence and pathogenicity are usually distant from the optimum fit of the association, partly because they are more likely to cause disease and eventually the removal (death) of their own individual hosts from the host population. High pathogenicity and virulence are partially responsible for the failure of several long-distance spillovers in the model (i.e. the upper dark area in the insert of fig. 5.3).

The most serious implication of our modeling is that colonization of the tails of the compatibility curve may be common when distant hosts are colonized. These regions of the curve of the new host species will overlap the fitness curve of the donor (original) host. The margins or edges of the fitness curve, however, may harbor pathogen variants that originate from more distant potential host species (fig. 5.3). This outcome of the simulations suggests that new host colonizations, when they are successful, may have a greater possibility of causing devastating epidemics. Thus, marginal (rare) variants must be sought in epidemiological searches for reservoir hosts, especially after the emergence of disease, as is now being done—belatedly—for the Ebola virus. Potential reservoirs may be host species distantly related to the newly acquired host, especially in the case of highly pathogenic and virulent epidemics. For instance, while the virus that causes eastern equine encephalitis appears to produce no obvious damage in its original host group, birds, it has devastating consequences in horses; these host groups are separated by 312 million years of evolution.

Darwin recognized the evolutionary significance of organisms' having the ability to survive under less than ideal circumstances. Pathogen populations may survive for long periods in association with a host species for which the variants that compose its population are only marginally fit (fig. 5.3). These phenotypic variants may—but need not—be highly pathogenic but exhibit low virulence. This explains two additional outcomes of the model, also supported by empirical examples. First, the variability of a pathogen is not limited to those variants that present the highest fit to the host species, allowing variability to increase far from the optimum fit value through the acquisition of novelties and recombination. Together with the importance of the edge of distribution for the arrival of the pathogen, these marginal variants also possess a greater potential to explore other available host species.

New pathogen populations can colonize a host for which they show low fit and can remain in the tails of the fitness distribution for a relatively long time (multiple generations) (fig. 5.3). The model does not require a high degree of fit between pathogen and host species. Conflicts between hosts and pathogens need not be readily resolved—increasing mutual fit between pathogen and host is not required for the maintenance of the association so long as both survive. That is, an association may persist indefinitely in a suboptimal portion of the compatibility curve. This was anticipated more than 30 years ago when the British scientists Roy Anderson and Robert May presented modeling results showing that we could not necessarily expect selection to favor decreasing pathogen virulence.[9]

Demographic and Variability Changes Following Colonization of New Hosts

Differentiation is greatly dependent on the level of genetic sharing between the original and newly formed populations. Given time and an accumulation of evolutionary novelties, therefore, a pathogen population established in a new host may (fig. 5.4e) or may not (fig. 5.4c) differentiate from the original population (fig. 5.4a). Pathogen populations may also increase their capacity to colonize new hosts. Each colonization of a new host by ecological fitting results in a pathogen composed of a subset of the original population; the immediate result is a change in the frequency of preexisting variants (fig. 5.4). Differences in available variants, influenced by the founder effect as well by selective pressure provided by the new host, can shift the frequency of variants in the new host rapidly. In the absence of strong genetic flow with the original population, this process promotes differentiation of the new pathogen population. Subsequent expansion of the new fitness space and further differentiation of the new population depends less on the fortuitous emergence of evolutionary novelties than on the level of genetic exchange between the two populations.

Predictions derived from the model provide a framework for explaining why the common assumption that mutation precedes host range changes is incorrect. If marginally fit pathogens occur in low frequencies in the original host, characterizing its population by standard sampling procedures—which aim to identify the most common variants—may miss those low-frequency variants and produce an incomplete picture of the pathogen's total variation (and thus its potential for colonizing new hosts). Correspondingly, if these marginal phenotypes represent the portion of the original pathogen population that can colonize a new host species, their frequency will likely be higher in the new host and easily detectable by sampling. When the original and the new host species are distant, only a few marginal variants of the pathogen associated with the original host species can successfully colonize the new host. In such cases, inadequate sampling of the original population could lead to the belief that newly evolved genetic variation was required in order for a successful host colonization to occur.

These results have special meaning for the control of pathogens, especially when their hosts are introduced to a new area. These pathogen populations can present low numbers of individuals of one or a few variants. They are difficult to detect in limited or occasional random sampling protocols performed by health inspectors, so they may pass undetected through checkpoints. While these marginal pathogens are more difficult to detect, they are likely to be more successful colonizers and potentially more capable of causing disease in new host species. Introduced host species bearing poorly fit pathogen populations may act as Trojan horses for local hosts. The implied threat ranges from asymptomatic international travelers and tourists, to introduced crops and livestock, to imported food, lumber, and flowers.

Propagule Pressure

Continuous colonization from the original population greatly facilitates increased variability of a pathogen population in a new host (fig. 5.4c). Termed *propagule pressure*, this process is well-known for invasive non-pathogen species, but the principle holds for pathogens as well. Multiple or continuous arrivals of propagules in a new ecosystem set the stage for introducing new variants. Ecological fitting gives the introduced species the ability to gain a foothold in a new place. This results in rapid increase in the variability of an invasive species, maximizing the capacity of the species to handle aspects of the new conditions that differ from those in its area of origin.

The capacities of an organism are inherited, and the rules of inheritance operate independently from the availability of fitness space. Therefore, more organisms are produced than can survive, and some of those that do survive wind up in less than ideal conditions. That host changes so often result in disease indicates that pathogens often end up in what are, at least initially, less than ideal living conditions. These results, with their unsettling implications for emerging disease, represent a reaffirmation of the Darwinian importance of survival of the adequate.

The scenarios provided by the computer simulations are compatible with epidemiological outcomes in the real world. They point to the possibility of distinct dynamics of host colonization that are fundamental for efforts of prevention, mitigation, and epidemiological control. If opportunity exists, EIDs may originate through one or a combination of three classes of host colonization dynamics.

In the first scenario, continuous colonization by a single pathogen variant allows the pathogen to be sustained by recurring colonization because transmission from the new host is impossible, infrequent, or greatly limited (fig. 5.4i). The pathogen is found in the host only because it is continuously introduced from the original host; the patho-gen is not capable of transmitting itself from the new host sufficiently to persist on its own. A well-known group of pathogens that exhibit this scenario are the hantaviruses. The hantaviruses to which people are most often exposed generally infect rodents (the role of other small mammalian hosts such as moles, shrews, or bats in zoonotic transmis-sion remains to be clearly determined). Rodents shed the virus particles through their body fluids, and humans may become infected through airborne transmission. Particular hantaviruses cause diseases such as hantavirus pulmonary syndrome (HPS) and hemorrhagic fever with renal syndrome (HFRS). Hantaviruses never need to become fully fit to humans in order to be potent pathogens—expansion of the virus's variability is unlikely to occur in human hosts. Among the consider-able assemblage of hantaviruses, human-to-human transmission has been recognized only in the Andean virus, from Argentina. In cases of HPS and HFRS, there is no transmission among this everlastingly novel host, and hence no pathogen population of these hantaviruses is ever continuously associated with humans, which for the most part repre-sent a dead end for these pathogens.

The second scenario includes colonization resulting from one or only a few exposures that nonetheless produce new pathogen popula-tions composed of variants capable of persisting in the new host species (figs. 5.4b, d, and e). Because of limited contact or low propagule pres-sure, the expected variability of the new population is initially reduced thanks to limited reproductive isolation from the original. The lack of continuous genetic exchange from the original population facilitates differentiation of each sister population from the original one; while this may occur simply as a change in the variability and frequency of phenotypes (or genotypes) (Figs. 5.4b and d), it may also occur through the accommodation of variability under the selective pressure of the new host resource, leading to the emergence of new variants by recom-bination and mutation (fig. 5.4e).

The third scenario (fig. 5.4c) involves multiple colonization attempts that increase the probability of establishing a new population of patho-gens by expanding the number of distinct variants colonizing a new host species. Intense and continuous propagule pressure may result in few successful colonization events followed by rapid expansion of the new population of pathogen, especially if contribution from the origi-nal population continues (a → c, c → d). Transmission among the new host species is possible, directly or via a vector, but the contribution from reservoir species is continuous and influences the variability of the pathogen population in the new host population. This dynamic reduces diversification (formation of independent species) despite a large host range. The rabies virus likely provides an excellent example of this scenario. Colonization of new hosts by RABV genotype I (the rabies virus) has been, and continues to be, common among bats—the original host group of RABV—and carnivores. Despite that, only two host colonization events have generated independently sustainable lineages of rabies virus circulating among North American carnivores.[10]

Directionality in Host Colonization

Flexible host colonization and continuing exploration of available fitness space leads inevitably to the conclusion that any change in host range need not be a one-way process. If host colonization occurs from host A to host B, the new population of the pathogen in B is still able to live in A. Because A is the original host species, there is no reason to think that colonization of pathogens from B to A cannot occur. Some term the colonization from A to B *spillover* and that from B to A *spillback*, but it is important to realize that they are both manifestations of the same phenomenon.

The introduction of the Zika virus (ZIKV) in new regions and human populations throughout the world is an impressive recent example. ZIKV colonized humans from African primates in the 1950s, although the virus had been known since the 1940s from a rhesus monkey that had been used as virus bait at a microbiological research station in Uganda's Zika Forest. In 1966 the virus was first found in the *Aedes aegypti* mosquito in Asia (Malaysia), and in 1977–78 the first report of human infections in Asia came from Indonesia. In 2015 the virus was detected in South America.[11] Whatever host—*Aedes* mosquitoes, humans, or any other species—was involved in the introductions of the virus, once the pathogen was introduced into a new geographic area it always seemed to find compatible vertebrate hosts and sometimes new insect vectors. In Asia horses, cows, carabao (water buffaloes), goats, ducks, bats, and mice have shown evidence of hosting ZIKV. In South America the virus has been detected in native monkeys and even in a different species of vector, mosquitoes of the genus *Culex*. This reveals that, given the opportunity, ZIKV has an enormous capacity to colonize hosts. While the involvement of many of these species in the circulation of the virus is yet uncertain, we cannot ignore the possibility that many of these cases are the result of colonization *from* humans—the B-to-A scenario suggested above. Host colonization in both directions is probably common in the evolution of host-pathogen associations and represents an additional risk factor for human, veterinary, and crop health.

Stepping-Stones

Host resource space emerges from the intersection of pathogen capacity and opportunity based on geographic distribution and ecology. And it is discontinuous, or patchy. In the short term, these discontinuities limit colonization of host species within a given biota. The expansion of fundamental fitness space seems limited. The modeling shows that pathogens, however, have ways of reaching host species that seem inaccessible. As pathogen populations increase their variability within the available fitness space represented by the host species they exploit, they accumulate evolutionary novelties. This leads to increased capacity; new hosts become accessible and can be incorporated into the pathogen host range (as illustrated in fig. 5.3 and fig. 5.4e). Thus, incorporating more hosts into the pathogen population increases both fundamental and realized fitness space.

As expected, it is easiest to add a new host if it is very similar to the original one (fig. 5.3). A pathogen, however, can reach host species that are very dissimilar to the original host by sequentially colonizing hosts whose fitness spaces overlap (fig. 5.4e). This is known as *stepping-stone host-switching*.[12] This process expands the reach of host colonization demonstrated in the simulations (fig. 5.2), allowing a pathogen eventually to colonize resources that are quite distant from the original host.

Stepping-stone host-switching explains the host distribution of species of the monogenean genus *Gyrodactylus*, ectoparasites of fishes throughout the world. *Gyrodactylus* species from Eurasia provided evidence for step-by-step host colonization that allowed a lineage of pathogens originally from Cypriniformes (carp and its relatives) to colonize phylogenetically distant fish groups such as Salmoniformes (salmon and trout), Esociformes (pike), Gasterosteiformes (stickleback), and Perciformes (perch) (fig. 5.6). From their original host stock, host range expansion to distantly related hosts apparently occurred multiple times, and in many instances, sequentially—from carp to salmoniforms, from salmoniforms to esociforms, from these to perciforms, and thence to sticklebacks. The example also suggests many instances of recolonization of ancestral host groups, with parasites of salmoniforms at least twice colonizing the ancestral host group, the cypriniforms.[13]

Once again we are faced with a plethora of terms for different outcomes of the same phenomenon. In the medical literature, stepping-stone host colonization is mediated by what is called an *amplifier host*. This term is not limited to stepping-stone colonization, however. Amplifier hosts represent various ways to expand the exploratory reach of the pathogen. They also act as what has been variously termed *samplers*, *selectors*, and *ecological bridges* (most commonly called *paratenic hosts* by parasitologists) between original and colonized hosts. These are all manifestations of the complex ways in which pathogens explore their surroundings, sometimes encountering viable fitness space (hosts) that could not have been predicted based on simple coevolutionary models.

Let's return to the hypothetical example of an insect colonizing a primate (fig. 5.7). Humans live mainly in cities and usually do not feed on a variety of insects. Our primate, however, lives in the forests and occasionally eats insects, by accident or intent. This allows insect viruses to explore the resource represented by the primate species. In the

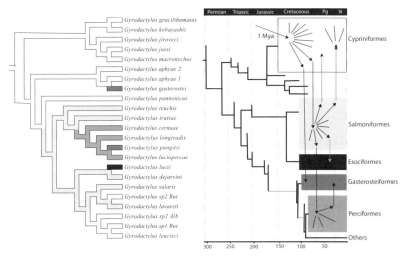

FIGURE 5.6 Phylogeny of species of ectoparasitic Platyhelminthes (on the left) and that of their host groups (on the right). The order of the host is optimized onto the parasite phylogeny (gray shades, corresponding to shades on the rectangles of fish orders, on the left phylogeny); diversification in association with a single host group is represented by black lines, and changes in host associations are represented by arrows among the boxes representing each host order. Species of *Gyrodactylus* diversified within distinct fish orders and expansion of host range followed by diversification is common. There are clear cases of recurrence, when the parasite lineage retrocolonizes an ancestral host group.

Expected compatibility association

FIGURE 5.7. A putative example of how an "amplifier" host may serve also as a sampler, selector, and paratenic host species. In this example, compatibility curves for three host species are presented (curves with different gray shades), with the frequency of phenotypes (genotypes) for the pathogen on each host species. Only marginal variants of the pathogen residing in the insect are capable of colonizing the tail of the primates' compatibility curve (hence the high frequency/low variability in the primate represented by the dark-gray bars). The "new" pathogen population "sampled and selected" by the apes will find humans a similar resource that is susceptible to infection. In this example apes, if hunted by humans, are transport (paratenic) hosts to humans.

insect, the original host species, the pathogen population evolved to accommodate the frequency of variants to the compatibility curve of the insect. Because the distance between insect and apes as resources is likely large, the insights from the models predict that only a few low-frequency variants of the insect virus may successfully colonize the ape species—hence the limited variability of this pathogen in the ape. Continuous ingestion of insects by primates is also continuous sampling of the primate by the pathogen living in the insect. This increases the chances that the primate will become exposed to marginal virus variants that can use primates as a resource. The process is a form of natural selection, because only the capable variants can survive and reproduce in primates. These specific variants may occur at very low frequency in the original population, but once they arrive in the primate, they may spread rapidly. Hunters entering the forest in search of bush meat may bring primates into a human settlement unaware that the meat contains large concentrations of the new viral variant. Colonization of a new host at this point is a far simpler matter, because humans are primates, a good indication that they represent similar resources for those pathogen variants that were sampled, selected, and amplified by primates in the natural environment. Apes, in this example, are samplers (they provide opportunity for new viruses to explore themselves as resource) as well as selectors (they impose selective pressure similar to humans on the variants of the virus population, selecting those variants that are likely more prone to colonize humans), and they are also amplifiers (they provide conditions for those selected variants to multiply among its host population, maximizing transmission to humans). Furthermore, they are paratenic hosts, acting as an ecological bridge that provides opportunity for host colonization that would otherwise be limited.

Does this story sound like science fiction? This is how we would use the results of the models to generate a hypothesis for the known dynamics of infection of EBOV in Africa. The reservoir of EBOV has been fundamentally unknown. The only evidence available points to fruit and insectivorous bats, but only fragments of the virus's DNA have been detected in these animals. The use of insects as the hypothetical reservoir of the virus in our example above was not arbitrary. Insects feed on primates and primates feed on insects, so insects are not an unlikely distantly related resource relative to primates.[14] Also, filovirus-like particles have been identified in some insect species, such as grasshoppers, a common food source in places where humans regularly eat insects. Perhaps, in order to identify the reservoir (or reservoirs) of EBOV in the endemic areas, we should undertake extensive sampling of all species that coinhabit the forests surrounding the areas in which epidemics have occurred in the past. Previous sampling efforts have been cursory, examining a limited number of readily available vertebrates. Without a doubt, the virus still exists in the places where outbreaks have occurred before. It may not be an agent of the EID in humans in any given place, owing to the extensive mortality it causes in native primates (or other hunted species), the necessary link to humans. In such cases, the amplifier (sampler, selector, paratenic) host may no longer be abundant or even present. If this scenario explains EBOV, we may have the ironic situation in which the virus may be transmitted to humans when infected animals are butchered for bush meat, while at the same time the reduction in game species due to bush-meat hunting might ultimately reduce exposure to EBOV. To further complicate this scenario, farmed animals may also act as amplifier host species. In fact, the recent suggestion that pigs might be involved in the June 2017 Ebola outbreak in the Democratic Republic of Congo indicates that we should heed the next example, the story of influenza A.[15]

Influenza: From Chickens to Pigs to Us

Phylogenetic studies from the late 1980s supported the earlier hypothesis that the influenza A virus originated in birds. Subsequently, swine acquired the virus, which differentiated into a swine form. Human influenza A is apparently a derivative of the swine form, showing that domestication can have costs as well as benefits. Swine remain suitable hosts for all three forms of the virus. When swine are infected with avian and human or swine and human influenza A at the same time, genetic exchange can produce viral strains that are highly pathogenic to humans. Such highly pathogenic strains are usually self-limiting on time scales that are short indeed for evolution, but that can be quite long if your family lives in an infected area. Evolutionary biologists are also intrigued by the number of times such pathogenic hybrid strains can be generated and the diversity of viral genetic backgrounds that can be co-opted. What intrigues the evolutionist in Budapest, however, terrifies the epidemiologist in rural China.[16] The ease with which influenza viruses engage in stepping-stone host colonization should make us very concerned about the largely unknown origin and potential threat to humans represented by the influenza virus responsible for the Spanish flu pandemic of 1918, which killed as many as 65 million people—roughly 10 percent of the human population at the time.

Back to Complexity: Oscillation Emerges Spontaneously from Ecological Fitting

The first modeling study showed how easily pathogens can gain a foothold in new hosts through ecological fitting. The second model added the more realistic element of repeated colonization efforts allowed by propagule pressure (fig. 5.8).

We now see the world in a new light, a galaxy of pathogens exploiting existing hosts while constantly exploring their surroundings, probing for additional viable fitness space in the form of new hosts. Colonization reduces the phenotypic and genotypic composition of pathogens in newly colonized hosts. Subsequent accumulation of variability, expanding fitness space, occurs through the acquisition of new features and recombination. The second model suggests that, given enough time, colonizing populations of pathogen species may even diverge and form new species. This cycle then repeats itself as long as new hosts are available in the simulation. Just as individual organisms oscillate between exploiting known resources and exploring potential fitness space, pathogen species oscillate between restricted host ranges (being specialized in fitness space) and broad host ranges (being generalized in fitness space). This oscillation dynamic is an emergent property of

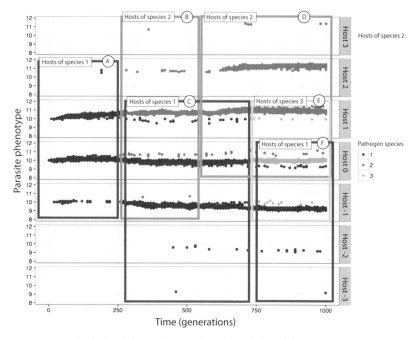

FIGURE 5.8   Single simulation of the second mathematical model that accompanied the pathogen populations in different host species (indicated in the right-hand column). Dots represent the variants of the pathogen, and distinctively colored dots represent different species evolving from the original population of pathogen in host 0. Since the first generations of these simulations, the pathogens increased their distribution (indicated by rectangles of distinct shades of gray) into distinct but similar host species (horizontal 2 is discontinuous, or "patchy," rectangles). Constant propagule pressure detected by the model apparently facilitated the colonization of new resources by increasing the variability of colonizers and reducing the randomness and low variability associated with founder effects. Modified from Braga et al. (2018).

the interaction between hosts and pathogens mediated by ecological fitting, the expected outcome of the oscillation hypothesis.

Oscillations in Real and Evolutionary Time

Oscillations emerge from pathogen-host interaction mediated by ecological fitting in sloppy fitness space, and this should be seen in both real and evolutionary time. Such oscillations usually are not easy to identify in real time. Humans, however, increasingly perform natural experiments that can allow us to test at least a portion of the process. Whenever we expose ourselves to host species with which we have not had continuous contact, we provide opportunities for their pathogens to explore us. Full exploitation of a new resource depends on the ability of the pathogen to colonize and maintain a new population in the new host, made possible by ecological fitting. Thus, the only distinction between pathogen populations in the original and the new host species is the overall variability and corresponding frequency of each variant in each population. The original population is, at least initially, more diverse, with frequencies of variants that reflect a longer exposure to the selective pressure represented by the original host. Subsequently, differentiation of both pathogen populations may occur, and the single population of pathogen may diverge into two distinct pathogen lineages, with unique characteristics and evolutionary fates.

Let's explore a well-known example of an EID, HIV (human immunodeficiency virus). SIV (simian immunodeficiency virus) is originally a nonhuman primate pathogen that broadened its host range four times to incorporate humans. At each of these points in time (indicated by a star in fig. 5.9), the ancestral virus species expanded its host range from one to two hosts. Subsequently, the virus specialized in the new (human) host species, resulting in five distinct lineages of HIV and a return to a one host–one pathogen association.[17] Overall, then, lineages varied from one to two hosts and, following putative isolation and differentiation of the populations into distinct lineages, one host again. This is the oscillation hypothesis in real time.

Oscillations are more evident on evolutionary time scales. They have not been reported in previous studies because the analytical methods used assumed that cospeciation and increasing specialization were the prevailing norms. Phylogenetically, associations were assumed to be mostly one pathogen occurring in association with a single species of host. This assumption leaves no room for generalized pathogens to be found, in the present or in the past. As we noted earlier, most cophylogeny studies assume that nonmatching between pathogen and host phylogenies is evidence of cryptic cospeciation. If your method of analysis cannot detect generalized pathogens, it naturally cannot detect oscillations in host range.

It is possible to analyze the evolutionary histories of pathogen-host associations in a way that is sensitive to the possibility of host range oscillations. Let's show this by reanalyzing what is known in the scientific literature as the classic case of cospeciation.[18]

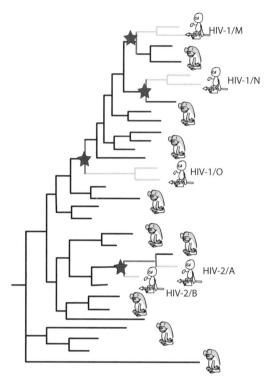

FIGURE 5.9 Phylogeny of SIV/HIV viruses showing respective host groups for each lineage. Addition of the novel host is indicated by a star.

The Classic Case of Cospeciation: An Example of the Oscillation Hypothesis

The family Geomyidae comprises an assemblage of rodents called pocket gophers, occurring from the Nearctic to Costa Rica. Geomyids have highly sedentary habits, considerable geographic range stability, and numerous species and subspecies characterized by strong genetic differentiation and ecological partitioning at local scales.[19] Pocket gophers' diversity represents an evolutionary radiation between 4.2 and 1.8 million years ago.[20] Contemporary diversity is concentrated in the temperate zone, and current geographic ranges indicate restrictions south of the Laurentide-Cordilleran during glacial maxima from the late Pliocene and Quaternary.

North American geomyids are parasitized by lice of the genus *Geomydoechus*. Transmission of the pathogens occurs in the nests, limiting opportunity for host range expansion. When carried by their original hosts, therefore, there is minimal chance of transfer even when the host expands its geographic range, in this case associated with episodic climate-change events. If a nest abandoned by one species is subsequently occupied by another, however, there is a chance that some of the parasites left behind will come into contact with novel hosts that are nonetheless closely related to their original hosts. A burst of diversification for pocket gophers, and presumably their louse parasites, coincided with a substantial regime of episodic variation in climate and habitat perturbation.[21] *Geomydoechus* species parasitize only pocket gophers, and divergence rates in hosts and parasites are highly correlated; consequently, cophylogeny assessment yields significant positive results. As we have suggested, however, this does not indicate cryptic cospeciation. Figure 5.10 compares the parasites and host phylogenies to show that half of the associations are due to host range expansion and subsequent differentiation. More important, this type of fine-scale examination, as opposed to the coarse-grained cophylogeny approach, shows clear evidence of oscillations in host range.

Reanalysis of additional published studies reveals support for oscillations, in groups as varied as viruses, parasitic flatworms, nematodes. and arthropods (e.g., lice, mites, and copepods).[22] Additionally, some researchers have documented portions of oscillations, such as the evolution of specialized species in groups of generalized species or of the reverse.

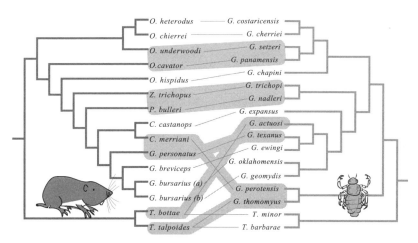

FIGURE 5.10 Phylogenies of pocket gophers and lice with episodes of host range expansion circled. Note alternating episodes of isolation (cospeciation) and expansion of host range (highlighted by shadows). Brooks, Hoberg, and Boeger (2015).

Evidence from real pathogen-host associations and from mathematical modeling support the core elements of the Stockholm Paradigm. Given a specified fitness space, pathogens have a substantial built-in capacity—ecological fitting—to oscillate between exploring that fitness space and exploiting each viable portion they discover. To the extent their capacity allows, pathogens will explore their surroundings, colonize viable new hosts, and exploit them. More importantly, ecological fitting allows ecologically specialized pathogens to expand their host range easily and without prior evolution of novel host-use capabilities, when the host resource upon which they are specialized is phylogenetically conservative and widespread. That allows generalizing in fitness space, taking advantage of opportunities that present themselves through ecological fitting in sloppy fitness space.[23] This leads to a critical insight—the more specialized a pathogen's host association is *locally*, the sloppier its fitness space is *globally*, and the greater the potential for establishing new associations, given the chance. We tend to think of changes in the nature of the conditions as a challenge to be fought and stopped, reversed if possible. But for pathogens, any changes in conditions represent opportunities for survival.

Likewise, any pathogen that is able to find a novel host in which it can survive even marginally may specialize rapidly in that host. Hence, in evolutionary time, oscillations should become evident whenever opportunity presents itself. Given a defined fitness space of arbitrary complexity, pathogens will oscillate within it until all available fitness space is filled with specialized pathogens.

The Next Step: Exploring and Exploiting . . . and Exploring Again

Our simulations allow us to understand how generalized species can diverge into more than one specialized species, but specialized species remain specialized. Once fitness space is saturated (i.e., each parasite has one host and each host has one parasite), subsequent evolution would involve cospeciation owing to a lack of opportunity for colonizing hosts. If that fitness space were altered by climate change, the only predicted outcome would be net extinction of pathogens and hosts. In other words, there should be only one oscillation cycle. The world of cospeciation would become a world of coextinction.[24] We know this cannot be the case, because we find little support for cospeciation in long time-scale studies and see evidence of rapid and rampant host range expansions—emerging diseases—in real time. This implies that oscillation cycles recur. What determines *that*?

If fitness space is sloppy, there will always be places where pathogens could exist that are not accessible. Inaccessibility can be due to local trophic constraints (the pathogen is transmitted by a vector or intermediate host that does not feed on, or is not eaten by, all susceptible hosts) or geographic isolation (susceptible hosts live in geographic areas where the pathogen does not). Pathogens may exist in a stable host community, but they continue to accumulate variability in isolation, limited to the realized fitness space, increasing the capacity to explore whenever opportunity space changes in such a way that new susceptible hosts become accessible. Even if specialized pathogens retain the capacity to survive in hosts other than the ones they "see," how do they find them, setting off a new round of generalizing that then leads to the emergence of new specialized associations? How do we get new diversity if the internal dynamics of the oscillation hypothesis are not sufficient? We think that repeating oscillations require an external catalyst—environmental perturbations beyond the control of evolving systems—that open up new opportunities for oscillation modules.

Summary

If fitness space is sloppy, even highly specialized pathogens have capacities greater than the opportunity of the moment. Any change in opportunity, therefore, can result in rapid emergence of new pathogen-host associations without having to wait for the evolution of new genetic information.

Those of you for whom alarm bells went off three chapters ago were right to be concerned. The ability to oscillate between generalizing and specializing—depending on the opportunities—is a built-in feature of the nature of the organism. Organisms really are inherently oscillators, and pathogens are very successful organisms. This is a critical issue for our understanding of EIDs and our decisions about what to do to mitigate their impact on us. We live in a heterogeneous and changing world. The capacity of pathogens will become evident as the changing environment provides the opportunity to connect fitness space that is currently discontinuous. The key to understanding what we should expect in the near future is linked to a full understanding of what has happened in the past. This is especially significant at a time when species distributions are changing as a consequence of climate change. After isolation, accumulation of variability increases the realized fitness space of a pathogen, allowing it to become an EID whenever the opportunity presents itself. We will show you in the next chapter that almost every case of EIDs in humans or domesticated plants and animals is a consequence of altered opportunity catalyzed by environmental change.

We next investigate how perturbations beyond the control of any single evolving system open up new opportunities for oscillation modules. We will show you what happens when the surroundings change and why we call this opportunity space. This will force us to confront the essential connection between climate change and emerging diseases.

There will be good news and bad.

SIX

Resolving the Parasite Paradox II: Coping with Changing Opportunities

Evolutionary diversification requires repeated opportunities to generalize and to specialize in fitness space. Changes in the nature of the conditions set the stage for both.

A biosphere in which fitness space was fixed and all that mattered was the nature of the organism would be simple. Pathogens entering that space would oscillate between exploring and exploiting until all viable hosts were inhabited to the greatest extent possible. They would then enter an existence in which time passed slowly and change was restricted to rare events that serendipitously enhanced their abilities to survive.[1] Such a world would have fewer pathogens infecting fewer hosts than the world we experience; pathogens would become locked into cospeciating with their hosts because they lacked the opportunity to encounter other hosts. This would not lead to the indefinite diversification that has characterized the history of life, including a seemingly endless supply of pathogens.

The history of life on this planet is one of selective diversification, not of selective replacement.[2] Evolutionary explanations should thus not focus solely on "conflicts of interest" arising from too many organisms wanting the same resources. More important to understand are resolutions of those conflicts that allow both of the groups involved to continue to coexist. Ecological fitting can lead to both generalizing and specializing in sloppy fitness space. And generalizing is the key to diversifying conflict resolution. But if fitness space does not change, it should be saturated with specialized species, each awaiting extinction the next time the climate changed. Without new generalizing, neither basic survival nor the possibility for novel specialized associations to emerge exist. We have yet to complete the story.

The final part of our tale involves understanding how pathogens cope with altered fitness space. Terrestrial life has been shaped by environmental changes that alter fitness space, ranging from constant fluctuations on small geographic scales lasting short periods of time to rare perturbations affecting the entire biosphere, and all combinations in between. Changes in the environment perturb living systems by increasing or decreasing connectivity in their fitness space. This alters the landscape of opportunity. Decreasing connectivity is an invitation to exploit locally isolated arenas, setting the stage for innovation and specialization. Increasing connectivity is an invitation to explore, using the capacity for ecological fitting embodied in the hereditary history of each species. This sets the stage for generalization. Perturbations that increase fitness space connectivity, therefore, should be the catalyst for generalization-biased evolution, which sets the stage for specialization-biased behavior when environmental stability decreases connectivity in fitness space. This leads us to the ultimate kind of oscillation, in which changes in the nature of the conditions lead to alternating episodes of generalizing and specializing, and those leading in turn to indefinite evolutionary diversification.

This is not a new notion. Formal ideas about oscillations in the nature of the conditions emerged in ecology more than a century ago. At that time, ecologists tended to think of dynamic changes as cycles rather than oscillations. On short temporal and limited spatial scales, they imagined an ebbing and flowing of a defined set of species, each with defined functional traits, in and out of a given area. Little net diversification occurs, even though organisms respond to changing environmental conditions.

Philip Darlington spent his career doing intensive natural history observations—mainly of carabid beetles—focused on an interest in how species came to live in the places where he found them. His 1943 monograph based on his encyclopedic observations of insect faunas on mountaintops and islands led him to a general framework in which new species as well as new ecological characteristics arose in "centers of diversification" with geographical ranges periodically fluctuating around a more stable, continuously occupied center as a result of environmental perturbations.[3] Darlington proposed that these biotic fluctuations might be interrupted by the formation of barriers to dispersal, producing episodes of isolation in which new species arose and ecological diversification might occur. Breakdown of those barriers produced new episodes of biotic expansion, setting the stage for yet more episodes of isolation. Populating the planet with species was thus a matter of sequential colonization emanating from the center of origin for any given taxon. Darlington's natural history observations also supported the long-held view that larger islands have greater species richness than smaller islands. He found similar patterns held for different-sized mountaintops.

Darlington's observations about the size of an area and the number of species living in it convinced another Harvard researcher, E. O. Wilson, to focus on the number of species living in a particular area more than on the manner in which they were formed. Wilson coined the term *taxon cycles* to refer to this extension of Darlington's framework.[4] Groups of species living together and affected by short-term, small-scale environmental perturbations disperse actively and colonize new areas when environmental changes allow expansion, then contract their ranges when environmental changes reduce the amount of suitable habitat. These taxon cycles occur without producing new species and without new species arriving in the stable, continuously occupied center of the geographic distribution of the species involved. Taxon cycles occurring on intermediate time scales and affecting whole communities are often associated with *succession*. Successional changes occur when species residing in an area alter their surroundings to such an extent that they can no longer live there; think about the transitions leading to old-growth forests. The way in which they alter their surroundings, however, leads to their being superseded by species that take advantage of the changes brought about by the previous residents. In its longest time-scale version, a series of successional events eventually brings the area back to the initial state. Thus, even on long time scales, there is still no net diversification inherent in the framework. Nor is there any thought given to what the excluded species do once they exclude themselves, and yet they must survive somewhere if they are to make a reappearance at the end of a successional cycle.

Collaboration with Robert MacArthur formalized Wilson's elaboration of Darlington's ideas. Their *equilibrium theory of island biogeography* was based on the view that dispersal from "source" areas to "islands" (actual or metaphorical), mediated by island size and distance, produces a linear log-normal relationship between species richness and the size of an island, called the *species-area relationship*.[5] Changes in the species-area relationship result from a dynamic balance between immigration—that is, colonization from a source area—and extinction. Any given "island" can support only a certain number of species, called the equilibrium number of species. When the island has fewer than the equilibrium number of species, it is open to colonization. When the equilibrium number of species is reached, no new species may colonize the area without displacing one already in residence. MacArthur and Wilson acknowledged that "local speciation" would confound the species-area relationship but suggested that for most cases it was probably safe to omit the production of new species within an area from the model, because its effect on the species-area relation is "probably significant only in the oldest, largest, and most isolated islands."

Within a decade, those pesky phylogeneticists began making trouble. A group of them suggested that it was not the dispersal of existing species, but the formation of new species, that created generalities in biogeography. They postulated that geographic fragmentation of entire biotas caused by the formation of barriers catalyzed geographic speciation in many species isolated at the same time in the same way. The patterns they envisioned reinforced geologists' newly rediscovered support for continental drift. Darlington, leaning on the opinions of geologists of his day, did not support the theory of continental drift and thus had to assume that all geographic speciation occurred as a result of populations dispersing across preexisting barriers. The form of geographic speciation advocated by the phylogenetics people that came to be called *vicariance* (and the research program vicariance biogeography), was a resurrection of David Starr Jordan's *law of geminate species* proposed half a century earlier.[6]

Island biogeography and vicariance biogeography are complementary theories, each one focused on what is neglected in the other. Island biogeography, as MacArthur and Wilson acknowledged, neglects assessments of the geographic origin of species, even though this is what distinguishes a source from an island. Vicariance biogeography neglects postspeciation movements, and yet this must be how ancestral species become widespread enough to be affected by vicariance. If each theory describes something valid, perhaps integrating both would provide a more complete framework.

Taxon Pulses and Species Distributions

In 1979 Terry Erwin, an entomologist specializing, like Darlington, in carabid beetles, proposed the *taxon pulse hypothesis*.[7] Erwin stated that his framework stemmed from Wilson's concept of taxon cycles and was intended to account for both dispersal and speciation in isolation, including ecological diversification, on phylogenetic time scales. In contrast with taxon cycles, taxon pulses occur over relatively long periods of time and are characterized by expansion into suitable habitat when previous barriers break down, altering previous source-island relationships. During an expansion phase, different species within a biota encounter additional geographical heterogeneity, including range contractions. Such heterogeneity may (1) stop the expansion of some species, resulting in species of restricted distributions; (2) affect only the rate of expansion for some species, producing widespread species; or (3) act as barriers to dispersal of sufficient magnitude to produce new species. Geological evolution, operating on longer time scales than biological evolution, may also produce relatively static or impermeable barriers, resulting in episodes of speciation by geographic isolation affecting members of these same biotas similarly (vicariance).

The taxon pulse was poorly discussed in the last 20 years of the twentieth century, even by those who recognized shortcomings in the vicariance biogeography program.[8] James Liebherr and colleagues, also entomologists who explored carabid beetles, proposed a set of phylogenetically based criteria for distinguishing taxon pulses from taxon cycles.[9] An explicit methodology for examining the phylogenetic context of geographic distributions in a way that could detect taxon pulses if they occurred, however, was still not available. That changed in the first few years of the twenty-first century. The noted paleontologist Bruce Lieberman published a series of landmark paleobiogeographical studies that expanded our view of diversification on the deep-time scales that confront paleontologists.[10] He noted that large-scale biotic diversification would likely occur over large enough spatial scales that historical biogeographical relationships would comprise a combination of general episodes of isolation such as, for example, pieces of continental land masses under fragmentation, as along with general episodes of connection as continental land masses collided and fused. By 2005 Lieberman's work had led to a unified methodology for detecting the historical signal of taxon pulses.[11]

Taxon pulses are historically repetitive, meaning that biotas resulting from them are made up of species that have been associated with each other for varying lengths of time and arrived under varying circumstances. Biotas assembled in this manner are complex mosaics resulting from the mixture of episodic expansion, isolation, and mixing during new expansion.[12] Diversification is driven by biotic expansion

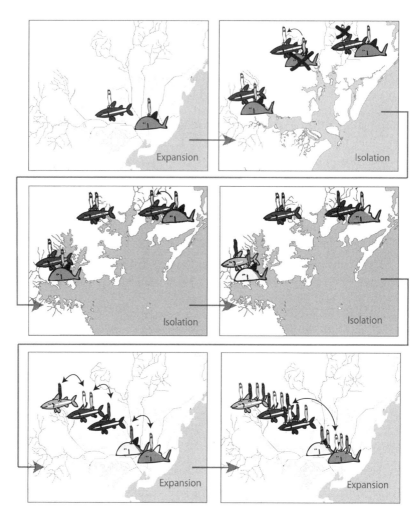

FIGURE 6.1 Depiction of the taxon pulse, a story of expansion and isolation over time. Stage 1: At the beginning, there are two species of freshwater fish and their monogenoid parasites with widespread distributions across a complex river system with near complete connectivity among a number of complex tributaries. Stage 2: Initial isolation episode. Sea level begins to rise, trapping fish and parasites in three different tributaries. Following isolation, we see exploitation and persistence by parasites, extinction of host species (X), and persistence by colonization (arrow). Extinction of a parasite species (X) and a subsequent colonization and speciation occur on a new host fish (arrow). An initial bout of isolation leads to increasing diversity for the parasite fauna. Sea level is falling again, and fishes and parasites reestablish connectivity and contact with this expansion phase, leading to exploration and new opportunity for colonization and new host-parasite associations. Episodes of expansion and isolation alternate between patterns of exploration and exploitation for parasites and increasing diversity over time. Based on distributional patterns revealed by Patella et al. (2017).

as well as isolation, so we expect to find general patterns associated with both phenomena. Episodes of biotic expansion, especially those involving large areas, will inevitably lead to complicated historical relationships among areas, characterized by biotas comprising species of different ages derived from different sources. Figure 6.1 depicts how taxon pulses work.

In this not-so-hypothetical example we view the coastal regions of southern Brazil during an extended history of changing sea levels, such as those during the glacial ages.[13] Initially, sea levels are low. This exposes an extensive and intricately connected river basin within which live—among many others—two species of fish, striped and gray, each inhabited by a single species of monogenoidean (a flatworm parasite living on the body or gills). Time passes, sea levels rise, and the accompanying marine transgression covers some coastal land with salt water, which decreases connectivity among the tributaries of the river system. The basin is transformed, with three separate rivers emptying into marine bays. The populations of our focal fish and their monogenoideans are isolated and eventually become different species. In one river system, the gray fish host goes extinct but its monogenoidean survives by colonizing the striped fish. In another river system, the monogenoidean associated with striped fish goes extinct and its place is taken by colonization of the monogenoidean from the gray fish. Time passes, sea levels fall, and a single large river system reemerges. Hosts and monogenoideans expand into this newly accessible fitness space, coming into contact. The parasites have the opportunity to expand their host ranges, and some of them do so, producing yet another mix of pathogen-host associations. When the next episode of isolation by marine transgression occurs, the parasites and hosts in each isolated area have already formed complex associations of different ages. Speciation again unfolds in isolation. Sea levels drop, the rivers are once again connected, and the complex of pathogen-host associations in each isolated river mixes, forming yet more pathogen-host associations. After only two iterations of this process, the pathogen-host associations that began with just two parasites and two hosts now make up a diverse pathogen-host biota.

Taxon Pulses and Emerging Disease

Climate acts indirectly by favoring other species.[14]

For us, the world is populated by two kinds of species—pathogens and hosts. Both are affected evolutionarily by the same abiotic and biotic variables in the nature of the conditions. The opportunities may be so specific that they affect only certain taxa, or so general they affect entire biotas. Sometimes those changes increase connectivity in fitness space, and sometimes they decrease that connectivity. Sometimes the changes are so ephemeral that they cause no permanent alterations in the composition or fitness space requirements of species making up those biotas; these are classic taxon cycles. These are also in the realm of short-term processes at local scales that influence the emergence of disease.[15] When the changes are more substantial in force and duration, micro- and macroevolutionary changes occur.[16] The taxon pulse is a label for the ways in which different taxa take advantage of opportunities provided by these latter kinds of changes, which oscillate between increasing and decreasing connectivity in fitness space. And when pathogens are given access to novel fitness space, they are capable of exploring it and exploiting all viable parts of it. Climate change is a primary source of current perturbations of biological diversity, and taxon pulse studies suggest this has always been true.[17] This is the basis for the triad of environment, host, and pathogen that remains central to understanding persistence, transmission, and the potential for disease emergence for all pathogens in all hosts.

At this point you the reader may be thinking that environmental perturbations are always a good thing. Time for us to remind you that in biology *it is never just one thing, and it always depends.*

It is true that oscillations in opportunity space have led to the indefinite biological diversification that characterizes the history of life on this planet and sustains our existence. But is it also true that technological humanity as we know it may not have arisen if the past 10,000 years had not been one of unusual climate stability. And that has given us a false sense of security. Yes, there were periodic volcanic eruptions and bursts of disturbances from storms, earthquakes, droughts, and floods. But until very recently the world was large and slow, and there was no Internet to spread the news of environmental perturbations in distant places, so for most humans such events were rare. The nature of our existence was fundamentally one of highly predictable environmental cycles, interrupted only intermittently by catastrophes.

As our world shifts into a period of constant and accelerating change, we are late to the party. Climate change in all its manifestations is neutral and dispassionate with respect to any given species, even us. And all species, including the 50 percent or more of Earth's biosphere that are pathogens, respond to environmental perturbations whether or not we are paying attention. In her groundbreaking book

Evil in Modern Thought, Susan Neiman suggested the Lisbon earthquake of 1599 was the catalyst for Western society to begin distinguishing between natural evils, which we now call natural disasters, and moral evils.[18] From that time forward, humans would increasingly see some environmental perturbations not as "messages from God" punishing some group of humans for their sins, but as natural phenomena that humans needed to deal with. We are still working on that last part, and this book is in a sense yet another exhortation for humanity to take collective responsibility for coping with natural phenomena.

We cannot cope with a future full of emerging diseases if we do not understand what is happening now. And we cannot understand that without understanding evolutionary origins. We perceive evolutionary history as a series of snapshots faded in proportion to their age. Their meaning becomes apparent only if we see them as part of a single complex but knowable history of life from which today's world emerged. The Stockholm Paradigm allows us to connect those snapshots into the real-time video in which we are immersed.

So, what are the implications of the taxon pulse dynamic for emerging diseases? First of all, periods of climate stability are usually times of isolation and specialization, of stasis and largely disconnected fitness space. Environmental changes that follow such episodes lead to biotic expansion events. Species that cannot tolerate the changes in the place where they have been living leave as fast as they are able, changing from exploitation to exploration mode as they seek familiar conditions. As they leave, they pass other species that have vacated their abodes and are in the process of discovering that they like the newly vacated territory. Biotic mixing is under way. A few species stay in place and survive, but they must cope with the loss of old neighbors and the arrival of newcomers. In some cases the new arrivals will be functional replacements for species that have recently left, maintaining preexisting trophic structures. In others the combination of departures and arrivals will produce novel trophic linkages. This produces an enormous amount of potential fitness space to be explored and exploited. And we now know how efficient pathogens are in that regard. The loss of the original host—by extinction or escape—will result in a pathogen's extinction if (a) the vacating host does not take it along when it escapes, or (b) another host does not become available when it goes extinct. Newcomer hosts from adjacent areas will often be closely related to the residents, increasing the ease of host range expansion by ecological fitting. Residents that were not in contact with the pathogen before the biotic expansion event may find themselves available; this is yet another source of potential for pathogen survival. The mixing process itself may provide opportunities for host range expansion by the stepping-stone mechanism.[19] This is why most host range expansion occurs in evolutionary bursts during episodes of increasing connectivity in fitness space associated with biotic expansion catalyzed by climate change. Pathogens are climate change's ultimate survivors, a lesson we are only beginning to learn.

Schistosomes are a group of trematodes living as adults in the circulatory system of various mammals and birds. They have a deep evolutionary history with vertebrates but are best known as a helminth group that causes great human suffering—schistosomiasis or bilharziasis. Seven of the 23 known species in the genus *Schistosoma* infect more than 240 million people in 76 countries, and as many as 700 million people are thought to be at risk of exposure in Asia, Africa, and South America.[20]

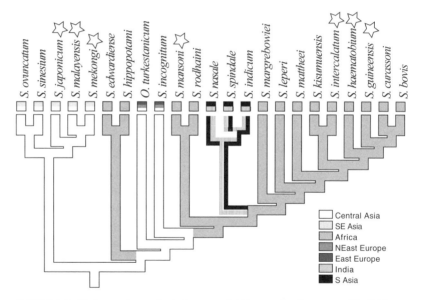

FIGURE 6.2 Phylogeny for species in the genus *Schistosoma* showing geographic distributions mapped onto the parasite tree to demonstrate the history of this group of blood flukes. Diversification demonstrates alternating patterns of expansion and isolation from an ancestral Asian distribution. From the Asian trunk of this tree, one event brought parasites into western Eurasia, and two episodes of expansion occurred bringing them into Africa. From Africa, geographic colonization occurred with expansion and isolation of the Indian subcontinent and southeast Asia. Humans are hosts for seven species in either Asia or Africa; this is associated with discrete events of host colonization from other mammals (stars) and in association with aquatic habitats.

Schistosome evolution is a story of snails, vertebrates and continents. Schistosomes have snail intermediate hosts, and the swimming infective stage called a cercaria directly infects a vertebrate host by penetration of the skin. Their evolutionary history clearly shows the taxon pulse signal (fig. 6.2).[21] The geographic distributions of schistosomes indicate repeated episodes of expansion and isolation events linking Asia and Africa beginning in the Miocene (fig. 6.3). The broad outlines of schistosome history reveal an origin in Asia (the *japonicum* group). Two episodes of expansion brought schistosomes to Eurasia and eastern Europe (*S. turkestanicum* and *S. incognitum*). Two more brought them to Africa (the *hippopotami* group, and the *mansoni* group + *haematobium* group), one of which was later the source of a fifth expansion event, from Africa onto the Indian subcontinent (the *indicum* group).

These geographic expansions involved host range changes. In Africa host range expansion involved both new mammalian groups, such as hippopotamus, and new snail groups; Asian schistosomes use members of a group of snails called pomatiopsids, while African schistosomes use members of a group of snails called pulmonates.[22] All seven species of schistosomes known to infect humans are the result of episodes of host range expansion at various times in Asia and Africa.

Schistosome evolution also seems to be part of a larger story. Two groups of roundworm parasites inhabiting Old World monkeys and great apes support an earlier, though still Miocene, episode of expansion from Africa to Asia.[23] From that point onward the roundworms and the schistosomes seem to be part of the same evolutionary story of expansion and isolation. The only difference is that one of the roundworm parasite groups infecting humans arrived in the New World from Eurasia with the original human colonizers of the continent, while schistosomes (*Schistosoma mansoni*) arrived much later, when infected Africans were enslaved and taken to the New World.

The schistosome example shows how taxon pulses can set and reset the stage for substantial complexity in pathogen-host associations through altering geographic and ecological connectivity.[24] We expect host colonization to be maximized during phases of biotic expansion, whereas geographic isolation promotes the emergence of specialization. In other words, host range expansion will be correlated with geographic expansion and opportunity; functional diversification and the potential for coevolutionary dynamics will be correlated with isolation events. Taxon pulses set the evolutionary stage for the ecological play involving geographic mosaics of coevolution that have outcomes rang-

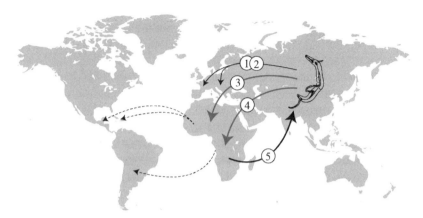

FIGURE 6.3 Geographic history for schistosome parasites showing episodes of expansion and isolation in a taxon pulse producing contemporary distribution and diversification of these flukes. *Schistosoma* originated in Asia, with subsequent episodes of expansion (solids lines and arrows) into Africa (3, *S. hippopotami/S. edwardiense*), western Eurasia (1, *S. turkistanicum*; 2, *S. incognitum*), Africa (4, *S. mansoni* and *S. haematobium* groups), and then from Africa back to the Indian subcontinent and southern Asia (5, *S. nasale* group). In the sixteenth century, *S. mansoni* was secondarily introduced (broken lines, arrows) from Africa to regions of South America and the Caribbean with the transport of infected African slaves.

ing from isolation on local landscape scales, with ephemeral emergence of local mosaics of disease, to the development of regional faunas.[25]

A recent estimate suggests that 58 percent of more than 1400 viruses, protists, worms, and arthropods known to infect humans were acquired from animal sources.[26] The taxon pulse provides the general mechanism for assembling such complex pathogen-host biotas, but if we want to understand how *we* came to have so many pathogens, we must understand something of human evolution.

One day between six and eight million years ago our ancestors dropped out of the trees. They remained in the forest, emerging three million years ago onto the expanding grasslands and savannahs of central Africa. There was likely nothing to indicate that they were special—they were just one of many lineages, existing, trying not to attract the attention of the big predators that shared their habitats. They had little influence on their immediate environment and did not comprehend larger-scale environmental changes. By the Pliocene they had no idea they were at the beginning of an ice age, though the climate was already cooling, producing fluctuations between wet and dry conditions and influencing the distribution of vital resources from water to plants.[27] Our ancestors somehow survived to the beginning of the Pleistocene, when a number of remarkable changes that had been in the making finally emerged. They had scrambled for grasses, seeds, fruits, nuts, insects, and the occasional slow and unwary rodent, but now their diets shifted. They began to scavenge actively, then turned to predation on antelope and other mammals, birds, and even fish and other aquatic animals.[28] The pointed sticks and sharp-edged stone tools they invented helped move them into competition with the lions, leopards, hyenas, jackals, and possibly remnant populations of African bears that once considered them a slow-moving food source. They may well have eaten one another at times.[29] As they turned the dining tables on the species with which they shared the savannahs, they began to expand across and then out of Africa. By 150,000–200,000 years ago they had become us.[30]

Our presence increasingly changed the rules of the game and the dimensions of the playing field. Early modern human populations may have remained in isolation for nearly 50,000 years, restricted by disconnected fitness space (permissive habitats) determined by prevailing climate conditions.[31] Genetic data suggest a single expansion event led to all non-African populations established at this time period, including the colonization of Australia about 65,000 years ago, although the signatures of earlier events may have been masked by small founding populations and extinctions.[32] It is possible that the first expansion of modern humans from Africa followed a path through the Arabian and Sinai corridors into southern Europe and Eurasia coinciding with periods of increased rains during the last glacial cycle, 107,000–95,000 and 90,000–75,000 years ago. Dry conditions that obtained 71,000–60,000 years ago blocked further expansion from Africa to Eurasia, but expansion toward southeastern Asia was open. Falling sea levels approximately 60,000 years ago established connections between Papua New Guinea and Australia, making additional human expansion possible. During a precession minimum about 59,000–47,000 years ago, monsoonal rainfall returned to North Africa. Expanding primary productivity and the reopening of corridors to the Arabian Peninsula facilitated a third major expansion of humans out of Africa. Later waves of expansion by Stone Age hunter-gatherers from Africa continued during the rest of the late Pleistocene, although with increasing variability in habitat connectivity resulting from dramatic climate events that marked the termination of the Northern Hemisphere glacial cycles.[33]

As humans began to occupy more of the biosphere, episodic climate changes, such as extremes of temperature during the Ice Age, influenced migration opportunities.[34] Modern humans arrived in the

Western Hemisphere by crossing Beringia about 15,000–23,000 years ago near the apex of the last glacial advances.[35] More archaic human populations were already present—Neanderthals especially in Europe and the enigmatic Denisovans in central and southeastern Asia. This resulted in a mosaic of modern and more archaic human lineages, often in contact, enhancing the potential for pathogen exchange during the Pleistocene.

That period also saw us emerge as a technological species, paving the way for ever increasing population and a broadening global footprint. This in turn dramatically influenced the arena for emerging infectious diseases. We have maintained some of the pathogens that infected our ancestors living in the African savannahs more than two million years ago, but we have acquired even more. For most of human history our populations were small and isolated, and our interactions with the environment were intimate, bringing us into contact with many potential pathogens. Africa may have been the initial source for many pathogens infecting humans and their crops and livestock, but wherever we have gone and having survived whatever episodes of climate change we have experienced, we have accumulated pathogens.[36]

Pathogen spread and sharing among early humans was a local phenomenon, limited by distance and population density. That began to change as our technological advancement found great comfort and security in a major climate shift at the termination of the glacial age around 12,000 years ago. We call the resulting extended period of highly unusual climate stability the Holocene or the Neolithic. It is at the beginning of this time that we find the first evidence of humans adopting a sedentary lifestyle.[37] Agriculture and the domestication of animals as food and companions rapidly emerged and took hold, then just as rapidly spread.[38] Domestications were complex events, often involving multiple geographic centers that framed the future trajectories for many parasites and diseases.[39] This included a rich legacy of pathogens; groups of roundworms in domestic cattle and sheep have origins in Africa where diversification extended over the past five to ten million years.[40] So, along with all the benefits of agriculture and domestication came a cost: an increased potential for sharing pathogens resulting from the increasingly interdependent associations between us and the species we had domesticated. Many of our contemporary pathogens originated in the increasingly intimate links among humans and food and companion animals emerging from domestication.[41] Even more, as sedentary living evolved into true urbanization, human living space provided viable fitness space for yet more animals, like rodents, that carried yet more pathogens happy to add humans to their repertoire.

A Tale of the Tape(worm)

Pathogens associated with our domestic animals tell stories of human-related food chains, centers of agriculture and urbanization and geographic expansion but are nearly always rooted in deeper history. A group of tapeworms that inhabits humans around the world exemplifies this. Species of the genus *Taenia* are parasites of terrestrial carnivorans that circulate in well-defined predator-prey systems. Predators, such as felids, canids, hyaenids, and ursids, among others, host approximately 50 *Taenia* species, with new species being found almost annually. Herbivores such as ungulates (hoofed mammals) and rodents harbor the infective larval stages, and the final, carnivoran, host acquires the tapeworm when it eats prey infected with such larvae. Parasite transmission dynamics are thus oriented around connections established in foraging relationships among particular groups of carnivorans and their prey. Host range shifts among carnivoran groups have had a prominent influence on diversification and distribution among most of these tapeworms. Three species of *Taenia* are obligate parasites in people—*Taenia saginata*, *T. asiatica*, and *T. solium*—and their larval or infective stages are found in the musculature or liver of domestic animals. *Taenia solium* has an especially broad intermediate host range that includes humans, other primates, domestic dogs, and swine.

For many years scientists believed that these tapeworm species had become associated with humans only since the domestication of food animals, such as cattle or swine. The story was that these parasites were originally found as adults in canids and as larvae in ungulates that the canids ate. When we domesticated dogs and ungulates and began eating the ungulates, we acquired the heretofore dog tapeworms. This would make the association between humans and *Taenia* spp. no more than 15,000 years old.[42] Phylogenetic analysis of *Taenia* established a novel ecological, geographic, and temporal setting for their association with humans as definitive hosts.[43] *Taenia* spp. that inhabit humans are older than previously thought, likely associated with early humans in the late Pliocene and Pleistocene (ca. three to two million years ago; see fig. 6.4).[44]

In addition to genetic estimates suggesting great age, the common ancestors of the *Taenia* spp. infecting humans were parasites not of wild canids or dogs, but of lions, hyenas, and bears. This leads us to

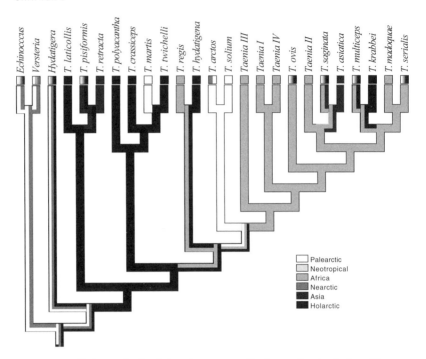

FIGURE 6.4 Phylogeny for *Taenia* based on a composite of current molecular-based trees depicting relationships among these tapeworms. Nakao et al. (2013); Terefe et al. (2014). We have mapped the geographic distributions of these tapeworms on the tree. *Taenia saginata* + *T. asiatica* and *T. solium* are not closely related to each, indicating that people have become hosts on two separate occasions, events associated with dietary shifts by our ancestors from scavenging to carnivory. The origin of the ancestor of *T. saginata* + *T. asiatica* appears to be associated with sharing scavenged or hunted prey that host larval *Taenia* with hyenas; for *T. solium*, colonization of hominins involved sharing scavenged or hunted prey that host larval *Taenia* with bears.

conclude that early humans likely acquired *Taenia* spp. as a result of more than one independent event of host range expansion by sharing scavenged or hunted prey infected with larval *Taenia* originally shared with carnivorans. *Taenia solium* is the sister species of *T. arctos*, a parasite of Northern Hemisphere bears in the genus *Ursus*, and that pair of species occurs within a broader assemblage of Holarctic *Taenia* spp. infecting felids, canids, and hyaenids.[45] Although *Taenia saginata* + *T. asiatica* are sister species, their most recent common ancestor was the sister species of *T. crocutae*, a parasite of hyenas.

What we know is consistent with the African origins of the ancestors of *Taenia* in humans (fig. 6.5). We can't be certain of African origins until we have a more complete phylogeny; the real story will certainly be more complicated. There is the potential that independent events of host colonization leading to the common ancestor of *T. saginata* and *T. asiatica*, possibly involving our cousin *Homo erectus*, may have taken place in Eurasia, where hyaenid and bovid diversity was well established in the Pliocene. The common ancestor for *T. solium* + *T. arctos* may have originated in Sub-Saharan Africa, with divergence of the sister species in the Palearctic. There are only two estimates of the age of divergence within this assemblage: for *T. saginata* + *T. asiatica*, divergence is placed at 0.78–1.71 million years ago;[46] and the Asian and African/European genotypes of *T. solium* have been dated similarly at 0.80–1.3 million years ago.[47] This means the original host expansion events by the ancestors of *T. solium* and of *T. saginata* + *T. asiatica* are even older. These conservative estimates seem to denote a common history of expansion out of Africa followed by isolation in western Eurasia

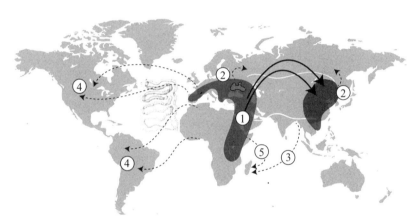

FIGURE 6.5 Major geographic episodes in the evolution of *Taenia*. Distributions reflect complex relationships of expansion and isolation across regions and continents. The initial radiation of human species may have occurred in Africa or the west-central Palearctic, with expansion across Eurasia (1). Isolation in the western Palearctic and eastern Eurasia (2) led to the divergence of *T. saginata* + *T. asiatica* and the recognized genotypes of *T. solium* in the Pleistocene. During Neolithic times and during the era of domestication of cattle and pigs, these parasites were acquired once they colonized our primary ungulate food animals. *Taenia asiatica* remained endemic to Asia, but *T. solium* and *T. saginata* became widely distributed, coincident with trade routes and the expansion of animal husbandry (dotted arrows). For example, *T. solium* became distributed in Madagascar following introductions of infected people or swine nearly 2000 years ago (3) and again during the slave trade in the past 200 years (5). Pig and cattle tapeworms were brought to the Americas with European exploration and occupation and the slave trade after the 1500s (4). These characteristic parasites that circulate in our primary food animals now have broad cosmopolitan distributions, with introductions throughout the world.

and in southeastern Asia near the middle of the Pleistocene, predating the origins and expansion of modern humans out of Africa.

No matter the details of place of origin, *Taenia* spp. infected humans long before domestication arose, at a time when the earliest humans left evolutionary traces of transitions from foragers eating primarily plant material to aggressive scavengers competing with animals such as hyenas, to active predators competing with species such as lions, tigers, and leopards on the African savannahs or the southern Palearctic. At the end of the Pleistocene, humans persisting in small groups of hunter-gatherers may, owing to our demographics—small groups, constantly on the move, having limited contact with other groups—have been remarkably free of pathogens that could circulate directly from human to human.[48] Nonetheless, they already hosted *Taenia* spp. as a result of their hunting success.

We acquired *Taenia* spp. from our carnivoran neighbors through these competitive acts of predation on at least two occasions during the transition by early humans from forest to savanna habitats.[49] Much later we domesticated the intermediate hosts of the tapeworms and took the pathogens with us on our explorations of the planet. Expansion across the western Palearctic and southern Asia may have set the stage for geographic isolation in southeastern Asia and the divergence of *T. saginata* and *T. asiatica* and the genotypes of *T. solium*. The timing suggests that *Taenia* spp. circulated among hominins prior to the origins of modern humans and well ahead of the domestication of cattle and swine.[50] Following their expansion from Africa, these *Taenia* spp. became widely distributed within but restricted to the Old World.

Our journey into the Neolithic is about the link between domestication and disease. The roots of agriculture and domestication of swine and ruminants had origins in the Middle East and in Asia between 9000 and 10,000 years ago.[51] If humans had already acquired *Taenia* by that time, our modern food animals, particularly cattle and swine, may have unwittingly become intermediate hosts during the process of domestication.[52] This would have involved *T. saginata* in cattle and *T. asiatica* and *T. solium* in swine, along with a possibility that the *T. solium* life cycle might have been maintained—at least to some extent—in some human populations through cannibalism.[53]

Taenia solium has discrete African and Asian genotypes reflecting isolation and divergence in the Pleistocene,[54] enhanced or maintained into the present through the discrete nature and regional history of swine domestication.[55] Genetic data reveal a more recent history of global introduction and establishment for this pathogen. Fifteenth-and sixteenth-century trade routes linking Africa and Europe, the Americas (Peru and Mexico), the Philippines, and southern Asia were especially prominent in the introduction and dissemination of *T. solium*.[56] In the Americas, populations of *T. solium* were derived from multiple introductions with people or swine from African and later European sources, but not from Asia. Asian and African genotypes occur in Madagascar, illuminating an unexpected and earlier history for the timing of human dispersal and colonization of the island from different sources.[57] Madagascar populations of *T. solium* with Asian affinities are related to those in northern Nepal and only distantly linked to insular southeastern Asia, which had been assumed to be the source. African genotypes arrived considerably later, only about 200 years ago, the result of the slave trade between Africa and Madagascar. Because early humans initially inserted themselves into African foraging guilds with carnivorans and ungulates, the initial colonization events for *Taenia* may not have been so strongly dependent on their population density, because other hosts helped maintain the parasite. Since becoming established in species capable of long-distance movement, however, their history has been one of expansion and mixing of early and later human populations and their domestic animals.[58] This served to expand the spectrum of human pathogens and also broadened the geographic and ecological distribution of pathogens among those organisms that we brought as food or companion animals.[59]

As our human ancestors expanded geographically, they would have been exposed to an array of vector-borne pathogens, each one initially localized by the vector distribution. Like *Taenia*, some of these pathogens could have been brought by humans to new places where susceptible but previously unexposed vectors waited, unaware that their lives were about to change. Scientists believe this kind of phenomenon has occurred often, leading to geographic breakouts of pathogens through the acquisition of new vectors (e.g., West Nile virus, chikungunya virus, Zika virus) or newly encountered susceptible domesticated or free-ranging host species.[60]

Many other pathogens of humans and livestock likely have an "out of Africa" origin and a complex history of sharing. The roundworm parasite that causes human river blindness in Africa, *Onchocerca volvulus*, was derived from a bovid host.[61] That particular parasite has not expanded into other parts of the Old World but was introduced to South America possibly as early as the 1500s, coincidental with the African slave trade. There the organism found a suitable vector in native black fly hosts, and the parasite was established through a classic case of host range expansion via ecological fitting. *Schistosoma mansoni* evolved in Africa nearly 120,000 years ago, derived from a host range expansion by an ancestral schistosome infecting other mammals. Like *Taenia* and *Onchocera, S. mansoni* was introduced to South America and the Caribbean about 600 years ago as a result of the African slave trade associated with colonization of the continent by Europeans. Schistosomes can only be acquired when water-borne larvae penetrate the host skin; this may have been associated with sedentary living near permanent water bodies or the emergence of fishing. A number of other pathogens of humans that are transmitted in an aquatic milieu seem to have been acquired as a result of increasing reliance on aquatic habitats.[62]

The bacterium that causes tuberculosis, *Mycobacterium tuberculosis*, emerged initially from Africa with our ancestors.[63] The bovine strain of this pathogen was derived much later as one of the unintended consequences of domestication. Human expansion carried this pathogen and disease throughout the Old World during the past 6000 years, an increasingly familiar theme. And, like *Taenia*, tuberculosis was thought not to have made the trip across Beringia and was unknown in the New World until arrival of the Europeans. When evidence of tuberculosis was discovered in Peruvian mummies predating European contact, that narrative seemed to fall into in doubt. It turns out, however, that the tuberculosis that plagued those ancient Peruvians was not *M. tuberculosis*, but another species more commonly found in local sea lions, an early case of an emerging disease waiting to happen.[64]

Of course, not everything that is in Africa began in Africa. Many gastrointestinal parasites in free-ranging ungulates have origins in Eurasia and more northern latitudes. Many of these parasite groups expanded from northwestern Eurasia into North America across Beringia in the past three million years, taking advantage of climatic conditions that precluded groups originating in warmer areas.[65]

Domestication of cattle and sheep was at the core of introductions of the liver fluke, *Fasciola hepatica*, which enjoys a global distribution.[66] Arising in the central Palearctic, this species was widely introduced throughout the temperate areas of the Old World coincidentally with developing trade accompanied by the spread of domestic livestock and invasive gastropod intermediate hosts. As with the history of the related giant liver fluke, *Fascioloides magna*, that we discussed earlier, establishment in new localities in western Europe involved host range expansions for both novel final (deer and their relatives) and intermediate (snail) hosts, facilitated by both climate and land-use change.[67]

Roundworms of the genus *Trichinella* have a unique life cycle—the same host is the intermediate and definitive host, and transmission to new hosts involves carnivory or scavenging infected carcasses (e.g., by rodents, carnivorans, other mammals, crocodilians, and some birds). Infective larval stages live in the musculature, for example in swine, and upon ingestion of raw or undercooked meat the parasites develop in the subsequent host, leading to successful transmission and persistence. The group has a history of association with vertebrates as much as 300 million years old. Our particular interest is in the members of a group within the genus characterized by the formation of a tissue capsule around larvae in host muscle that originated in Central Asia approximately 30 million years ago. The group comprises nine species whose radiation involved multiple expansion events from Eurasia into North America with mammalian hosts across Beringia in the Pliocene/Pleistocene, one into the Neotropics from Eurasia with the first small cats nearly 12 million years ago, and two or three events into Africa (fig. 6.6).[68]

Trichinella distributions in Europe and in southeastern Asia reflect

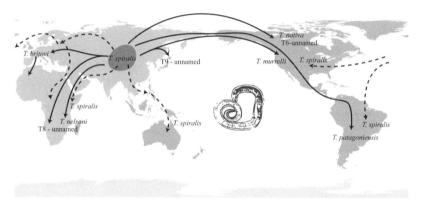

FIGURE 6.6 Global geography and history for species of the encapsulated group of *Trichinella*. The initial radiation began nearly 30 million years in central Asia, where the greatest diversity for *T. spiralis* is observed. Independent events of expansion occurred in the Pliocene (solid lines and arrows) into Africa and isolation with the distribution of *T. nelsoni* (the sister of *T. spiralis*) and for T-8 (an unnamed species). T-8 is the sister of the Palearctic *T. britovi*, which also expanded into Africa about 100,000 years ago. Geographic colonization in the Pliocene through the Nearctic (*T. patagoniensis*) to South America, and during the Pleistocene into North America (events for *T. murrelli* and for T-6 + *T. nativa*), was the basis for the origins of Western Hemisphere fauna. *Trichinella spiralis* became widely distributed in central and western Europe from central Asia, associated with swine domestication and expanding husbandry in the Neolithic (dotted lines). These populations were eventually distributed into the Western Hemisphere after the 1500s with the transport of infected pigs and rats in association with humans (they were synanthropic).

the history of swine domestication.[69] The species of greatest health and economic importance, *T. spiralis*, has a relatively short association with humans, swine, and rats. Asia is one of the centers for swine domestication, and the extensive genetic variability of *T. spiralis* there suggests an association with swine in the Neolithic or even earlier. European lineages of the parasite, in contrast, exhibit genetic uniformity consistent with a single origin about 6000 years ago. European trade and colonization introduced the parasite assemblage to North Africa and the Americas over the past 600 years.[70]

The "barber-pole nematodes" of the genus *Haemonchus* are stomach and abomasal pathogens in ungulates.[71] Although most of the diversification of the 12 species in this genus was confined to Africa, the group appears to have originated in Eurasia and diversified from there (fig 6.7). Most of the diversification encompassed a series of within-Africa taxon pulses associated with a considerable amount of host range oscillations involving many ungulate groups, including the Caprinae, Bovinae, Antilopinae, Giraffidae, and Camelidae. The pulses correspond to identifiable episodes in the Middle Miocene (from 14 to 15 million years ago), the Pliocene (between 3 million and 2.6 million years ago) and the Quaternary (less than 2.6 million years ago).[72] This included multiple cases of stepping-stone host colonization.[73] Four species expanded out of Africa back into Eurasia in the Holocene: *H. longistipes*, associated with domestication of the dromedary and with the subsequent expansion along trade routes for incense and spices from Africa to India;[74] *H. placei* and *H. similis*, primarily associated with domestic cattle; and *H. contortus*, primarily associated with domestic sheep and goats (Caprini). Global introductions to every continent expect Antarctica for the latter three species were coincidental with European colonization after the 1500s, which has seen the host range for *H. contortus* explode to include ungulate hosts of 40 genera.

These examples show us that many pathogens, as well as their associations with us and with the plants and animals upon which we depend for survival, are far older than we imagined. And their histories indicate they have not only persisted through many previous episodes of climate change, they have often taken advantage of those episodes.

Anthropocene Impacts

Domestication and global dissemination of pathogens infecting humans and their crops and livestock are recurring themes in human

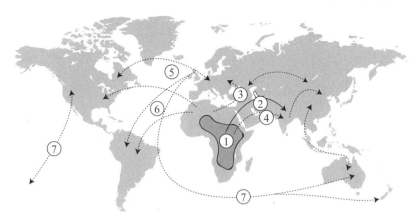

FIGURE 6.7 Geographic history for the species of *Haemonchus* nematodes living in the abomasum of domesticated ruminants. There are 12 species, and this group is primarily limited to Africa, where it has radiated among antelopes and other bovids since the Miocene (1). The complex radiation unfolded with sequential arrival and host colonization of different ungulate groups from Eurasia into Africa. Only four species of these nematodes escaped Africa, an event associated with domesticated cattle, sheep, goats, and camels. Here we show the routes and history of expansion for those species in domesticated hosts. Solid lines are historical events that may not be directly linked to human involvement; broken lines denote anthropogenic expansion and introductions connecting regions and continents; and bidirectional arrows indicate recurring exchange. *Haemonchus longistipes* expanded from Africa across southern Eurasia into India coincidental with host colonization and domestication of the dromedary (2). Neolithic domestication of cattle, sheep, and goats led to subsequent expansion into Asia for *H. contortus*, *H. placei*, and *H. similis* (3). Later trade routes such as the Silk Road, which linked Africa, Europe, and India, further disseminated the fauna (4). European contact and exploration introduced these three species to South and North America, with recurring or overlapping introductions reflecting an exchange of stock late into the twentieth century (5,6). *Haemonchus* was introduced to Australia and New Zealand from European and other sources in the nineteenth century (7). Global introductions led to broad dissemination of *H. contortus* through ecological fitting and host colonization among a considerable number of ungulate species in the Western Hemisphere.

evolution since the Neolithic, especially during the past 600 years.[75] Pathogen circulation for modern humans began in the Pleistocene with early mixing of African and Eurasian biotas.[76] Most pathogens have histories based on transmission patterns that originated prior to the advent of agriculture and domestication and are likely to involve links among Neanderthals, Denisovans, and modern humans. A suite of distinct disease packages from the Pleistocene (some viruses, bacteria, and helminths) and Holocene (plague, cholera, arboviruses, malaria, and helminths, among others) define or characterize this history. Significantly, such diseases as measles, smallpox, pertussis (whooping cough), and a range of other viral pathogens in humans originated via host colonization from domestic stock from Europe in the Neolithic.

As the Neolithic waned, our world became increasingly small and fast, with access to every corner of the planet. Our relationship to climate underwent a transformation. We had been among the passive bystanders to climate, often looking for a cave or later a stone hut in which to shelter. As we marched fitfully into the late Holocene, we saw a growing frequency of military campaigns, invasions, occupations, and population displacement. Long-distance trade along sea routes and land connections grew. Our travels required us constantly to contend with new pathogens and diseases. During this time transmission that originated through exposures to infected livestock became enhanced by crowding but limited by immunity, and both factors increasingly characterized conditions up to the Middle Ages.

As the Middle Ages arrived, the human condition—particularly in Europe—was no less tumultuous and certainly not improving to a great degree. The burden of human disease increased. Expanding trade, religious pilgrimage, and warfare spurred geographic expansion, influencing the spread and emergence of many of the pathogens limited to humans, such as leprosy, smallpox, measles, tuberculosis, and typhus.[77] And to make matters worse, the frigid circumstances of the Little Ice Age jeopardized human health and food security and ended 500 years of Norse occupation in Greenland.[78]

Plague has an especially convoluted history of expansion and pandemic explosion linking Eurasia and western Europe. Caused by the bacterium *Yersinia pestis*, plague is a well-known disease with ancient populations widespread in Central Asia and more restricted occurrences in northeastern Africa, the Arabian Peninsula, southern China and Mongolia, and northwestern India.[79] Plague is one of the iconic human diseases spread by rodents and their fleas. During the past 5000 years this pestilence has swept across the planet, including at least three pandemics that killed more than 200 million people. Plague may have initially spread out of Eurasia into Europe in the Late Neolithic or Bronze Age between 4500 and 3700 years ago with the westward expansion of steppe-pastoralists.[80] The disease returned to Central Asia with subsequent dispersal by human populations, although that history is obscure. The Plague of Justinian swept through Constantinople in the sixth century CE. That emergence seems to have spread with rats, fleas, and people moving northward on cargo boats navigating the Nile from the African Great Lakes region north to Alexandria and on to the Bosporus.[81] The second plague pandemic, known as the Black

Death, unfolded between 1347 and 1665, eventually infecting half the population of Europe. The overlapping trade routes, religious pilgrimages, military conflicts, and human migration due to the recurring crop failures, famine, and torrential rains that characterized the Little Ice Age became threat multipliers for the disease.[82] Rather than being a reemergence of endemic plague following an earlier establishment from Asia, the second pandemic in Europe may be linked to multiple events of geographic expansion into western Europe.

Plague pandemics are attributed to climate and weather patterns, especially in Eurasia.[83] Climate cycles involving a minimum of 16 events catalyzed the episodic recurrence of plague in Asia and expansion into Europe; these were taxon pulses in near contemporary time. In Central Asia warm springs and wet summers synchronize vegetation growth and the increased rodent and flea populations required for transmission. Environmental changes reducing vegetation growth can cause rodent populations to crash within one to two years, whereupon fleas colonize alternative hosts, including urbanized rats and humans. Infections of humans, rats, and camels in trade caravans on the Silk Road requiring a 10-to-12-year journey from Asia to European ports on the Mediterranean facilitated the westward pulse of expansion. A subsequent three-year spread among rats and humans involved connections among the major port cities, leading to 61 recognized maritime introductions around Europe and the Mediterranean between 1346 and 1859. In a slow world dominated by land caravans and maritime trade, substantial lag times were involved between outbreaks in Asia resulting from climate oscillations and episodes of introduction and emergence in western Europe.[84] Recurrent expansion from Eurasia would result in a complex mosaic of plague genotypes, which may account for the persistence of plague in a now-extinct focus identified in France indicating emergences in the 1400s and 1700s.[85] Plague arrived in North America for the first time from southern China and Hong Kong, coming ashore at San Francisco in 1899.[86] All exposures and subsequent outbreaks in North America originated from this introduction. The establishment of plague in North America was part of a third global pandemic in the early 1900s.

Plague is still very much with us, though there has not been a global pandemic in more than a century. Cholera, by contrast, provides a chilling view of recurrent pandemics in recent times and on short time scales. Caused by the bacterium *Vibrio cholerae*, the disease exploded onto the world stage with what is called the first pandemic, sweeping out of India and into China and Asia during 1817–20. There have been six additional global pandemics, all linked to climate and globalization.[87] We are currently in the grip of the seventh pandemic, which originated in 1961 and has circled the globe three times, infecting three million to five million people annually. Cholera emerges most often along the coastal regions of the Bay of Bengal in association with increasing sea surface temperatures during ENSO (the El Niño Southern Oscillation, commonly known as El Niño) events that influence near-shore production cycles. The bacteria become amplified and enter marine trophic structures that connect to the human food chain. ENSO events also warm ocean waters, and cholera requires temperatures of at least 15 degrees C to develop and cause illness. The bacteria are carried along established shipping routes to urban centers in near coastal areas.[88] International shipping and tourism hubs become sites where multiple populations of cholera meet and mix. There, in a manner analogous to the situation with influenza A, recombination among members of different populations increases the production of pathogenic genotypes. Poor personal hygiene and contaminated water supplies allow the bacteria to spread. Drought focuses surface water resources and increases concentrations of bacteria, whereas flooding challenges water security and sanitation infrastructure.[89] Within current risk areas, warming oceans will extend the length of time water temperatures remain above 15 degrees C. Latitudes currently free of cholera may begin to experience water temperatures above the 15-degree threshold for at least some period of time; coastal Alaska and the North Sea could see expansion of a number of *Vibrio* species, including those that are highly pathogenic, within this century.[90]

Some of our most serious disease threats come from pathogens that ravage food crops.[91] Wheat, for example, is the world's largest food crop and along with barley represents 26 percent of the global food supply. It is also one of the most widely distributed and genetically homogeneous food plants in cultivation, and its existence as a global monoculture creates particular vulnerabilities to rapid pathogen emergence. Foremost among these are the rust fungi, three species in the genus *Puccinia* that have long been documented to wreak havoc and that represent an expanding threat today, capable of causing more than $5 billion in losses annually. The genus *Puccinia* includes about 3500 species, most of which are pathogens of wild grasses (family Poaceae). Wheat is one of those species, so it is no surprise that it hosts rust fungi. Each of the three wheat rust species seems to have originated in different parts of Eurasia, where they remained isolated and in obscurity for most of their existence.[92] *Puccinia striiformis*, the yellow leaf rust, for example, originated in the Himalayas and experienced an early separation in China and Pakistan, independent of the radiation for the other wheat rusts.[93]

Rusts are excellent explorers, with broad host ranges and an evolutionary radiation involving extensive host-plant colonization, geographic expansion, and ecological diversification.[94] Humans acquired wheat rusts by domesticating wheat in the Fertile Crescent and then spread it into the Trans-Caucasus region more than 12,000 years ago.[95] Connectivity among farmlands linked previously isolated rust species with susceptible wheat crops in environments favorable for transmission.[96] Like all good oscillators, they took advantage of the invitation to explore. The wheat leaf rust *Puccinia triticina* seems to have colonized wheat from *Aegilops speltoides* in the Middle East. The evolution of the yellow rust, *P. striiformis*, involved episodes of geographic and host range expansion from China and Pakistan into Europe. The origins of the stem rust, *P. gramini*, are not so clear.[97] But it is likely all three of the species met wheat in the Fertile Crescent and Trans-Caucasus regions, where they created problems for early agriculturalists.

European farmers recognized that barberry was somehow connected to stem rust epidemics by the seventeenth century. The French city of Rouen banned the planting of barberry near wheat fields in 1660. Laws banning the importation and planting of European barberry were enacted in several New England colonies in the mid-1700s—to no avail, since a native barberry grows throughout the wildlands. And when that species was planted near wheat fields, infections stemming from rusts imported from Europe broke out with a vengeance. In 1865 de Bary had established that the fungus had a complex life cycle that appeared to require development on one host to produce spores capable of infecting the other. Researchers in the early 1900s recognized that de Bary's belief about rust life cycles was not absolute. These fungi are capable of great feats of ecological fitting as a result of extreme plasticity in their reproductive modes. All of them are quite capable of surviving only on wheat plants, or only on wild grasses, or only on barberry, undergoing indefinite episodes of asexual reproduction. So it appeared that the hope that barberry eradication would eliminate the pathogens was forlorn. The only good news was that forcing the fungi to persist in asexual modes for extended periods of time would decrease the odds that new virulent strains would emerge as a result of recombination.

The wheat rust fungus had a large potential host range among wild grasses, making it a constant threat for infection and reinfection of crops of other domestic grasses. You might eradicate large numbers of barberry, even ban it from being imported or planted close to agricultural lands, but you cannot eradicate grasses, which make up 12,000 species of plants covering 40 percent of Earth's land surface. Despite those warnings, wheat continued to be grown and shipped throughout the world, and the rust fungi always hitched a ride, reaffirming that geographic and host colonization are critical issues in global disease emergence.[98] Yellow rust fungus, for example, has had extensive global spread along trade routes during the past 100 years (fig. 6.8). The recent history of rust dissemination illustrates the role of pathways facilitated by people. Everywhere that humans have introduced wheat-based agriculture, rust pathogens have been transported, introduced, and established as pathogens along with the grain.

The Western Hemisphere and India have long experienced seasonal outbreaks and crop losses, but new pathogenic strains of *Puccinia* have recently emerged in Europe and Africa.[99] Most outbreaks occur on a regional scale, such as the northward seasonal migration of black rust that occurs in the central region of the United States and Canada. Sea-

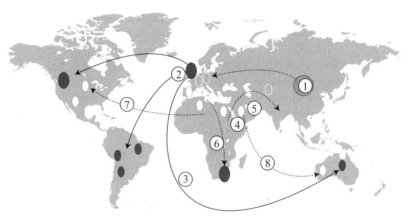

FIGURE 6.8 Long-distance introductions and migrations of yellow wheat rust fungus linked to human activities since 1900. (Ali et al. [2014]). (1) An association of yellow wheat rust and the wheat host have origins in the Himalayan region, serving as a precursor for global expansion. (2) In the early 1900s a pathogen strain was introduced and established from the United Kingdom to the Americas. (3) United Kingdom to Australia in 1979. (4 and 5) Northern Africa to the Middle East in 1986 and continuing to India in 1998. (6) North Africa to South Africa in 1996. (7) Africa to the southern United States in 2000. (8) The Arabian Peninsula to Australia in 2003. Patterns of lines and color symbols that are the same show shifts for single strains of fungi at different times over the past century. Episodes of introductions of single fungal strains have occurred on multiple occasions, for example into the Americas in the early 1900s and into Australia about 30 years ago. Multiple strains have also been established, for example in the United States and Mexico and in Australia.

sonal shifts also tell the story of episodes of local extinction and re-invasion in those regions whose climates do not permit persistence of the pathogens. Epidemics are driven by climate envelopes—hot days (25–30 degrees C) and mild nights (15–20 degrees C) with ample precipitation; for example, in North America these interactions determine the climate- and weather-related northward expansions that occur seasonally from the southern United States and Mexico northward toward Canada. New strains of yellow rust occur over large areas in Europe, North Africa, East Africa, and Central Asia where they had previously been considered eradicated. Genetic analysis shows that the strains most successful in colonizing new geographic areas are most closely related to the virulent strains that caused earlier epidemics from North America to Afghanistan.[100] These fungi clearly illustrate the threat of pathogen dissemination as an unintended consequence of accidental introductions through expanding agricultural lands and opening trade routes, ultimately damaging food security on all continents.[101] In a similar manner, movements of domesticated livestock in agricultural lands and along trade routes introduced nematode parasites worldwide.[102]

The examples we have discussed in this section have a common theme: disease emergence associated with climate change in recent and contemporary times. Hantaviruses and plague circulating in southwestern North America and hantaviruses infecting voles in Fennoscandia are sensitive to even minor fluctuations in climate. In North America these are linked to multiannual to decadal shifts in ocean-atmosphere circulation.[103] Notably, seven of the nine outbreaks of yellow fever in the United States during the nineteenth century can be linked to these events, in which eastern North America is subjected to high temperatures and humidity. Global climate warming has a disproportionate influence across northern ecosystems that are a bellwether for planetary change.[104] Lungworm nematodes in caribou and musk oxen on the tundra of the central Canadian Arctic provided the first snapshots of climate-based responses in pathogen-host systems anywhere on the planet.[105] Adult protostrongylid nematodes live in the lungs of ungulates, and their larvae develop in slugs or snails whose development and viability are influenced by temperature and moisture. If temperatures are too low, parasites do not develop, transmission does not occur, and infections do not persist. Prior to the 1970s temperatures on the Canadian mainland limited the life cycle and transmission to a two-year process for the lungworm *Umingmakstrongylus*. Rising temperatures since 1970 caused a shift from a multiyear to an annual infection pattern. This led to larger parasite populations involving ever greater numbers of hosts. Initially the parasites remained excluded from the adjacent islands across the Northwest Passage, where temperatures were too low to permit completion of the parasite life cycle. This changed by the early 2000s, when two unrelated species became established on the low Arctic islands in response to permissive new environments. Warming has also led to a rapidly expanding geographic range for unrelated filarioid nematodes among cervids (reindeer and introduced white-tailed deer) and blood-feeding arthropod hosts in Finland.[106]

Summary

The Stockholm Paradigm provides a direct link between climate change and emerging diseases. Climate change does not cause emerging diseases, but it provides opportunities that set the stage for pathogens to explore new fitness space. Pathogens expand geographically, leading to new distributions of established diversity, which in turn sets the stage for the emergence of new specialized diversity. Each episode of generalizing (expansion) and specializing (isolation) is initiated by environmental perturbations that increase connectivity in fitness space. There is abundant evidence that pathogen diversification has followed this evolutionary pattern, and that each episode has been catalyzed by some form of climate change. Oscillations—from host range changes to taxon pulse dynamics—are generalities that do not show strong periodicity because the triggers are contingent. We cannot predict with certainty that a Stockholm Paradigm episode will be initiated on any given date, but we can anticipate that whenever climate change occurs, a Stockholm Paradigm episode will be initiated. The structure and distribution of the complex pathogen-host systems we see now are much more strongly influenced by events in the past than we had thought, and sometimes more than by things happening now. As a result, their regularities may emerge only on the scale of global climate episodes. Such oscillations take time and space to unfold. This is an excellent example of our recurring theme of complex systems: *it is never just one thing, and it always depends.* Only a few of the pathogens that humans host are relics of our distant ancestry. This is because we largely left the African forest environs where they still live. Nonetheless, we have an enormous number of "new" pathogens acquired in the course of history since we moved into the savannahs and beyond. Simply bringing our old habits into new habitats exposed us to some pathogens we had never encountered. But we also changed our habits, first by adding more animals to our diet as we actively scavenged the remains left behind by carnivorans, then developing the tool-making abilities that allowed us to become predators in our own right. New pathogens were a major cost of that dietary bonanza. About 12,000 years ago, climate stability broke out across the globe. Humans quickly changed their habits again, and agriculture, animal domestication, and sedentary lifestyles led to new interfaces with our world. These changes brought us into contact with many of the pathogens that still affect us, our crops, and our domestic animals. At first, the costs of newly acquired pathogens seemed to be outweighed by our increasing success in improving our living conditions. But those improvements led, inevitably, to unprecedented population growth, and with that came geographic expansion. Humans bringing the lifestyle of sedentary pastoralists and agriculturalists into places new for them came into contact with other humans, either hunter-gatherers or other sedentary pastoralists and agriculturalists. And while those contacts laid the groundwork for cooperative activities, such as trade, too often they produced conflict. Conflict, in turn intensified the movements of human populations, bringing humans and their domestic plants and animals into contact with yet more pathogens.[107]

By the end of the eighteenth century it appeared that our technological innovations would allow us to survive our burgeoning population, our movements, our diseases, and our conflicts. There were warning signs, however. Yellow fever emerged in Philadelphia during 1793 when infected French refugees displaced by the Haitian slave rebellion arrived at the same time that increasingly warm conditions amplified populations of the *Aedes aegypti* mosquito vectors.[108] Jenner's initial vaccination experiments in 1796 were a response to recognizing that humans were beginning to have trouble surviving disease without help, and in 1797 the Reverend Thomas R. Malthus warned that food production could not keep pace with population growth. At the time, however, these events were never connected.

Largely undetected by us, the era of climate stability was ending. This ushered in the modern era (the Anthropocene), characterized by an ever smaller and faster world, with a tightening web of threat multipliers: climate change and climate fluctuations, increasing population and population density, globalization of economies and travel, urbanization, and reliance on highly polluting and nonrenewable energy sources. These factors create problems with water and food supplies and security, poverty and social and political instability, conflict, and migration.

There is no doubt that in the twenty-first century, the web of threat multipliers for emerging disease is tightening rapidly: climate fluctuations today mediate the incidence of a multitude of pathogens, from malaria in South America to Rift Valley fever in East Africa, dengue in Thailand, [109] and a cholera outbreak in Yemen that threatens to destroy the country. All the examples we have discussed so far, and a host of others to which we have only alluded, are the complex landscapes for pathogens, diseases, and us. In each circumstance these pandemics arose during periods not just of climate change, but also of climate fluctuations that destabilized human societies. People were on the move, responding to patchy food security, famine, socioeconomic and political instability, and conflict. Previous episodes of pandemics subsided with the return of climate stability, but we are now not certain when climate stability will return.

Throughout the world, climate change increases the threat of emerging disease. This is not a new insight.[110] What we bring to the discussion is the recognition that the pathogens affected by climate change today have long evolutionary histories that are important for us to understand. The historical record shows that everywhere we go, and everywhere we take our domestic plant and animal companions, pathogens are shared, rapidly and enthusiastically. Furthermore, the capacity to thrive and diversify as a result of coping with environmental perturbations is a built-in feature of pathogen biology. What is happening today has happened before. Biotic expansion and isolation phases of taxon pulses can be catalyzed by external perturbations ranging from minor local climate fluctuations to global catastrophic events such as asteroid impacts. And those capacities are on display globally, right now. This means that as climate change becomes more pronounced we should expect more, rather than fewer, emerging diseases.

We suggested earlier that we see the world as being made up of two groups of organisms, pathogens and hosts. If the Stockholm Paradigm explains the evolution of pathogens, does it also explain the evolution of their hosts?

Let's find out.

SEVEN

A Paradigm for Pathogens and Hosts

In adaptive radiation and in every part of the whole, wonderful history of life, all the modes and all the factors of evolution are inextricably interwoven. The total process cannot be made simple, but it can be analyzed in part. It is not understood in all its appalling intricacy, but some understanding is in our grasp, and we may trust our own powers to obtain more.
—GEORGE G. SIMPSON, *THE MAJOR FEATURES OF EVOLUTION* (1954)

Vernon Kellogg wrote that neo-Lamarckians, orthogeneticists, and neo-Darwinians all agreed that the core question in evolution was understanding how the right adaptation always arose at the right time. He dismissed the traditional Darwinian that what was necessary for survival under changing conditions must already be present in a species population—in low numbers, in the background or on the margins of fitness space, but preexisting—as absurd. We now know that he was wrong on that issue, but we can understand his perspective. For Darwinians, evolution was effective, not heroic. At the end of the nineteenth century, however, a love of the heroic dominated the scientific zeitgeist. What could be more heroic than the vision of a species, faced with an incoming asteroid, or maybe just encroaching desert on the margins of a rainforest, bravely standing its ground and averting extinction by magically producing "the right stuff" needed to survive? What about entire groups of species resulting from such heroic efforts? A single species, having survived an environmental onslaught by coming up with the right stuff, then goes on to a long existence, living happily ever after by producing lots of descendant species? Just a few years after Kellogg's book appeared, the race was on to provide the most heroic account of evolution.

In the Beginning . . .

We met Henry Fairfield Osborn as a key leader of the North American orthogenesis movement. Coping with environments was not important to most orthogeneticists, but a particular sense of it was crucial for Osborn. With prose appropriate to the heroic feeling of the time, Osborn proposed that

it is a well-known zoölogical principle that an isolated region, if large and suffi- ciently varied in its topography, soil, climate, and vegetation, will give rise to a di- versified fauna according to the law of adaptive radiation from primitive and central types. Branches will spring off in all directions to take advantage of every possible opportunity of securing food. The modifications animals undergo in this adaptive radiation are largely of mechanical nature, they are limited in number and kind by hereditary, stirp, or germinal influences, and thus result in the independent evolu- tion of similar types in widely separated regions under the law of parallelism or homoplasy. This law causes the independent origin not only of similar genera but of similar families and even of similar orders. Nature thus repeats herself on a vast scale, but the similarity is never complete or exact.[1]

Each unique set of characteristics was an *adaptive mode*, and all spe- cies exhibiting the same adaptive mode occupied an *adaptive zone*. Osborn believed that every lineage traveled through various adaptive zones, each marked by increasing ecological specialization, eventually becoming so specialized that it went extinct. He believed that adap- tive radiation was driven by internal principles of diversification rather than responses to external demands or opportunities, so he rejected both Lamarckian inheritance and natural selection.[2] Osborn's sense of adaptation was similar to Darwin's use of adaptation to mean "func- tional ability." It was not, however, connected to Darwin's notion of adaptation as "selective fit to the environment."[3] Osborn noted: "All that we can say at present is that Nature does not waste time or effort with chance or fortuity or experiment, but that she proceeds directly and creatively to her marvelous adaptive ends of biomechanism."[4] In accordance with other orthogeneticists, Osborn attached special significance to associations between different species, including pathogen-host systems. His "seventh law of adaptive radiation in the external body form" was *symbiotic adaptation*, in which

vertebrate forms exhibit reciprocal or interlocking adaptations with the form evo-lution of other vertebrates or invertebrates. It is these two principles of too close adjustment to a single environment and of the nonrevival of characters once lost by the chromatin which underlie the law that the highly specialized, and most per-fectly adapted types become extinct, while primitive, conservative, and relatively unspecialized types invariably become the centres of new adaptive radiations.[5]

Neo-Darwinians quickly joined the scrum. George Gaylord Simpson, also a paleontologist, co-opted Osborn, proposing a framework in which natural selection played a central role. Simpson argued that the environment could be divided into "a finite and more or less clearly delimited set of zones or areas."[6] He envisioned adaptive radiation as a three-step phenomenon, beginning slowly as small, isolated popula-tions accumulated mutations (evolutionary novelties) that would grad-ually make them increasingly more poorly adapted to their environ-ments. This would allow a population (in a random yet heroic way) to make a "quantum leap" into a new adaptive zone. In essence, the novel traits would "preadapt" (remember our discussion of problems that particular term has created) the population for survival in the new zone. Of course, if no new zone were available, this would seem to pre-adapt the population for extinction, but at this point Simpson is on a roll, so let's go with it. Once in the new zone, the leaping ancestor and its descendants would diversify along one or both of two differ-ent pathways: (1) geographic expansion + speciation, producing many species with the same *evolutionarily conservative ecologies* in different places, and (2) ecological diversification + speciation, producing many species with *divergent ecologies* in the same place.

Simpson's ideas about adaptive radiations emerged at the same time physicists were popularizing quantum mechanics. Quantum mechan-ics is based partly on the proposition that the double life of photons (energy/wave versus matter/particle) introduces a fundamental dual-ity into the universe. Simpson built a phylogeny/ecology duality into his adaptive zones, which could be specified simultaneously by the group that occupied them (e.g., the "felid zone") and the adaptive trait that matched the environment (e.g., the "carnivore zone"). Different zones could thus have different degrees of environmental and taxo-nomic complexity, which made it difficult to determine exactly what constituted a zone. Simpson's ideas were later modified by a variety of authors, including himself, but his basic components remained: an initial major change (which became known as a *key innovation*) moved a group into a new zone, where the group diversified to an extent limited only by the amount of available space in the zone.[7]

Simpson successfully shunted orthogenesis aside by co-opting the concept of adaptive radiation. But his efforts to offer a grand vision for adaptive radiations within neo-Darwinism fragmented evolutionary studies of large-scale evolutionary diversification.[8] And all formulations produced by that fragmentation foundered. The term *zone* could never be defined objectively. Basically, if you shut your "ecology" eye, adaptive zones appear as phylogenetically coherent groupings that may not be ecologically cohesive; and if you shut your "phylogeny" eye, adaptive zones appear as ecological groupings that may not be phylogenetically cohesive. Neither has it been possible to define objectively what qualifies as a key innovation—or, more important, how that innovation caused a population to "leap" from one zone to another. Traditionally, there was no necessary link between the key innovation and speciation, only between the key innovation and the degree of distinctiveness, which in turn was related to the distinctive nature of the new adaptive zone. So, if many of the higher taxa grouped in such a manner were not phylogenetically cohesive, their purported distinctiveness might be illusory, as well as many of their putative key innovations.

Most biologists studying evolutionary diversification have been quite happy to go through life with one eye firmly closed. As a result, we have two research programs among the academic descendants of Osborn and Simpson. Researchers in one focus their attention on the *radiation of adaptations* associated with invasion of the new zone. Although speciation may be incorporated into this program, it is not a critical component. The only requirement is that some unit of biological diversity must show substantial functional diversification. The other group of researchers focuses on the *radiation of species*. This research is triggered by an old observation that some evolutionary groups appear to be unusually species-rich compared with others. Rosa, another orthogeneticist, proposed that sister groups often were the products of two different "stems," the "precoce" (or smaller stem) and the "retarded" (or larger stem). Using orthogenetic reasoning, he argued that the smaller stem would be more phylogenetically conservative (archaic), would reach its apogee of large and exaggerated forms earlier, and would go extinct sooner than its more species-rich sister.[9] This theme was later adopted in modified form by Hennig as the *progression rule*, in which he proposed that the asymmetry was due to the sequential dispersal by members of a species into new areas, where they were subject to different selection pressures and became new species.[10] The goal of that research was to identify groups of "unusually high" or "unusually low" species number, then try to uncover the processes responsible. Not surprisingly, arguing from the present into the past beginning with a single group produced no general explanations. Differences in species number between sister groups can arise for at least three different reasons. First, it is possible that asymmetry develops stochastically rather than due to the influence of any forces extrinsic or intrinsic to the biological system. For example, say you have two sister species, A and B, existing quite contentedly for eons. Then one day, purely by chance, A is subdivided, producing daughter species C and D. If speciation occurs at random, then the monophyletic group comprising C and D has a higher probability of being subdivided again than does its monotypic sister group (species B), and so on. In other words, the more bets you place on the speciation roulette wheel, the greater your chance of winning purely by chance. It is also possible that an asymmetry in species richness could arise due to factors extrinsic to any particular group. Joel Cracraft suggested that the distribution of biodiversity was affected by two major environmental factors that contribute to the asymmetric distribution of groups of species.[11] The first factor was *environmental harshness*.[12] The observation that diversity in the tropics is higher than diversity in temperate or arctic regions is often attributed to differences in speciation rates. It is possible, however, that extinction rates in temperate to arctic habitats have been higher than extinction rates in the tropics owing to historical increases in environmental harshness in the colder areas without any particular differential rate of speciation by latitude, which limited the number of terrestrial and freshwater species living at higher latitudes.[13] Cracraft's second factor was the history of geological change and geographically mediated speciation. Biological diversity tends to be clumped in "hot spots" corresponding to areas with historically high rates of geological change, rather than being uniformly distributed across a given habitat or zone. For example, tropical diversity is clumped in South America, the Indo-Malaysian region, the South African Cape region, Madagascar, and the Rift Valley lakes of Africa, areas whose geological histories are extremely complex. In other words, a group may be unusually species-rich because it exists in an old and persistent geological hot spot, where it has been subdividing on a regular basis, producing and accumulating many species via repeated episodes of geographic isolation. Finally, it is possible that asymmetry in species richness is due to intrinsic factors, which brings us to the concept of a key innovation. Shifting the evolutionary focus from degree of distinctiveness to the number of species in a group allowed the key innovation concept to be reintroduced to the study of radiations.[14] Both changes in competition pressure by movement into a new, unoccupied space (either ecological or geographical) and improved competitive ability within the ancestral space have been cited as possible mechanisms by which the origin of a key innovation might have a positive effect on net diversity. It is important to emphasize here that using phrases such as *reduced competition* and *ecological release* (e.g., release from competition or predation pressure) as an explanation for the effect of a key innovation on patterns of radiation is too imprecise. Both of these terms imply absence. If movement into a new area removes competition, what causes the population to change? If the population remains static in the new zone, you have evidence of a range expansion but not of a radiation. Surely the evolution of the population in the new zone does not occur because of the absence of old conditions of life, but because the population has encountered a new opportunity space.[15] Species do not leave ancestral fitness space without entering new fitness space. A key innovation may influence net diversity in two different, but not of necessity mutually exclusive, ways. It may increase the likelihood that members of a group will undergo episodes of rapid speciation.[16] A variety of factors, including adopting a specialist foraging mode, or changes in breeding systems (sexual selection) and in population structure, have been postulated to have a positive effect on speciation rates. Alternatively, it may influence net diversity by increasing population size, expanding the range occupied by the species, or both, decreasing its chances of going extinct. The interaction between increased rates of speciation and decreased rates of extinction is often difficult to disentangle. For example, a comparison of species richness between phytophagous (plant-eating) insect groups and their nonphytophagous sister groups showed that, in 11 out of 13 comparisons, the phytophagous lineage was more species-rich, leading to the conclusion that adoption of phytophagy was significantly associated with an increase in net diversity.[17] Yet, not addressed was how that net diversity had been produced or whether only a single factor was involved. While we agree with the basic idea of a causal coupling of species richness and functional diversification, we believe the mechanism should be viewed with some caution. First, the idea that speciation produces species that always and progressively partition the environment into smaller and more specialized subunits is not Darwinian, though it predated Osborn.[18] And it was this aspect of Osborn's adaptive radiation scenario that Simpson, and many later authors, were never able to successfully remove from their research programs. Secondly, the concept that the environment is somehow divided into predetermined niches in the absence of the organisms that inhabit those niches suffers from the same problems as did the concept of adaptive zones. Organisms impose themselves on their surroundings—metaphorically creating their own niches—existing where they occur and doing what they do to survive because of their particular natures and history.[19] Organisms are not pulled into unoccupied niches—there is no external force drawing species to their destinies. Richard Lewontin eloquently wrote that the "error is to suppose that because organisms construct their environments they can construct them arbitrarily in the manner of a science fiction writer constructing an imaginary world . . . Where there is strong convergence is in certain marsupial-placental pairs, and this should be taken as evidence about the nature of constraints on development and physical relations, rather than as evidence for pre-existing niches."[20]

An organism first moves to the zone/niche/resource because of its own unique biological properties (ecological fitting); only after that does the interaction between the new zone/niche/resource/selective regime and the biological system begin. The appearance that the surroundings are constructed as niches is analogous to a kind of optical illusion that Hume warned about. The more historically conservative organisms are, the more likely it is that each generation will be characterized by many organisms all preferring the same resources, those resources being the ones preferred by the previous generation. The repeated exploitation of the same resources could fool some into believing the environment was doing the structuring when in reality evolutionary history was responsible. Early in the twenty-first century, a protocol containing the steps needed to corroborate a claim of adaptive radiation for any given group emerged. Its effectiveness was argued by examining possible examples of adaptive radiation on a case-by-case basis; as a result, it did not achieve the goal of a general explanatory framework. Indeed, each step in the protocol was accompanied by a list of multiple qualifiers that would plague any attempt to pin down even one example to being considered a well agreed upon case of adaptive radiation. The researchers recognized this deficiency and lamented the lack of a comprehensive null hypothesis for studies of evolutionary radiations.[21] More recent efforts have reiterated that framework or provided an even more comprehensive list of possible kinds of evolutionary radiations, but a unified research framework has not emerged.[22]

So, here we are trying to cope with what many consider a core feature and success story of neo-Darwinism, and it's a mess. A century of studies of evolutionary radiations managed to learn a lot and accomplish nothing with respect to a general explanation. We believe the search for a general explanatory framework for evolutionary radiations has failed because we have been trying to adapt an orthogenetic framework to a neo-Darwinian one. All is not lost, though, so long as we go . . .

Back to the Future . . . Again

Organisms are lazy; species are even more lazy.[23]

Let's go back to Darwin. Evolution is not heroic; it is quietly effective, perhaps even lazy, at least as lazy as conditions will permit. And in keeping with the long history of evolutionary radiations' being the preoccupation of paleontologists, we introduce the work of Alycia Stigall.[24] Like all paleontologists, Stigall has documented that external perturbations, stemming from major climate-change events, are responsible for changing the nature of the conditions. But unlike many of her fellows, Stigall has focused not on the species that went extinct, but on those that survived. They are the ones that were so widespread and numerous that some populations either were not affected by the perturbation or were capable of moving away from the changed conditions to areas where their preferred conditions still occurred. Through luck or ability, those that survive are the ones that take advantage of changing opportunities, based initially on inherited traits that existed before the perturbation. This should sound familiar.

Throughout the history of life, most species, when faced with climate change, have moved to places that are most similar to the conditions in which they lived previously and in which they already possess the means for survival, or they die trying. They do not "stand their ground" heroically, coming up with the right adaptation at the right time. For hosts as well as for pathogens, it is critical to remember that *perturbations in climate do not cause innovations—they allow exploration, which sets the stage for innovations to emerge.* Isolation during periods of climate stability is the context that allows innovations to emerge. This is why innovations cannot be predicted, although they can be explained after they emerge. Once again, this perspective rests on a solid Darwinian foundation:

Indefinite variability is a much more common result of changed conditions than definite variability.[25]

The direct reaction of changed conditions is biotic expansion, the extent to which and manner of which is determined by each species' capacity, that is, ecological fitting in sloppy fitness space. This sets the stage for indefinite variation to be expressed, upon which selection acts,[26] because

natural selection is the indirect action of changed conditions.[27]

We make a bold claim: that the Stockholm Paradigm, which explains the evolutionary diversification of pathogens, also explains the evolutionary diversification of their hosts. What evidence do we have to back that up? (Spoiler alert: there is a lot.)

Capacity for Ecological Fitting—Functional Traits Are Specific and Conservative

In chapter 3 we identified three classes of qualities of inherited traits that promote the capacity for ecological fitting in pathogens. In chapters 4 and 5 we showed how those traits functioned for pathogens trying to survive in a world that is largely indifferent to them. Do these categories exist for hosts as well? The short answer is a thunderous "yes."

The first capacity is what we call *contextual flexibility*, also known as phenotypic plasticity. In the first few years of the twenty-first century Mary Jane West-Eberhard wrote a book about this kind of capacity.[28] She provided such comprehensive coverage of the concept that it immediately became its own research discipline, and her book became the "industry standard." The second class of capacities is *situational flexibility*, better known as co-option. As we noted in chapter 3, Darwin recognized this kind of capacity. It was first documented in a phylogenetic context in 1983, based on a study showing that the deciduousness of oak trees, which helps temperate oaks cope with seasonally cold climates, originated evolutionarily in seasonally dry climates.[29] A beautifully written contemporary assessment of this concept was published in 2008.[30] And finally, the capacity that is the foundation of Darwin's nature of the organism is *historical flexibility*, or evolutionary conservatism. A comprehensive treatment of this class of traits, and its pervasive occurrence and influence in biological diversity, appeared in the early years of this century, and subsequent studies have reinforced that book's conclusions.[31]

These categories stem from Darwin's view of the "nature of the organism" and represent the general ways in which different organisms cope with their surroundings, especially with changes in their surroundings. They indicate that there is great capacity for ecological fitting among all life, not just pathogens. An excellent example of the contributions of all three categories to the structure of biodiversity is a study of ectomycorrhizal communities occurring in association with different trees. According to the authors of that study, the communities are structured by specific and conservative elements of microhabitat preferences and transmission dynamics. These are precisely the findings we discussed previously for the evolution of frog parasite communities.[32]

Fitness Space: Is It Sloppy?

It is tempting to say that if Darwinian evolution applies to all species, and if pathogen fitness space is sloppy, then host fitness space must be sloppy. But this is a good opportunity for us to discuss a critical phenomenon showing that fitness space is fundamentally sloppy for hosts in a way that facilitates host range changes characteristic of emerging diseases. That phenomenon is the observed success of invasive and introduced species and our ability to predict such occurrences.

If native species are in exploitation-biased mode and invaders are in exploration-biased mode, it is likely that if there is conflict between invader and native, there will be substantial flexibility in the system, leading to conflict resolution in which both species persist. We can assume that, relative to each other, any given invader is in exploration-biased/generalizing mode, while a resident with similar functional requirements is in exploitation-biased/specializing mode. At the time and place of introduction, all other things being equal, priority effects do not play a role. But at some point the invader will shift into exploitation-biased mode, and if that leads to direct competition with a resident, any prior advantages enjoyed by the native species will come into play. This could explain situations in which an invader initially expands quite rapidly in the new habitat, seemingly putting a native species at risk, yet, after a period of time, the invader fades into marginal habits and the native makes a comeback. Multiple outcomes are possible, one of which is that the invader will become "naturalized" in habitats that are marginal for the native species.[33] This is also the process that produces pathogen pollution.[34]

Ecological niche modeling has been very effective in understanding and predicting the places that are at risk of inadvertent introductions and the chances of success for deliberate introductions of species.[35] The reason for this success is twofold: first, introductions and invasions are easy and common because the capacity for ecological fitting is great, fitness space is sloppy, and fundamental fitness space is always larger than realized fitness space;[36] there is thus massive opportunity in the world. Secondly, the capacities possessed by every species that are relevant to taking advantage of opportunities are both specific and conservative. The "predictive power" of ecological niche modeling is what philosophers call *retrodiction*: you predict that the future will be like the past. The more specific and conservative such traits are, the better we can anticipate where the species possessing them might be able to find viable fitness space.

Fitness space is so sloppy and ecological fitting capacities are so extensive that many species are capable of being introduced all over the world, given the opportunity, and often we cannot prevent such invasions and introductions even if we wish to. This is equally true for hosts and pathogens.

Functional Diversification: Oscillations between Generalizing and Specializing

Simpson saw half of the picture when he asserted that adaptive radiations resulted from diversification accelerated by ecological opportunity.[37] As with the nineteenth- and twentieth- century researchers studying the evolution of pathogen-host systems, the foundational studies of evolutionary radiations lacked a coherent mechanism for generating generalized species. The Stockholm Paradigm suggests that generalizing in fitness space requires only the capacity for ecological fitting, realized fitness space that is a subset of fundamental fitness space, and a means by which species may encounter fitness space that had not been accessible previously. This complex set of connected capacities should lead to evolutionary oscillations between generalizing and specializing in fitness space. What would that look like?

We may not have convinced all readers that phytophagous insects are actually pathogens, despite the fact that all of them have specific host preferences and at least some of them cause "plagues" even in their preferred hosts. So, examples using phytophagous insects and their host plants are a good bridge between "real" pathogens and "real" nonpathogens. Ecological fitting in sloppy fitness space sets the stage for two major categories of evolutionary transitions.[38] The first is *taking advantage of opportunities permitted by the rules of the game*. In general terms, such evolutionary transitions are the easiest (i.e., the least costly in terms of time and genetic change) to achieve because they require only a change in the nature of conditions, not in the nature of the organism. These transitions require only altered environmental conditions to bring species into contact with novel sources of evolutionarily conservative resources. They may also be called *opportunity-limited*, but this does not preclude a role for natural selection in the outcome (i.e., they may also be *selection-limited*). The second is *changing the rules of the game*. The evolution of new capacities may permit a species to occupy new fitness space. This kind of evolutionary change is more difficult to achieve than simply moving in fitness space representing ancestral resources, because it requires a change in the nature of the organism. These transitions may also be called *information-limited*, but this does not mean natural selection plays no role in the outcome (i.e., they may also be *selection-limited*).

Let us examine the two categories of transitions based on an example using some leaf beetles in the genus *Ophraella* (fig. 7.1).

The figure should look familiar. Much of the speciation in the insect group (think "pathogens") took place within the context of the plesiomorphic resource (host colonization among members of the same tribe). Many of those events evidently occurred in a burst during the Plio-Pleistocene, a time of substantial environmental change.[39] This implies that potential for speciation may have increased during that period of environmental crisis because the beetles were given the opportunity to colonize new hosts. Increased sympatry of previously allopatric hosts in rapidly shrinking refugia would have increased the chances of host colonization through ecological fitting if many of the newly sympatric hosts shared the same specific and conservative resources preferred by the insects. Both colonization of more distantly related host plants and movement between tribes encompass fewer events, and as yet there is no indication that these events were correlated with each other or with any particular episode of environmental change. These host colonizations are more difficult to understand.[40]

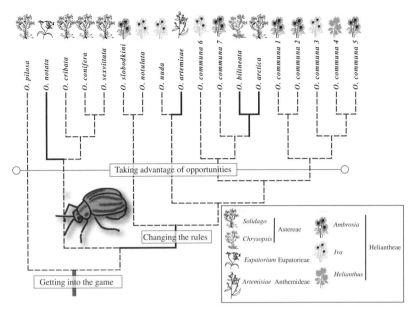

FIGURE 7.1 Two different types of transitions in coevolution mapped onto the phylogeny for *Ophraella*. Dotted lines = taking advantage of an opportunity; bold lines = changing the rules of the game. Redrawn and modified from Brooks and McLennan (2002).

In this example, only two pairs of recently evolved sister species of *Ophraella* (*O. communa* 1 and 2, and *O. conifera* + *O. sexvittata*) show possible cospeciation. The system nonetheless has evolved in a context highly influenced by phylogenetic conservatism, so there is a strong cophylogeny signal despite minimal cospeciation. Many species form easily during periods of isolation, when fitness space is disconnected and climate is stable. Many of them go extinct when climate changes lead to both local alterations of and new connections in fitness space. They lack the capacity to cope with changes or to take advantage of the opportunity to explore permitted by the new connections in fitness space. Constraints on the ability to explore new fitness space can be—perhaps usually are—the result of adjacent fitness space being occupied by a relative already in control of the resources required by a potential colonizer.[41] There is abundant evidence of ecological fitting in sloppy fitness space, with oscillations between generalizing and specializing among hosts as well as pathogens. Thus far, it seems that host evolution does not differ from pathogen evolution. Next, do hosts respond to external perturbations with taxon pulses?

Response to External Perturbations:
Do Hosts Experience Taxon Pulses?

The larger ecological explosions have helped to alter the course of world history and . . . can often be traced to a breakdown in isolation of continents and islands.[42]

Darwin's metaphor of the tree of life portrayed evolution as a process of diversification, an indefinite increase of species over time. Everything we know about the fossil record and the phylogeny of life on this planet supports that view: many species have gone extinct, but none has been replaced. The taxon pulse provides a natural means by which species form in isolation, then mix together during periods of biotic expansion. Multigroup studies in Mexico and in the Hawaiian archipelago and the Caribbean islands show clearly that the taxon pulse dynamic underlies the evolution of entire biotas across broad geographic scales.[43] But this leads us to a conundrum. How can accumulation occur if the specific functional traits that determine species' fundamental fitness space are phylogenetically conservative? Won't that keep related species from occurring in the same areas? Evidently not. In the multigroup studies of Mexican terrestrial biotas and Hawaiian Islands biotas, assembly has involved approximately one-third sequential (forward) colonization, one-third in situ speciation, and one-third back-colonization. Thus, as many as two-thirds of those biotas comprise species that coexist with "close relatives." How can this be?

Only in politics does "conservative" mean "no change ever." In evolutionary biology, conservative means that ecological change occurs less often than the formation of new species. As we noted earlier, changes in functional traits in many different groups of insects living in the Caribbean occur about once in every 30 speciation events (3 percent).[44] This supports the notion that the nature of the organism is to be characterized by traits that are specific and conservative. But is that enough to allow accumulation of diversity in biotas? The answer is a qualified yes. In the simplest case, if only one of the 30 species produced from 30 isolation events evolves novel functional traits, it would have the potential to colonize as many as 29 other biotas already containing close relatives without coming into conflict with any of them. Does that mean the other 29 species in the group would be unable to colonize other biotas without coming into conflict with resident relatives? Not necessarily; ecological fitting could allow some older species and newer colonizing species to live together—the result of conflict resolution leading to coexistence of both colonizer and resident, even if they began with similar fitness space requirements.

Documenting the full story of biotic diversification and assemblage by taxon pulses is complicated, and a truly robust study would require analysis of even more taxa than the few multitaxon analyses published thus far. Even in those studies, not all groups participated equally in all the historical episodes, though they were all consistent with one complex history. Nonetheless, the signature of the taxon pulse as an overarching process is always clearly legible. Most groups in each study agree on the episodic highlights but differ in the details. Each species plays a role in any given biota, but to truly comprehend it, we must know the story of the species with which it lives and from which it evolved. This is why studies in which multiple taxa have been analyzed together show the taxon pulse dynamic so clearly. One particular study helps us connect hosts and pathogens. Analysis of the historical biogeography of hominoids, proboscideans (elephants and their relatives), and hyaenids (hyaenas) since the Miocene epoch (approximately 12 million years ago) revealed a shared history of taxon pulses (fig. 7.2). More significant was that the pattern of repeated episodes of expansion, isolation, and new expansion was the same as the one involved in the evolution of schistosomes and of the two groups of roundworm parasites that we discussed earlier. A portion of this history of multiple taxon pulses may refer to the host range expansion that led to the common ancestor of *Taenia saginata* and *T. asiatica* possibly involving *Homo erectus* in Eurasia, where hyaenid diversity was well established in the Pliocene.

To date, eight studies of hosts using methods capable of detecting taxon pulses have been published, and all have found clear evidence that diversity has been shaped by taxon pulses.[45] Those studies are only the tip of the iceberg; a large number of recent studies in historical biogeography, using methods analogous to the cophylogeny methods, have identified alternating episodes of expansion and isolation, leading to mosaic patterns of biotic assemblage within any given area. Just as cophylogeny studies should no longer be considered evidence of cryptic cospeciation, these biogeographic studies should not be considered cases of cryptic vicariance. Rather, we should assume they are evidence of taxon pulses unless proven otherwise.[46] Taken together, this sampling encompasses an extraordinary representation of Earth's biodiversity, sufficient for us to believe that the evolutionary history of species' geographic distributions is a history of taxon pulses.

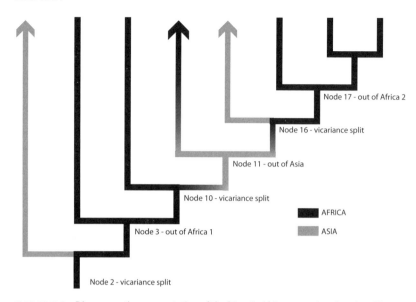

Node 17 - out of Africa 2

Node 16 - vicariance split

Node 11 - out of Asia

Node 10 - vicariance split

AFRICA

ASIA

Node 3 - out of Africa 1

Node 2 - vicariance split

FIGURE 7.2 Diagrammatic representation of the historical biogeography of proboscideans (elephants, mammoths, and mastodons), hominoids, and hyaenas since the Miocene (the last 12–15 million years). "Vicariance split" nodes refer to episodes of isolation between Africa and Asia; "out of" nodes refer to episodes of biotic expansion. The evidence shows clearly that all three groups (and presumably what they ate and what ate them) experienced significant diversification as a result of participation in multiple taxon pulse episodes. Folinsbee and Brooks (2007).

Back to the Pathogens

The principles that govern the structure, life cycles, habitats and activities of free-living and parasitic animals are really the same.[47]

The most extraordinary adaptive radiations on earth have been among parasitic organisms.[48]

The elements of the Stockholm Paradigm underlie the dynamics of evolutionary diversification for hosts. Let us now turn full circle and look at evolutionary radiations of pathogens. If, as we suggested earlier, high reproductive rates in pathogens have to do with increased capacity to explore sloppy fitness space, we should see a strong correlation between parasitic groups in which boosted reproductive capacities have evolved and evolutionary radiation in both functional diversification and species numbers. Among the flatworms, most pathogens belong to a group called the Neodermata ("new skins").[49] This group includes three groups of "health" significance: the Digenea (we have already discussed schistosomes and liver flukes); monogenoideans, which are major concerns for commercial production of both marine and freshwater fish; and tapeworms (we have already discussed *Taenia*).

The immediate ancestors of the Neodermata were themselves either pathogens or opportunistic hangers-on. They had a simple transmission mode, each generation moving from the outside of one arthropod to the outside of another (fig. 7.3).[50]

The pattern became more complicated in the common ancestor of the Neodermata; a vertebrate host was added, and the adult parasites became endoparasitic. Do either or both of these changes count as key innovations? They conform to Simpson's original scenario for a radia-

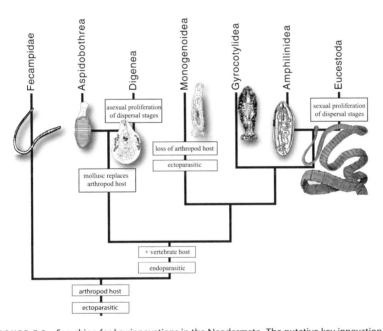

FIGURE 7.3   Searching for key innovations in the Neodermata. The putative key innovation (the plesiomorphic life-cycle pattern for the species-rich Neodermata) is a life-cycle pattern involving an arthropod intermediate host, a vertebrate definitive host, and endoparasitic habitats for the adult parasites living in the vertebrate. The Gyrocotylidea, Amphilinidea, and Eucestoda (tapeworms) have retained that general pattern. The trematodes (Aspidobothrea + Digenea) display one variation: a molluscan host was substituted for an arthropod host in their ancestor. The Monogenoidea have a secondarily simplified life cycle (a return to the plesiomorphic state "ectoparasite with only one host," although that host is a vertebrate, not an arthropod). The Digenea and the Eucestoda have independently evolved different mechanisms to proliferate larval dispersal stages. Redrawn and modified from Brooks and McLennan (2002).

tion because they involve movement into and diversification in new fitness space, in terms of both hosts and places to live within the host. At this level of analysis, it appears that the origin of these two habitat-shifting traits is associated with a significant change in net diversification (Fecampiidae [fewer than 20 species] versus Neodermata [approximately 14,000 species]).

Paradoxically, if we expand the phylogenetic scope of our study, we find that the intuitively pleasing notion that being endoparasitic (living *in* the host) is really significant evolutionarily may not be correct. This is because the Mongenoidea have both shifted back to the ancestral ectoparasitic (living *on* the host) and have not experienced a decrease in net diversity. By the same reasoning, the hypothesis that there is something inherently important about adding a vertebrate host is also refuted, because the Aspidobothrea, Gyrocotylidea, and Amphilinidea all have very few species, despite having a vertebrate host. What is the next step in this quest? Diversity is asymmetrically distributed among groups within the Neodermata. This mirrors patterns for other very old groups such as spiders and flowering plants.[51] Let's search for possible key innovations among the species-rich groups of neodermatans as compared to their species-poor sister groups.

From a pathogen's perspective, hosts are resource patches clumped in fitness space. Furthermore, pathogens are usually less mobile than their hosts, so being able to move from host patch to host patch is important. This is why transmission dynamics are so specific and evolutionarily conservative: if you find something that works, stick with it. Pathogens that counter this relative immobility have evolved some form of one of two general strategies:[52] (1) direct development or autoinfection (or both); and (2) production or amplification of dispersing larval or juvenile stages, by either (a) cloning (asexual) or (b) increasing egg production (sexual). The benefits of autoinfection are obvious; there is a substantial decrease in the risk of your offspring not finding a suitable host. The benefits of producing or amplifying larval or juvenile dispersal stages are also obvious. Such stages, often aided by overwhelming numbers or host-seeking or host-behavior-altering capabilities, function to (1) decrease intraspecific competition by partitioning resources among different developmental stages, (2) partition the parasite gene pool among different developmental stages in different hosts to buffer against problems with transmission, and (3) increase the likelihood of transmission from one host to another. Increasing reproductive output is the sexual equivalent of amplification of larval or juvenile stages via asexual cloning. Both of these strategies have a direct, positive effect on the reproductive output of their bearers. That alone should imbue them with enhanced capacities to explore their surroundings, increasing the odds of finding new viable fitness space. Are either, or both, of these traits associated with an increase in net diversity in the groups where they have emerged?

The Eucestoda (tapeworms) and the Amphilinidea (ca. 3000 versus ca. 10 species) show the greatest disparity in species richness between sister groups. The eucestodes have used strategy 2a above to cope with the patchiness problem; the vast majority of adults are segmented, and each segment (proglottis) contains at least one complete set of male and female reproductive organs—testes, ovaries, and a uterus. The increase in reproductive output by adult tapeworms is dramatic. The average daily output of 100 eggs per worm in the Monogenoidea, and even the exceptional 24,000 eggs per worm in some of the larger digeneans,[53] pales by comparison to the 720,000 eggs expelled per day by the human beef tapeworm, *Taenia saginata*.[54] The lifetime fecundity of a tapeworm is enormous compared with the lifetime fecundity of species of other groups of neodermatans. The proliferation of dispersal stages created by the new strategy ought to increase the likelihood that some larvae will find suitable hosts, decreasing the chance that a given population will go extinct.[55] This has an indirect effect on speciation because it allows a species to persist long enough to encounter a variety of speciation-causing factors. The sister group of the eucestodes, the Amphilinidea, possesses the persistent ancestral trait of lacking repeating reproductive units.

Support for the effect of repeating reproductive units on net diversification can be found by investigating groups within the Eucestoda that have secondarily lost repeating reproductive units. One such group is the Aporidae, parasites of ducks and swans. As predicted, the aporids are significantly less diverse than their sister group within the Cyclophyllidea (six versus approximately 300 species).

The second-largest asymmetry in species richness is between the Digenea and the Aspidobothrea (5,000 versus 50 species). Both aspidobothreans and digeneans share an ancestral life-cycle pattern involving a molluscan and a vertebrate host. Aspidobothrean larvae hatch from eggs, develop directly into juveniles, and are ingested by a mollusk-eating vertebrate, where they develop to adulthood. Hence, each aspidobothrean embryo can potentially give rise to only a single adult. Digeneans, on the other hand, are characterized by a series of complex developmental stages in the molluscan host, at least one (and usually two) of which produce a large number of cloned larvae or juveniles (depending on the species and the stage). The reproductive potential of a single digenean embryo ranges from 1000 to 10,000 adults. This is significant: of 100,000 *Fasciola hepatica* (the liver fluke that infects both sheep and humans) eggs deposited in a pasture from the middle of winter to the end of summer, only 17 hatch and infect a snail host.[56] The reproductive potential of those 17, however can compensate for the 99,983 that did not hatch if each of the 17 is capable of producing 5883 cercariae asexually. Like the eucestodes, digeneans have solved the patchiness problem by increasing the number of dispersing stages but have evolved a different mechanism (strategy 2a above). The effects of increasing dispersal stages on net diversification should be similar to those described for the tapeworms; however, it is difficult to test this hypothesis because, unlike the tapeworms, no digeneans are known of in which the capacity to produce cercariae has been lost.

Finally, we are left with strategy 1: autoinfection. Autoinfection occurs in only one clade, the Monogenoidea, and that clade is not significantly more species-rich than its sister group (the Gyrocotylidea + Amphillinidea + Eucestoda). Does that mean the trait in the Monogeneoidea is not associated with their species richness? Well, not necessarily. The absence of a significant asymmetry in species number following the origin of a putative innovation can be the result of two different phenomena. The first is that our hypothesis of a causal relationship between the innovation and net diversification is, in fact, incorrect. The second explanation is less obvious and more difficult to approach. It is possible that the positive effect of a key innovation in one group was countered by the positive effect of a different key innovation in the sister group. For example, the sister group of the monogeneans contains a species-rich clade (the Eucestoda). If the eucestodes had not developed sexual proliferation of dispersal stages, then we would not expect them to be significantly different (in terms of species number) from their sister group. Let us use the basal members of the Eucestoda, the Caryophyllidea, as our guide.[57] There are only about 111 species of these tapeworms, which are unsegmented, that is, they lack the putative "key innovation" of multiple reproductive units. In our alternate universe, then, there are only 111 tapeworm species (and fewer dietary restrictions in many of the world's cultures), and the comparison between the Monogenea and their sister group now becomes significant (5000 versus 129 species). Back in our universe, we believe that it is a useful exercise to reduce the confounding effects of a species-rich group (in this case, the tapeworms) in the sister group.

Autoinfection has a positive effect upon reproductive output in or-

ganisms possessing the capacity for it.[58] Is the association between monogenoidean diversification and this novel reproductive ability more
than just a spurious correlation? From a theoretical perspective, the
answer to this is yes, because the subsequent changes in life-history
characteristics have a direct effect on the genetic structure of groups
of organisms that spend so much time together, they are more likely
to mate with each other than with any others. Parasite *demes* (local
breeding aggregates; in the case of pathogens, demes are the reproductive individuals living at the same time in the same host) are ephemeral. In the vast majority of cases, they are reassembled each generation on or in hosts by the relatively random immigration of larval or
juvenile stages, usually from a larger gene pool than that represented
by the members of the original deme. Parasites that autoinfect their
hosts and inhabit long-lived hosts, however, can produce more than
one generation on or in the same host organism. This reduces the
ephemerality of the deme structure, increasing the potential that differences appearing within a deme could be maintained by inbreeding. Such inbreeding might "promote rapid increase in gene frequency
and potential fixing of beneficial variation."[59] Monogenoideans are
thus characterized by the appearance of a novel reproductive strategy
that increases the opportunity for relatively rapid speciation to occur under a variety of conditions. The hypothesis that the evolution
of autoinfection influences the rates of speciation is supported by an
additional piece of evidence. Within the monogenoideans, one of
the most species-rich groups, the Gyrodactylidae, takes autoinfection
one step further. Instead of laying eggs on their hosts, they reproduce
parthenogenetically.

Taking into account the appropriate qualifiers for incomplete knowledge and data, it seems that the traditional idea that any evolutionary
boost in reproductive capacity in pathogens leads to increased diversity is true. There is nothing magical or heroic about this; it is simply
a consequence of increased propagule pressure leading to enhanced
exploration of potential fitness space. But each such evolutionary innovation is highly significant, because the increased ability to explore
when the opportunity presents itself sets the stage for everything to do
with pathogen diversity and the emergence of disease during climate-
change events.

We see some readers with furrowed brows. Why is it so complicated?
Why not be like viruses and bacteria and just maximize asexual reproductive output? That's a great question, and the answer is indeed
a little complicated. First of all, asexually reproducing lineages evolve

instantly into new species each time a mutation arises. And viruses and obligately asexual bacteria cope with the world by brute force—very rapid reproduction with many strains, most of which appear and disappear rapidly. In addition, it turns out that even among bacteria and viruses there is more genetic mixing than you might expect; many of them, especially bacteria, are *parasexual*, possessing amazing ways to stir up the gene pool, so to speak. Remember the case of influenza. But, interestingly, there is something we call the paradox of the cost of sex. No, this has nothing to do with emergent sexually transmitted infections (though that is a global problem). Simply put, the cost-of-sex problem is this: reproducing asexually, meaning that all offspring are clones, is a really good thing if you've got a great genome. However, cloning lineages can be wiped out by a single bad mutation. Sexual reproduction provides some protection against the impacts of bad mutations, which can be eliminated by recombination. Some theoretical biologists have noted that the ideal compromise would be a combination of some of each, a phenomenon called haplodiploidy.[60] And it turns out that there are a lot of different versions of haplodiploidy in the world. Humans even have a small remnant of the haplo part of a haplodiploid system—we call it *meiosis*. Now, back to the Neodermata. Each of the special reproductive modes we identified in the species-rich groups involves an alternation of sexual and cloning reproduction; in the Digenean, larvae are cloned; in the Eucestoda, adult reproductive units are cloned; and in the most species-rich Monogenoidea, entire adults are cloned by parthenogenesis. So, while the actual mechanisms differ, each is likely a developmental modification of a common ancestral reproductive system, and they all have similar functional impacts in terms of evolutionary persistence and diversification.

The Scope of the New Paradigm

Stigall's results can be summarized thusly: environmental perturbations causing a change in the nature of the conditions lead to biotic expansions based on existing capabilities (ecological fitting); this leads to generalizing (ecological fitting in sloppy fitness space). When climate stability returns, local isolation reduces connectivity in fitness space, leading to specialization with or without strong coevolutionary interactions (the geographic mosaic theory of coevolution). This sets the stage for survival when the next environmental perturbations hit (the oscillation hypothesis). Finally, new perturbations lead to a repetition of the sequence, this time with new players, which produce mixing leading to complex ecosystems (taxon pulses).

There seems to be no fundamental difference in the evolutionary dynamics of pathogens or their hosts.[61] There is simply evolution, and it is fundamentally Darwinian. Not heroic, but effective. Darwinian evolution is the history of species persisting, sometimes thriving and sometimes just surviving, but always coping with whatever conditions present themselves, in whatever manner is possible given their inherited capacities, oscillating slightly out of phase with environmental changes.

An Integrated Research Program

The Stockholm Paradigm proposes, mathematical modeling of the proposal predicts, and the empirical evidence supports the following characterization of the world of evolvable life that includes a universe of pathogens: The evolutionary potential of pathogens is immense, and they can take advantage of opportunities rapidly under a variety of conditions. Ecological fitting in sloppy fitness space results in oscillations between exploiting and exploring that lead to a maximum amount of available fitness space being occupied. That leads to a maximum degree of fragmentation of fitness space, with a maximum degree of specialization in pathogen-host associations. Once that occurs, the system as a whole experiences exploitation-biased evolution. When the system is affected by external perturbations, the initial effect is the establishment of new connections in fitness space, leading to exploration-biased activities for some period of time determined by the size and power and duration of the perturbation. During that time species generalize as much as possible in the new larger fitness space. When the perturbation ceases, we return to exploitation-biased oscillations, except this time the pathogens, their distributions, and their hosts are different species in different associations in different places than their immediate ancestors. No matter how intense the local interactions are, capacity will always be greater than opportunity. Species will always have what appear to be hidden capacities allowing them to take advantage of novel opportunities (and thus escaping a turn of events altering their previous conditions of existence). Paradigms comprise multiple interconnected hypotheses that combine to suggest a coherent concept. Because of their composite nature, paradigms per se do not have predictions, but the hypotheses stemming from them do. Next we discuss some of them.

Sloppy Fitness Space

At any given time, most species will be specialists (a fact that includes the prediction that most pathogens will have restricted host ranges), but this will reflect opportunity more than capacity. Fundamental fitness space (including potential host range) is much larger than realized fitness space (including actual host range). Fitness space (including host range) changes may be stimulated by any external factor (from tectonic events to climate change to anthropogenic activities) that alters accessible fitness space.

Ecological Fitting in Sloppy Fitness Space

Cophylogeny studies will show lots of host range changes within a constrained host group, indicating that host range changes reflect changes in available portions of fundamental fitness space. Phylogenetic comparative studies will show alternating episodes of host range expansion and host range reduction, indicating large-scale patterns of changes in host range. Successful invasive species will occupy the same or similar fitness space as in their source area (niche modeling studies will be largely successful in predicting successful introductions).

Generalizing and Specializing (Includes the Geographic Mosaic Theory of Coevolution)

Generalizing in fitness space (including geographic/host range expansions) will be associated with phylogenetically conservative specialized microhabitat preferences and transmission dynamics. Specializing in fitness space (including geographic fragmentation and host range reductions) will be the source of innovations in specialized microhabitat preferences and transmission dynamics.

The Oscillation Hypothesis

Fitness space (including host range) changes alternate between specializing (exploitation-biased) and generalizing (exploration-biased). Such fitness space changes may occur rapidly enough that we miss the transition from specializing to generalizing to specializing (including cases of host range changes from an original host to the original host + a new host to only the new host).[62] The large-scale phylogenetic pattern of host range changes appears as oscillating increases and decreases in host range over phylogenetic time scales.

The Taxon Pulse Hypothesis

Geography: pathogen and host geographic distributions will alternatingly expand, then fragment, then expand; but they will not coincide precisely. Host range: geographic expansion events will be associated with host range expansion, and geographic fragmentation events with host range reductions. Geography and host range: episodic geographic and host range expansion and isolation will result in the assembly of complex faunal mosaics.

Summary

Wanderer, there is no road.
The road is made by walking.
By walking, one makes the road
and, upon glancing back, one sees
the path that will never be trod again.[63]

Realizing that the Stockholm Paradigm resolves long-standing paradoxes and confusions in evolutionary biology concerning the nature of pathogen-host systems and of evolutionary radiation should be a cause for celebration. But the intellectual satisfaction we feel moves to the margins in light of the implications of our findings.

At the height of its technological achievements, humanity now lives in a time of environmental perturbations of ever-increasing intensity and duration that are associated with disease emergence. This threatens our ability to survive as a technological species with the potential to achieve better lives for future generations. We believe that understanding the relationship between pathogen evolution and evolution of the biosphere as a whole is one way to help in that effort. The various oscillation cycles we have identified are "reset to repeat" by external perturbations of a magnitude great enough to cause extinctions in those species caught in the "can't leave/can't cope" (no opportunity/no capacity) trap of Darwinian extinctions. This is a good news/bad news situation. The good news is that such perturbations set the stage for extinctions to be overcome by new evolution. The bad news is that pathogens will participate in the same way as hosts. We cannot "extinct' our way out of the emerging disease crisis.

The Stockholm Paradigm indicates—paradoxically—that something potentially reassuring is occurring. Remember that it is the nature of fitness space to be sloppy, and always to be a source of evolutionary potential, complementing the indefinite variation produced by inheritance that is part of the nature of the organism. The paleontological studies by Alycia Stigall in particular concerning evolutionary dynamics during past major episodes of climate change give us hope that the lessons of the past can help inform proposals for how humanity deals with all the implications of the current episode of global climate change. The crisis of emerging diseases is one particular manifestation of the general way in which the biosphere copes with massive external perturbations.

The good news is that the massive biotic expansions caused by climate change are an indication that the biosphere is beginning to respond evolutionarily in the way it has to all previous such perturbations. And the result of that has been the generation of new biodiversity, forming new, complex biotas. The bad news is that the evolutionary ability of the biosphere to cope with large-scale climate change and other perturbations means that there will be a spike in emerging diseases as a by-product of the way the biospheres copes with climate change. This reinforces the notion that we are facing an existential crisis, in which the biosphere is—to use Darwin's terminology—indifferent to our fate. Therefore, is it incumbent on us to decide how we are going to deal with it.

The emerging disease crisis is one indication that the biosphere is beginning to cope evolutionarily with climate change in the way it has always. But it is also an indication that the biosphere is not asking our permission or waiting for us to decide to cope and survive.

The emerging disease crisis is yet one more threat multiplier associated with climate change and will continue so long as there are climatic perturbations, which will continue for the foreseeable future. And climate change is not the only source of perturbations leading to emerging disease, as we will see next.

So,

Fasten your seat belts; it's going to be a bumpy night.[64]

EIGHT

Emerging Diseases: The Cost of Human Evolution

Today is the slowest day of the rest of your life.
—IAN GOLDIN, PUBLIC LECTURE, STELLENBOSCH INSTITUTE FOR ADVANCED
STUDY, STELLENBOSCH, SOUTH AFRICA, APRIL 19, 2017

Now we shift gears, taking our new insights about pathogen-host systems from basic science to a world of accelerating climate change and emerging diseases. The concerns we voice are the concerns of other scientists as well, some of them echoing down from more than a century ago. Most of us believe humanity faces a precarious future, and we must act with great urgency if we are to find a hopeful outcome. As scientists we are trained to deal with problems with a certain sense of detachment, putting our true feelings aside in an effort to draw rational conclusions. In the following chapters, we will try to combine scientific detachment with our true concern for the welfare of humanity, but we will not please everyone.

Detachment is a rare virtue, and very few people find it lovable, either in themselves or in others.[1]

The bad news is on us, all of us. We are biologists, so we will begin by calling ourselves out.

Conservation Biology

Between Stasis and Oblivion, There Is Evolution

In most people's minds, conservation biology has taken center stage for discussions about humanity and a changing climate. It brings together a galaxy of concerned individuals, all with different motivations and backgrounds. Evolution is the only natural process by which the biosphere has recovered from all previous episodes of climate change. And yet, major policy decisions in conservation biology do not fully integrate the evolutionary nature of the biosphere. *We criticize only the advocates of conservation biology who promote a narrative that all climate change and habitat alteration indicates something is wrong, when the unifying theory of biology says that is not always true.*

Stasis Is an Illusion, not a Goal

Many policies in conservation biology revolve around a perceived need to preserve existing diversity where it is and as it is; or, in a more extreme version, to try to return the biosphere to a state presumed to exist before the Industrial Revolution. We believe these are inappropriate and unreachable goals, given the evolutionary nature of the biosphere.

In a Darwinian world, species are the children of time—specifically, of the interaction of the nature of the organism and the nature of conditions through time.[2] What appears to be stasis in biological systems is an artifact of our attention and life span. In reality, the biosphere is in constant motion, renewing life that carries a load of historically conservative structural and functional traits. This does not mean the biosphere teeters "on the edge of chaos."[3] In fact, life occurs in a robust "window of vitality."[4] Such robustness stems from the nature of the organism, but it imbues organisms with a certain degree of insensitivity to their surroundings, especially when it comes to reproduction. And that produces natural selection. Biological stability is thus long-term persistence as a result of constant environmental change. Phylogenetically conservative and persistent features of form and function maintain continuity in that change. This has fooled many people into thinking that the proper "conditions of existence" are a state in which nothing changes. Worse, it has fooled some of those people into nostalgia for a past that never existed, the belief that things were better in the past.

It is a feature of complex systems that external perturbations of great magnitude will cause the system to respond in unpredictable ways—ways that could well be contrary to the intentions of whatever (or whoever) is applying the external perturbing force. This should be understandable to all readers: human history is littered with people who died as a result of unintended consequences stemming from attempts to control the world. We have evidence that efforts expended trying to force the biosphere into compliance with human aspirations can backfire horrifically. This is not just true of warfare, but also of the building of cities and nature reserves. If the biosphere is a complex functional system, we must have a light touch in our interventions. If we wish to actively preserve any part of the biosphere, we should focus on preserving the evolutionary process in general rather than particular places, species, or species associations.[5] The biosphere is not a collection of variable yet static entities; it is an evolutionary system with an indefinite ability to survive even massive external perturbations. This leads to a simple-sounding policy—save as many places as possible, link them together as much as possible, and then let evolution take over. This is the way the biosphere has responded to previous mass extinction events, with remarkable success each time. We may lose some species to which we have formed attachments, but what remains will be capable of sustaining a transitional biosphere that will evolve into a new (meta)stable one.

Evolutionary radiations are evolution's way of coping with major external perturbations. As Stigall's work shows so clearly, most species that cannot move when the climate changes go extinct. Some that are able to move go extinct as well, but the survivors are among those that move, either finding a new place where they can live as before or buying enough time while on the move to evolve the ability to survive, somewhere, somehow. Every radiation thus begins with biotic expansions catalyzed by external perturbations that decrease species fitness in their area of origin, but that also link previously unconnected pieces of fitness space, allowing each species a chance to flee for its life. Species that are capable of leaving areas where their preferred conditions are deteriorating for areas where they can still find preferred conditions not only survive, they become generalized in fitness space. This diversifies their portfolios, increasing the odds of survival in an uncertain future, and setting the stage for evolutionary innovations to emerge in the changed conditions to come. How has that been working out?

Some perturbations have been of such a magnitude that a significant proportion of the Earth's species were unable to cope with the

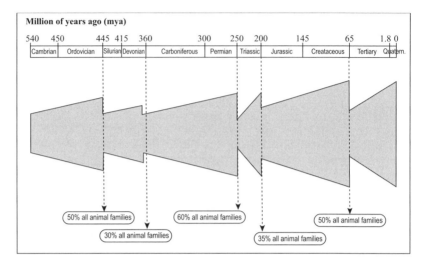

FIGURE 8.1 Diagrammatic representation of the impact and aftermath of the five great mass extinction events. The x-axis indicates time and the y-axis indicates the diversity of animal families estimated from the fossil record. Regardless of the severity of the extinction event, evolvable life has enough capacity—in the form of indefinite variation—to rediversify. Summarized from sources on the Internet.

changes and became extinct. Five of these terrestrial convulsions are called the *great mass extinction* events. But mass extinction events are also mass evolutionary reset events, because every mass extinction has been followed by relatively rapid and widespread new diversification of hosts and pathogens.[6] This shows that there is enormous evolutionary capacity in the biosphere, which allows evolution to generate new and abundant biological diversity (fig. 8.1).

Evolution is not merely capable of producing new complex biotas after every major perturbation event; it is the *only* mechanism by which that occurs. This fundamental truth must be better internalized in the research activities and policies of conservationists. Every day, in many parts of this planet, bright and energetic people are out collecting and counting and measuring. Everywhere they record information, finding populations of many species—plants and animals, large and small—changing in a manner correlated with environmental change. Those changes are too often interpreted as something bad—reasons to intervene to stop whatever is causing the changes. This means, however, that the research protocols and explanatory frameworks used by these researchers have no criteria by which correlated biological and environmental changes can be explained as the result of the evolutionary change that allows species to cope with environmental changes. What if a reduction in the population size of a certain species is temporary? What if the reduction is an indication that the population is responding selectively to climate change and, if left alone, will begin to increase in size as older variants that could not cope with climate change disappear? After all, one of the core principles of evolutionary biology is selective response to local environmental conditions.

This distinction is critical. Our well-intentioned interventions to impose stasis could conflict with the ongoing evolutionary process. Certainly, we do not want to counter the only mechanism known that allows the biosphere to cope with climate change, neither by hastening nor by inhibiting potential for adaptive changes. Conservation biology measures that corral portions of the biosphere in situ risk blocking the only known mechanism for coping with climate change available to evolvable life. Evolution is a never-ending story. Change is an essential and inescapable element that we neglect at our peril, and our current problems are a testimony to that fact. We can be certain that pathogens have not stopped evolving just because we think about them in contemporary time.

Pathogens Are Neither Villains nor Victims

Conservation biologists have not achieved consensus about pathogens, but they have accumulated strong opinions. Some consider pathogens to be natural enemies exacting a terrible cost on their hosts and requiring extermination if possible. That runs afoul of the more general notion that we should be preserving as much of the biosphere (and its evolutionary dynamics) as possible. Others reason that surely, if pathogens exist, they must provide some essential ecosystem service. This leads to ideas that pathogens perform important eugenics roles, "regulating" host populations in a way that somehow makes the hosts more or less vulnerable to extinction, depending on your assessment of the value of the host species. Within this group, a small minority—mostly parasitologists—consider pathogens to be special victims of climate change requiring nurturing and care lest we lose them. All of these groups tend to use colorful descriptive language, such as "arms races," "attacks," "plagues," "infestations," "defense," "enemies," "aliens," "enemy-free zones," "coextinction," and "invasive species," when justifying a particular point of view.[7]

Let's apply some professional detachment to this emotion-laden topic. First of all, from a scientific standpoint, pathogens have *func-*

tions, not *purposes*. Pathogens did not magically appear on Earth for the purpose of causing disease or regulating host populations. They are simply members of the biosphere, species made up of individuals preoccupied with survival in the conditions of life in which they find themselves. And as they oscillate between exploiting and exploring their fitness space in such preoccupied states, they will contact their surroundings—including hosts—in a variety of interesting ways. Most humans don't care about this seething mass of pathogens except when some of them have a negative impact on host species for which they have a particular fondness or socioeconomic need—or, of course, on themselves.

Secondly, evolution does not eradicate pathogens or coeradicate pathogens and hosts. Evolution is not a story of hosts being attacked by enemies, the enemies either wiping out the hosts or the hosts wiping out the enemies, or both wiping out each other as the result of reciprocal intensification of conflict. Nor is it a story of hosts running away to a mythical pathogen-free part of the planet. It is a story of conflict resolution: evolution eliminates pathogen-host associations that cannot coexist.[8] During biotic expansions, species mix together and pathogens capable of infecting hosts to which they have never been exposed have a chance to enlarge their host range. The new hosts, having never been exposed to the pathogen, have no resistance and thus some disease outbreak often accompanies the early stages of the new association. After selection has increased the proportion of resistant and tolerant hosts in the population, and acute disease is not an issue, the pathogen remains as an element of *pathogen pollution*. This is how evolution produces persistent pathogen-host associations—lots of them. Pathogen pollution is a euphemism for the expected fate of pathogens that can coexist with their hosts, even if they do not get along that well. To return to an earlier observation: pathogen-host associations are neither random nor optimal; they are simply functional. This is why EIDs are an expected outcome of the biosphere's coping with climate change based on its tremendous evolutionary capacity.

What Lessons about Survival Can We Learn by Studying Species That Are Going Extinct?

Whether it is Darwin's prosaic "survival of the adequate" or Spencer's poetic "survival of the fittest," evolution is about survival and persistence. We have abundant evidence that many species are unable to cope with ongoing climate change. What can we learn about our own survival from them? If we want to survive, we need to learn survival lessons, not death lessons, from events happening around us. Why, for example, if we live in a world in which 50 percent of the current species are predicted to go extinct as early as 2050, are we seeing massive expansions of pathogen host ranges? Are they the only species on the planet participating in biotic expansion in the face of perturbations in the nature of the conditions?

At some point we must prioritize survival in allocating research resources and effort. Elton's *Ecology of Invasions* (1958) was not a call for biology to focus on failure, becoming a scientific chorus chanting the names of species as they disappear. We have wasted more than half a century and now must begin asking some difficult questions about research funding priorities. Can we afford to continue to spend money and expertise on species that are not surviving? Can we afford to spend money trying to keep them from going extinct? When do we begin paying attention to nonpathogen species that are thriving—who they are and why they are doing so well? By saving as many of the areas in which the species that are thriving live, will we be saving more of the at-risk species than if we target those at-risk species individually?[9]

The three topics we addressed in this section have a common connection, suggested by the emotional reactions they are likely to elicit. They are elements of a persistent non-Darwinian view of nature. Eighteenth-century Europe was alive with brilliant natural philosophers, including Jean-Jacques Rousseau, from Geneva. He inspired a tradition of looking to nature for validation of what he considered to be key principles of natural philosophy. The concept that nature maintains itself in an unchanging state is ancient, reaching back as far as Herodotus in Western philosophy. Rousseau was inspired by this sense of order and predictability in biological systems, which he concluded was a good thing. He also reasoned that anything disrupting this "balance of nature" must be bad.

The twentieth-century notion of the balance of nature claimed that "healthy" ecosystems exist in a stable equilibrium. The Gaia hypothesis is a balance-of-nature theory based on the assumption that the biosphere acts as a set of coordinated systems maintaining a homeostatic balance. The concept was a core principle of ecological research and management of natural resources until it was abandoned in the last quarter of the twentieth century for lack of evidence. Then, in the ricochet fashion of academic research, ecology bounced from homeostatic stasis to chaos, claiming that the biosphere is governed by random events that can be described by chaos theory, but never fully explained or predicted. Cooler heads began to exert their influence in the early twenty-first century, when it became apparent that the biosphere was neither random nor optimal, neither static nor chaotic, but evolutionary and functional.

Most of ecology and evolutionary biology now accepts that there is no static balance of nature. Too many involved in conservation biology, from concerned members of the public to policy makers to researchers, seem not to have gotten the emails about that. Indeed, parts of conservation biology serve as a reactionary repository for some of the last vestiges of Rousseauean biology on the planet. It is the source of the persistent belief that the natural state of the biosphere is timeless and changeless. Or at least it would be if humans weren't trying to destroy it. Or perhaps at some point in the past it existed in a better state, so we should impose stasis before things get worse. Maybe we should begin now, using genetic engineering to bring back long-dead species such as the wooly mammoth and quagga, and reintroduce big cats to North America to restore some mythical balance. So humans should either do nothing or actively intervene to return nature to something called its "natural state," which is always the result of someone's nostalgia for a past that never existed.

These views are contrary to the recognition that the biosphere is an evolved biological system in which change is an essential part of survival and persistence. The notion of a static natural equilibrium is also contrary to our understanding of the great upheavals that have wracked our planet in the past, causing mass extinctions that are then followed by bursts of evolution producing innumerable new species. Many species form easily during periods of isolation, when climates are stable and fitness space is disconnected; but many of them go extinct when climate change leads to local alterations and new connections in fitness space. The ones that go extinct lack the capacity to cope with the changes and to take advantage of the opportunity for exploration provided by the new connections in fitness space. Of course, this can be the result of a species' being forced by circumstance to explore fitness space that is already occupied by a relative with similar resource requirements and lacking the capacity to co-opt or change. A call not to impede evolution is not the same as a call to accelerate extinction.

We must not lose the love of nature espoused by von Humboldt and Goethe and Rousseau—*biophilia*, as E. O. Wilson called it.[10] But our emotions cannot rule our decisions about the biosphere as it pertains to the survival of humanity. We must not advocate policies that place limits on evolution. Either such policies will be immediate failures, or we may experience some short-term gains followed by losses far worse than if we had done nothing at all. The more we try to impose stasis, the more we take away the biosphere's ability to respond to perturbations, which is the essence of survival. Indefinite sustainability requires indefinite variation; in other words, without evolution we are sunk. But fortunately, we get evolution for free: life on this planet is evolvable, and evolving, life. If experience unclouded by emotion has shown us anything, it is that all policy decisions that are contrary to evolution—anything that attempts stasis and control—work against us and become part of the problem.

Evolution has taught us that the old will inevitably pass away; every human knows this lesson personally, and we all resent it. But there is a positive note. Evolution has also taught us that the passing away of the old sets the stage for the arrival of the new, and humans also experience this personally. Global climate change has happened before and has had a significant impact on the biosphere. This means we can compare the biological effects of this episode with previous ones and, potentially at least, learn the lessons of history.

Health

The challenges are not underappreciated; the responses, well-intentioned as they are, have been inadequate.

We are not clinicians or health care providers. But we understand that when diseases pop up faster than we can cope with them, something is wrong. When a cholera outbreak threatens to bring down an entire country, when a single outbreak of Ebola depletes the resources of the World Health Organization, when yellow fever and plague outbreaks exhaust global emergency vaccine supplies, and when the global cost of disease exceeds the GDP of all but 15 countries, thinking outside the box becomes a necessity rather than a luxury.

Knowing what they knew, thinking what we all thought in the twentieth century, health care professionals could not anticipate the emerging disease crisis fully. There were warning signs, but they did not catch our attention. The sheer number of emergent diseases since the turn of the century has given many researchers pause, but for those who control policy, little has changed. Their conceptual framework has not permitted them to comprehend fully what is happening. For those reasons, it is not surprising very little thought has been given to proactive measures for coping with the crisis. If you don't believe that what is happening is only the beginning of something that could, and should, have been anticipated, you will tend to believe the situation can be handled by reaction alone. And yet—again—in other areas of health practice, we have internalized the message that crisis response is not only more expensive than prevention, it is unsustainably more expensive. Early changes in diet and exercise patterns are always preferable to open-heart surgery.

For too long, emerging disease was viewed as if some malign influence were creating the capacity to seek out targets of opportunity. Some believed that selection would always be there to counterattack, and that if humans intervened on the side of the angels to help selection fight off the enemy, victory would surely be ours. No one took it seriously when modeling efforts showed that selection could not eliminate pathogens or even necessarily reduce their pathogenicity.[11] We have had cause and effect backward. Disease is an indication that a pathogen is exploring the limits of its fitness space by ecological fitting and has encountered viable (if suboptimal) fitness space—not that it has mutated into an enemy attacking a new host.

We now have no excuse for being repeatedly and unpleasantly surprised.[12] The Darwinian concept of preexisting capacity taking advantage of new opportunities is the general explanation for all emerging disease. Capacity is always greater than opportunity, but opportunity drives the bus. Perturbations caused by climate change increase connectivity in fitness space, inviting humans and pathogens to explore. While we have planning conferences, they explore.

Medicating the Ill and Symptomatic Medicine

Pathogens are specific and phylogenetically conservative in their transmission modes and microhabitat preferences, which are linked to signs and symptoms. Physicians, veterinarians, and crop and wildlife disease specialists have all relied on this feature of evolution in treating disease. The word *malaria* brings to mind a range of signs, symptoms, and treatment options for all species of *Plasmodium*, which makes diagnosis and treatment more rapid and efficient. But this is a two-way street. Similar symptoms are produced by multiple pathogens (how many produce "flulike symptoms"?). How often has your family physician taken blood to screen for a pathogen before reaching for the prescription tablet? Do no harm, indeed!

Vaccinating Those at Risk and Genetic Load

Genetic load refers to the reduction in the mean fitness of one population relative to a population composed entirely of individuals with optimal genotypes. It is an inescapable yet unintended by-product of the success of medical technology.[13] When we intervene to help a host-selective response to pathogens, we counter selection from the host and the pathogen's perspective. Annual vaccinations are an easy illustrative example.

Let's imagine a host population made up of some individuals who are resistant to a given pathogen and others who are not. The smaller the proportion of susceptible hosts, the lower the genetic load. We expect selection to favor resistance, so we expect to find that most host populations have a distribution like the one shown in figure 8.2, in which the majority of hosts are resistant. Vaccination programs do not target those individuals directly, because they are already resistant. Therefore, vaccines target and selectively preserve hosts that are not genetically resistant. In this way, mass vaccination programs—if they are successful—will actually increase the genetic load in susceptible host populations (fig. 8.2). An increasingly larger proportion of the host population will require the vaccination to survive, leading to herd immunity policies. Mass vaccination programs are no longer just a good idea; they are essential. When that point is reached, if we stop mass vaccinations, a far larger proportion of hosts will be in danger of disease than before we began the vaccination program.

Annual vaccinations have a complementary unintended cost. Consider a pathogen with a range of variation in its ability to infect a given host, or in pathogenicity. Our modeling shows clearly that selection will tend to match pathogen genetic variation with host genetic variation. If a seasonal vaccine targets the most common pathogen variant and the vaccine is successful, the next year that variant will be less common. Eventually, the vaccination protocol will make common variants less common while allowing rare variants to become more common (fig. 8.2); at some point, it will be impossible to predict accurately what next year's most prevalent variant of the pathogen will be (this may explain the failures of annual influenza vaccines in 2013 and 2018). Vaccination programs based on providing annual protection against the anticipated most common variant of the pathogen risk becoming efforts in pure guesswork. Furthermore, when hosts and pathogens respond selectively to pharmaceuticals rather to than

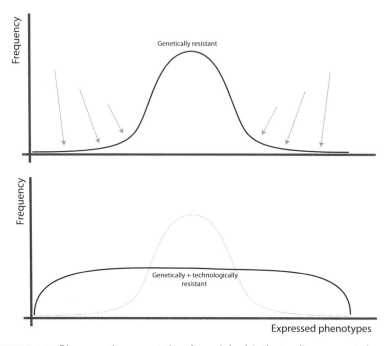

FIGURE 8.2 Diagrammatic representation of genetic load. In the top diagram, most of the organisms in a population are genetically resistant to diseases the population has been exposed to. The bottom diagram indicates the impact of mass vaccinations on the genetic makeup of the host population—an increasing proportion of the population is not genetically resistant, meaning mass vaccinations become increasingly important. If we consider the diagrams to be representations of the pathogen population, the top diagram indicates that there is generally only a small proportion of the phenotypes that make up most of the organisms in the pathogen population. If an annual vaccine program targets the most common pathogen phenotypes, genetic load produces a pathogen population in which an increasing number of phenotypes become more equally represented. Vaccination programs targeting the most common pathogen phenotype become less effective.

to each other, we risk adding another dimension to the problem of increasing genetic load.

Eradicating Reservoirs and Biological Control

Rachel Carson's *Silent Spring* led many to understand that eradicating biodiversity to control disease might be too costly. Nonetheless, this mode of coping with disease persists. One application is *biological control*. The term refers to the practice of attempting to control one introduced species by introducing another. When put in those terms, it sounds a bit like "two wrongs make a right." A common form of biological control involves an introduced pest insect that is inflicting damage on a native or an introduced plant of commercial value. The introduced pest insect is not a pest in its native area. Why? In its native range, the pest is attacked by a parasitoid wasp, which lays its eggs in members of the pest species, which die as the parasitoid larva eats its way out of the host. So, the solution is clear—import the parasitoid species to the place where the pest has been imported, and let nature takes its course.

You, the reader, now understanding ecological fitting in sloppy fitness space, will begin shouting, "Don't do it!" Sure enough, within a couple of years the parasitoid discovers a native species of insect it likes and is busy using it. Sometimes this is just collateral damage. But too often the parasitoid not only uses a native species but actually prefers it. As a result, the introduced pest is granted an evolutionary reprieve. And in the worst-case scenario, the native species that is now the focus of attention for the introduced parasitoid is itself an economically important species—for instance, an essential pollinator for other crops or protected wild plants. And to make matters worse—if that is possible—establishing pathogens in suboptimal hosts through the introduction of the pathogen or the host may increase the risks of sampling rare and pathogenic genotypes, increasing the pathogenicity of existing pathogens. Likewise, the evolution of resistance to pharmaceuticals or pesticides in one population of a pathogen simply adds to the pathogen's ecological fitting repertoire, making additional spread of the pathogen more likely.

And we still alter landscapes and species distributions on large scales with scarcely a thought about the long-term, widespread consequences. Parasitology students are taught one of the most notorious episodes as a cautionary tale. Between 1960 and 1970, Egypt built the Aswan High Dam. The United States refused to help fund the project, citing concerns about disease. Parasitologists warned that when large amounts of flowing water were turned into large amounts of standing water, there would be an explosion in the populations of certain snails, whose specific and conservative biological traits include a preference for standing rather than flowing water. Seems like no big deal: snails eat algae, so if we have lots of them, they may control the spread of algae in the newly formed reservoir. But it *is* a big deal if those snails happen to be the vector for *Schistosoma haematobium*, which live in the blood vessels surrounding the urinary bladder. In addition to blocking circulation and flooding the host's bloodstream with its own metabolic waste,

S. haematobium produces a prodigious number of eggs, each of which has a tiny spine that digs through host tissue until it arrives in the urinary bladder and is expelled into the environment—along with blood from its traumatic migration through the host tissue. It is a monster of literally biblical proportions: an outbreak of *S. haematobium* is thought by some to have been the causative agent of one of the plagues—the one that caused the waters of Egypt to turn red with blood—that Moses called down upon the Egyptians to encourage the pharaoh to grant his people freedom. This was truly a known issue. Since nobody wanted to pass up an opportunity during the cold war, one side agreed to do what the other would not. The dam was built, and schistosomiasis rates in the area skyrocketed, augmented today by a high rate of hepatitis B infections in the inhabitants around the Aswan Dam. Conflicts and military deployments being temporary, ultimately there was no net geopolitical gain to either external power—just an increase in human misery inside Egypt. Is it possible that climate change itself could eradicate biodiversity in such a way that disease transmission is interrupted, perhaps even leading to the extinction of pathogens? Might we find that extinction has a silver lining? Uh, no. Coping with climate change is a matter of ecological fitting in sloppy fitness space, so we would predict that more hosts than pathogens might go extinct as a result of large-scale and prolonged climate change. Parasitologists know that some hosts are "parasite hotels," hosting a disproportionately large diversity of parasites compared to their relatives. This has not yet been investigated directly, but we suspect that those hosts are the descendants of the survivors of extinction events that eliminated other relatives, the pathogens of those that went extinct moving to the survivors, who expanded their geographic ranges into areas where their relatives were going extinct. Pathogens are the ultimate evolutionary survival machines.

Agricultural Health and Food Security

Loss of environmental resources affecting food security and food safety will shortly be substantial and prolonged.[14] The emerging disease crisis is by no means limited to pathogens that infect humans.[15] Climate change influences food security and safety in part through changing opportunities for pathogen dissemination over small and large scales, driving the potential for disease emergence.[16] Diverse assemblages of pathogens brought together by climate change represent critical impediments to global food security, in food sources and industries ranging from domestic ungulates to aquaculture, fisheries, and crops.[17]

Physical and biological perturbations caused by climate change are threat multipliers for emerging infectious disease that influence non-pathogen factors affecting health and food security.[18] How water is distributed—too much here and not enough there, with extreme fluctuations in precipitation—is the most critical of these, because it links direct threats to food production with ease of pathogen transmission. The availability of potable and agricultural water is affected by land use, ocean warming and acidification, rising sea levels, and desertification. Attempts to capture freshwater sources may dramatically change aquatic systems and the potential for disease transmission, causing fisheries to collapse. Climate change limits the amount and distribution of well-watered arable land, leading to conflict and migration. Waves of human displacement increasingly challenge global infrastructure for food and health resources as refugees introduce pathogens to, and receive pathogens from, sources new to them.[19] Pathogen pollution exerts an ever-greater cost on public/agricultural and veterinary/wildlife health services. The direct impacts of accelerating climate change on health resources and capacity are critical threat multipliers for EID among people whose health is already affected by environmental disruption.

The Stockholm Paradigm leads us to a sobering conclusion: no matter what we do in recognizing and treating new outbreaks, new infectious diseases will continue emerge so long as the external perturbations associated with climate change persist. We must assume we will be living in a world of EIDs for the foreseeable future.

Density and Connectivity

More than 100,000 years ago the initial waves of human population expansion were catalyzed by habitat changes, as our ancestors left the forest trees for savannah grasslands, then by dietary changes accompanying the invention of tools that allowed us first to scavenge and then to compete with top carnivorans for animal prey. We were already clever and aggressive; the new diet led to larger and more fecund females, fueling a population boom that created the need for geographic expansion as humans sought water and food. Both dietary changes and geographic expansion brought those earliest humans into contact with pathogens they had never encountered before. Emerging disease was likely not the key factor in keeping their populations small and on the move—that was due to a biotic expansion event catalyzed by extensive climate oscillations and glaciation from 90,000 to 20,000 years ago—but it certainly played a role.

The last ice age began to abate about 20,000–18,000 years ago and was associated with incremental warming. Around 14,000–12,000 years ago the Pleistocene breathed it last gasp amid rapid, abrupt, and tumultuous shifts between cooling and warming, eventually ushering in our current era of unusually stable climate conditions beginning about 11,700 years ago. It was during that time, at the inception of the Holocene, that domestication and agriculture, along with sedentary lifestyles, left their earliest traces. Some of that amplified population growth propelled geographic expansion. But the rest of it produced population densities in sedentary settings never before seen in human history. Humans, like many other species, were entering a period of biotic stability characterized by isolation, exploitation, and innovation. At the same time, increasing population density in isolation increased opportunities for contact with pathogens already known from the area. Domestic plants and animals suffered the same fate as humans, leading to a demand for more and more food, as well as higher and higher densities of domestic plants and animals.

For the next 5000 to 10,000 years, humans living sedentary lifestyles tended to cope with short-term but often intense climate fluctuations, and associated plagues, famine, floods, drought, and conflict, by abandoning their abodes to start again elsewhere. The fate of the Khmer in Angkor, the Tamils in the inland areas of Sri Lanka, and the Mayans in southern Mexico and northern Central America during the Middle Ages are notable recent examples. We can imagine countless conversations: "If this [drought, flood, famine, plague, conflict] goes on much longer, we're going to have to leave." We do not know that disease was ever the primary reason ancient humans abandoned their abodes, but we think it likely that disease was often the final insult: "and now the kids are all sick."

Growing populations are exploratory, so humans entered new habitats daily as their local populations expanded. That increased human realized fitness space but made it disproportionately sloppier: as soon as humans made a new kind of habitat part of their realized fitness space, all examples of that kind of habitat throughout the world became part of their fundamental fitness space. Migrating humans found fewer new places in the world to be completely novel. Human population expansion thus mimicked the biotic expansion phase of taxon pulses, most likely with predictable results in terms of disease (lots of ecological fit-ting leading to rampant host range expansion—in other words, a sharp increase in EIDs). Species do not fill fitness space optimally or maxi-mally, but functionally, according to how much and which parts are available to them at any given time. Human beings followed this path; We did not expand smoothly across the globe, like some kind of gentle but inexorable tidal wave of life. Rather, we had a clumped distribution owing to the high population densities in populations adopting seden-tary lifestyles, some members lurching away when that density became more than they could bear. This produced two major innovations rel-evant to the emerging disease crisis: urbanization and globalization.

Urbanization and the Density Trap

More than half of humanity now lives in cities, compared to 14 percent a century ago. And within a decade, that figure is expected to climb to 65 percent. For most of the past 12,000 years, sedentary humans coped with climate change, and with the increasing risk of disease associated with increasing population density, by leaving their settlements in search of better conditions. As a result, the number of cities that have been abandoned is greater than that of the ones that are still inhabited. We are, however, a clever species. And in the past 5000 years we began to learn how to endure environmental changes that would previously have overwhelmed us and caused us to move.

Part of that learning experience involved recognizing certain con-nections between urban life and disease. These included such obvious things as poor sanitation, poor nutrition, inadequate water, and prox-imity to certain habitats—for example, foul-smelling swamps full of mosquitoes. But they may have included more obscure things, such as the presence of royal ibises along the banks of the Nile. Like most Egyptian mummies, human and animal, royal ibises had their innards removed as part of the mummification process. What is intriguing, however, is that at least some of those ibises had their gizzards returned to their abdominal cavities, packed full of the snails that transmit *Schistosoma haematobium*. Is it too far-fetched to believe that humans could reason from the presence of royal ibises in places where there was a reduced threat of schistosomes to a belief that whatever they were eating was associated with the disease?

As we recognized connections between certain aspects of urban life and diseases, humans produced technological advances, mainly means of decreasing the possibility of outbreaks and of responding rapidly to mitigate outbreaks when they did occur. Part of the success of such measures was due to the fact that these early urban dwellers were coping with known diseases, not with new ones encountered as a result of colonizing a new geographical area. When such measures were successful, humans were capable of staying longer in their cities. That marked the true emergence of urbanized humanity: we no longer felt we needed to flee a challenge. But the technological innovations were costly, and their success was translated into ever-increasing population density as fewer people died of disease. In addition, for reasons not understood until 1797, city dwellers seemed to be constantly running out of food. The cost of successful urbanization, in addition to new disease patterns, was decreased self-sufficiency—increased reliance on flows of essentials from outside. Trade between cities and the countryside, or between different cities, was shifting from a luxury to a necessity.

Three generations ago Elton warned that climate change and environmental disruption would produce massive migration. And he was correct. By the second half of the twentieth century, however, the technological advantages of urbanization had changed the traditional dynamic. Rather than spreading into new territory, humans began migrating to fewer locales. For all its benefits, urbanization became a kind of *density trap* for humanity, although we did not recognize the perils of increasing dependency on flows of essentials from outside. Not understanding this aspect of the history of urbanization, many today believe that highly technological cities are buffered from disease outbreaks. Alongside the various benefits of living in urban settings, however, are a number of threat multipliers for emerging disease. Five of them are especially significant.

Firstly, modern cities produce almost none of the basic essentials needed for their inhabitants' survival. They require a constant flow of energy, water, and material goods in and out. That is, they have an extraordinarily high level of connection to an enormous amount of fitness space containing an enormous number of actual and potential pathogens.

Secondly, cities are places of high human population density and contact. Remember that epidemiological network studies distinguish pathogens that establish themselves readily but spread slowly from those that establish themselves with difficulty but spread rapidly once established. High-density cities create conditions conducive for the second category of emerging disease. Many pathogens are more likely to become established in rural areas than in cities; but, once there is a disease outbreak in an urban setting, the higher the population density, the greater the likelihood of exposure to and spread of the disease. Bear in mind what happened the first time the Ebola virus entered an urban setting, in Guinea, Liberia, and Sierra Leone in 2014.

Thirdly, highly technological cities function through extreme division of labor with extreme interdependency. This means that disabling a relatively small proportion of the workforce by disease can wreak havoc with a city's regular operations.

Fourthly, large technologically advanced cities allow their human inhabitants to live and walk through landscapes that permit pathogens to persist on a daily basis. You might not think of cityscapes as permissive environments for pathogens. Take a stroll across the tundra at −40 degrees C in a stiff breeze rumbling off the Alaska Range, or a short walk in the shimmering heat on the Nullarbor Plain in the depths of summer, both places that harbor pathogens, and you might want to reconsider. So, large technologically advanced cities, especially cities that have attempted to "go green" by creating parklands, provide nurturing environments for animals that carry diseases that can be transmitted to humans. This includes domestic dogs and cats, but also rodents, rabbits, skunks and other mustelids, coyotes and foxes, raccoons, even hedgehogs and songbirds. In the autumn of 2016 a rabid raccoon from somewhere in the wildlands of New York hopped a ride in someone's pickup truck and crossed the international border into Hamilton, Ontario, where it disembarked and infected local raccoons, which then posed a severe threat to all other susceptible species—including humans—in the city. West Nile virus was introduced into North America when a single infected tourist took a commercial flight from Africa to New York, where local mosquitoes fed on him and transmitted the pathogen to birds living in New York City's Central Park. Mosquitoes that fed on those birds infected humans living in the city and wild birds living outside it, so the disease spread rapidly to encompass most of North America, where in a short time it established itself as part of the continent's pathogen pollution. And then there was monkeypox, introduced into the United States in 2003 by African rodents imported for the pet trade, although it was fortunately recognized and quickly thwarted, just as in the case of the rabid raccoon in Ontario. In early 2018 city parks in São Paulo, Brazil, were closed amid concerns about yellow fever and malaria.[20]

Fifthly, and most significant, a population of underpaid, undereducated, undernourished people who are virtually invisible to public health services supports the wealthy standard of living that attracts people to large technologically advanced cities. They are the backbone of urban centers, where they handle the food, water, garbage, laundry, and hospital waste; they clean the toilets, handle and cook the food, and change the linen in restaurants and hotels. In the cities of many developing countries, a distressingly large number of people subsist by scavenging the refuse and garbage of their more fortunate neighbors. In 1842 Edgar Allan Poe published a short story called *The Masque of the Red Death*. It told an apocryphal tale of a town in which smallpox broke out. The privileged rich people in the town took in as many supplies as they would need to wait out the outbreak, then closed the gates to their castle and settled down to party until the plague was spent. When the plague passed and the surviving poor people in the town opened the castle gates, they found that all the rich people were dead. That could never happen to us, right? We're much smarter than they were in 1842, after all.

What if we had a recurrence of the Spanish flu pandemic of 1918?[21] What would happen and what would we do? Table 8.1 summarizes an apocryphal thought experiment, projecting from the 1918 epidemic to a world of 2030 in which there are 4.5 times as many people, and 4.6 times as many people living in cities, as there were in 1918. If urbanization increases the risk of infection by that same amount—4.6 times—virtually every human on the planet will fall ill, and up to nearly 2 billion people will die. If urbanization's benefits in terms of living conditions and health care offset the increased risk created by high-density human living conditions by 50 percent, "only" half the humans on the planet will fall ill and "only" about 1 billion will die. Even in that scenario, global commerce and civic functions will be greatly disrupted.

The cost of human disease associated with urbanization does not reside simply in the number of people that die in a given outbreak. In terms of humans alone, the costs include the people who get sick and cannot work, the health care systems that expend human and material resources caring for the sick, the cost of the inexperienced workers who replace those who become ill, lost worker productivity as parents stay home to nurse sick children, and lost educational opportunities as teachers and students fall ill. Analogous arguments hold true for disease outbreaks in plant and animal resources of economic importance to humanity.

As we began to recognize connections between certain aspects of our lives and diseases, humans nonetheless continued migrating into new areas, coming into contact with novel pathogens and introduc-

Table 8.1 Depiction of an apocryphal thought experiment

1918	
World population	2,000,000,000
Urban population	280,000,0000 (12 cities of more than 1 million)
Infected (25% urban/25% nonurban)	500,000,0000 (25% of humanity)
2030	
World population	9,000,000,000 (4.5 × 1918)
Urban population	5,850,000,000 (20.8 × 1918)
Scenario 1: infected (75% urban/25% nonurban)	5,175,000,000 (57.5% of humanity)
Scenario 2: infected (50% urban/25% nonurban)	3,712,500,000 (41.2% of humanity)

Note: What would happen if the world of 2030, with 4.5 times as many people and 20.8 times as many people living in increasingly high-density habitats (cities), were subjected to a pandemic of the influenza variant that infected the planet in 1918? What would Spanish flu 2.0 look like? We consider two scenarios. In scenario 1 the risk of infection increases from 25 percent to 75 percent in the highly dense urban areas, while the risk of infection remains at 25 percent elsewhere; in scenario 2, the benefits of urbanization with respect to health decrease the risk of infection resulting from higher population density in urban areas from 75 percent to 50 percent.

ing ones they carried with them to new places. Urban-dwellers added the risk of amplified levels of local pathogens resulting from increased population density. They produced technological advances pertaining to health but at the cost of decreased self-sufficiency, because the technological innovations were themselves costly and because their success was translated into ever-increasing population density. That, as we mentioned before, converted trade from a luxury to a necessity, setting the stage for the emergence of hyperconnectivity and interdependence among human groups.

Globalization and the Hyperconnectivity Trap

Urban-dwellers coped with the population density trap by developing more and more complex trade networks to connect the cities with suppliers of necessary resources living outside the urban boundaries. As trade shifted from a luxury to a necessity, groups of mobile and sedentary humans in different places became connected. As cities grew in size, they required ever-greater amounts of resources from external sources. And that led to even more connectivity; this was the beginning of *globalization.*

True cities emerged about 7000 years ago, but only within the last 3000 years have we developed the ability to move large numbers of people and large amounts of goods around the entire planet. Globalizing trade connections were limited by time; each trade adventure lasted extended periods of time, often years. The development of rapid transportation has thus been key to advancing globalization. That began at the end of the fifteenth century with oceangoing voyages of discovery to every part of the world, but even as late as the 1950s shipping was still the major factor limiting intercontinental connectivity. By the 1960s air travel began creating the trade and travel connections that now link every continent every single day. Real-time communication, based first on radio and now the Internet, amplified and coordinated by global satellite systems, made the world go faster.

As a result of these activities, humanity has been able to create an enormous amount of technological fitness space for itself. Our global technology coverage means we can move things faster and farther than ever before. This has changed our world from the large, slow one of as recently as a century ago into one that is small and fast. As humanity becomes increasingly urbanized, it is not surprising that the greatest concentration of technological fitness space occurs around large cities. Human population is much higher than previously thought possible, and at least part of the reason must be that globalization has broadened the distribution of food and resources sufficiently to permit a larger number of people to be accommodated.

We have never before had this much technological capability. And yet, we are unusually fragile in some ways. Even a small drought can hurt humans and agriculture—just ask California, or São Paulo, or Cape Town. Droughts were at the core of the Syrian conflict, which has displaced millions, and now ravage central Africa, where famine and disease expand. A drought followed by torrential rains can strain infrastructure, as one of us learned recently when traveling through the Johannesburg airport, listening to public announcements apologizing for the damaged infrastructure resulting from recent floods. Most universities grind to a halt if connections to the Internet are interrupted for even a short time. Globalization also allows food and people and technological support for health to be distributed more equitably around the world, so even though birth rates are declining, the human population continues to grow because people are living longer and infant mortality rates are dropping. In a hyperconnected world, this leads to ever more people concentrated in ever-growing cities, exposed to more pathogens.

Each major transition in human civilization has brought with it an increased risk of disease. We are still encountering new pathogens as a result of expanding into new areas, and we still have known disease issues in our population centers; now globalization is bringing new pathogens to us. Globalization introduces nonhuman hosts and their pathogens to novel fitness space, with potential disease consequences for human and nonhuman residents of that fitness space. So, global-ization is more than just humans moving into new fitness space; it is humans bringing many species of plants and animals—and many of their pathogens—with them. Humans have a long history of moving plants and animals around the globe in any event, and globalization simply amplifies the distance and frequency of these introductions. Such speed and ease of distribution increases the odds of pandemics.

Surely large technologically advanced cities should be able to offset the risk of new pathogens' being introduced as a result of trade, travel, and migration by detecting incoming pathogens at their point of en-try. In theory, yes. But inspection and quarantine programs cost time, money, and human resources and slow the flow of goods. Globalization means increased risk, meaning increased costs of risk assessment and intervention. Thus far, the economic cost of slowing the flow of goods, services, and people has outweighed the potential cost of a major dis-ease outbreak.[22] But humanity does not need to be plagued by a low-probability, high-impact pathogen—think *The Andromeda Strain*—that threatens to kill all humans. All we need is a steady drip, drip, drip of high-probability, low-impact pathogens introduced as an unintended consequence of globalization, each persisting as *pathogen pollution* fol-lowing its initial acute outbreak;[23] never disappearing, always having the potential for a new outbreak. This is how the connectivity trap amplifies disease risk; efforts in cities to keep pathogens out became ever more costly as cities required or demanded ever more connectivity with the rest of the world. Today a pathogen can be disseminated glob-ally on a time scale of days, if not hours. No matter how well we come to cope with them after they happen, they will keep coming. And even if we never face a pathogen as deadly and destructive as that which caused the 1918 influenza outbreak, we will always face the health sce-nario of "death by a thousand cuts."

Gambler's Ruin: A gambler who raises his bet to a fixed fraction of his bankroll when he wins, but does not reduce it when he loses, will eventually go broke, even if he has a positive expected value on each bet. In evolutionary terms, Gambler's Ruin means that no matter how well a species or population seems to be coping with a current environment, it can still go extinct if conditions change too rapidly or in ways that do not allow it to cope.

John Maynard Smith introduced the Gambler's Ruin metaphor to evo-lutionary biology, and it is a good cautionary tale for the circumstances in which humanity now finds itself. We live in a highly constructed technological niche, and we are living beyond our means. The emerging disease crisis is a wake-up call, an indicator of how easily the tapestry of civilization can be frayed.

Limits to Growth

As of April 2019 the human population was estimated to be at an all-time high: more than 7.7 billion and growing, despite globally decreasing birth rates. For 10,000 years we did not realize there was a problem. That changed in 1797, when the Reverend Thomas Robert Malthus published *An Essay on the Principle of Population*.[24] That publication is remembered mostly for the startling, and depressing, conclusion that human population growth was outstripping agricultural growth. Although Malthus's intention was to point out a looming humanitarian crisis and urge action to stave off his dire predictions, he is caricatured today as someone who was at best indifferent to the plight of the poor, at worst an early advocate of cultural genocide.

Darwin used Malthus's basic assumption, which we paraphrase thusly:

Reproduction is part of the nature of the organism, while food availability is part of the nature of the conditions. The nature of the organism being by far more important than the nature of the conditions, it seems that there must be some general mechanism operating on all living systems that limits populations.

Applying this principle to humanity, biologists conclude that there must be some maximum sustainable number of humans on this planet. Beyond that number, whatever it is, biologists believe there will be what we euphemistically call a "population correction."

We do not view Darwin's inferences about Malthus's insights as some kind of social program, but rather as a material reality that can be dealt with or ignored via various social programs. In the real world, there is no free lunch; living in technological fitness space has costs as well as benefits. Malthus becomes a problem not by advocating this truism, but by ignoring it. And we have ignored it for a very long time. We call ourselves Man the Hunter,[25] but evolutionarily we are Man the Hunted. At issue is not whether we eat meat, or whether males were the primary meat getters (human females have become larger over the past two million years, so they might have been the primary beneficiaries of adding animal protein to our diets). The critical issue is that we are descended from prey animals, not predators. While we have made ourselves into predators, our basic reproductive biology remains that of prey. We reproduce easily, but now we can defend ourselves from predators (other than ourselves). As a result, our population grows. Increased population density initiates biotic expansion events driven by the relentless reproductive urge common to all life forms. So, we explore and exploit new kinds of fitness space, surviving by eating almost anything that is available. We combine an exceptional flexibility to utilize and create fitness space with a high reproductive rate.

There is growing evidence that, as a result of climate change, crop and livestock biotechnology is falling behind in its experiment to show that we can boost food production to feed an indefinitely growing human population.[26] Experiments asking whether technology can prove Malthus wrong has been answered with a resounding no. The Green Revolution, for example, is a term given to a global agricultural program, based solely on boosting food production, imposed on the developing world by well-meaning people who ignored the Malthusian reality.[27] The Green Revolution increased food production, and many who would otherwise have died did not. But there were no population control programs linked to the Green Revolution, and in Sub-Saharan Africa alone, the Green Revolution led to a baby boom called the African youth demographic bulge.[28] That, in turn, has increased conversion of wildlands into agricultural lands, reduced water supplies, led to greater conflict and migration, and increased disease risk.[29]

The 1972 book *Limits to Growth* summarized the results of a computer simulation commissioned by the Club of Rome. It created an immediate sensation with this conclusion:[30]

If the present growth trends in world population, industrialisation, pollution, food production, and resource depletion continue unchanged, the limits to growth on this planet will be reached sometime within the next one hundred years. The most probable result will be a rather sudden and uncontrollable decline in both population and industrial capacity.[31]

Scientists who had seen troubling evidence in their own work on biodiversity and climate were alarmed at the suggested magnitude of the problem. This led quickly to the emergence of what is now commonly referred to as the environmentalist movement. The book also raised the ire of all policy makers who were devoted to the notion that we can grow our way out of any socioeconomic problems. And because that encompassed virtually all political and economic thinking of the day, every part of the political spectrum felt insulted and demeaned by the findings. But the authors of the analysis did not make any system-specific policy recommendations. They simply pointed out that this was a global problem and would require a global response.[32]

The world has dithered for nearly half a century since *Limits to Growth* appeared. None of the world's countries entertain policies based on limiting growth—that is for other countries with inferior socioeconomic philosophies. Those who denied the need to take action were joined by advocates of a school of thought in the humanities called postmodernism. Postmodernism emerged after World War II, when a number of intellectuals, notably Jean-François Lyotard, began to doubt the view of itself that humanity in Western culture had promoted since the Enlightenment. In particular, postmodernists rejected human understanding and objectivity. This attitude in general goes against meta-narratives derived from earlier ages—the belief that humanity is good, or that human morality and goodness can be reached through rational thought, the belief that science can support humanity, the belief that science is objective. Needless to say, postmodernism has been received badly by the scientific community.

Postmodernism, the school of "thought" that proclaimed "There are no truths, only interpretations" has largely played itself out in absurdity, but it has left behind a generation of academics in the humanities disabled by their distrust of the very idea of truth and their disrespect for evidence, settling for "conversations" in which nobody is wrong and nothing can be confirmed, only asserted with whatever style you can muster.[33]

Many postmodernists strongly reject Malthusian concepts; they see them as a manufactured excuse for maintaining certain political and socioeconomic power asymmetries. They disagree that Malthus's fundamental assertion is an inescapable truth. Ironically, those in charge of the existing power asymmetries who have been attacked by postmodernists also deny the Malthusian reality. For these people, the fact that human civilization has not yet collapsed casts doubt that it ever will, so they continue business as usual. They ignore the fact that the authors of *Limits to Growth* stated categorically that the collapse their model predicted might not occur for as much as a century—half of which has passed—and, even worse, that whenever it happened it would likely occur suddenly and massively, giving us no time to react.

Others believe that *Limits to Growth* underestimated the impact of globalization on redistributing people and food in ways that are far more equitable than in the past, so the absolute number of humans that can survive on this planet is higher than anticipated. They also believe no one should be dull enough to think that human population density is not a problem and that food is going to become equally available and abundant to all humans if our population continues to grow. The vast majority of scientists, therefore, maintain a strong belief that no matter what technology we throw at the problem, human population cannot grow indefinitely.

The business-as-usual projections in *Limits to Growth* were remarkably close to reality as of 2010.[34] Any advantages brought about by technological developments not anticipated in *Limits to Growth* were offset by the accelerating pace of climate change. As we have noted before, climate change has a history of accelerating rapidly and on a global scale, which has often led to abrupt changes in ecosystem structure and species diversity as the biosphere crosses tipping points that were not obvious beforehand.[35] There is concern that environmental changes may occur far more rapidly than our technological capacities can cope with them in the near future. In addition, while birthrates are dropping, health technology has decreased infant mortality and extended life spans in enough of the world to keep us on a precarious route to a global population maximum in 2100, just beyond the century mark indicated in *Limits to Growth*. The 2014 assessment of the predictions in *Limits to Growth* was summarized thusly:

The dotted line shows the Limits to Growth "business as usual" scenario out to 2100. Up to 2010, the data is [sic] strikingly similar to the book's forecasts. . . . Resources are being used up at a rapid rate, pollution is rising, industrial output and food per capita is rising. The population is rising quickly.

As pollution mounts and industrial input into agriculture falls, food production per capita falls. Health and education services are cut back, and that combines to bring about a rise in the death rate from about 2020. Global population begins to fall from about 2030, by about half a billion people per decade. Living conditions fall to levels similar to the early 1900s.

Global pollution measured by CO2 concentration is most consistent with the BAU [business as usual] scenario . . . but this ten year data update indicates that it is rising at a somewhat slower rate than that modeled.

Peak oil could be the catalyst for global collapse. Some see new fossil fuel sources like shale oil, tar sands and coal seam gas as saviours, but the issue is how fast these resources can be extracted, for how long, and at what cost.

Regardless of what role oil constraints and price increases played in the current GFC [global financial crisis in 2008], a final consideration is whether there is scope of a successful transition to alternative transport fuel(s) and renewable energy more generally. . . . To transition requires introducing a new transport fuel to compensate for possible oil production depletion rates of four per cent (or higher) while also satisfying any additional demand associated with economic growth. It is unclear that these various conditions required for a transition are possible.

In other words, scientists studying climate change since Arrhenius at the end of the nineteenth century have been fundamentally correct. The years 1896–2070 span 174 years, 123 of which have passed.

Mike Stuart, a colleague of ours, teaches students about the emerging disease crisis using the analogy of the U.S. real estate bubble that burst in 2008, nearly destroying the American economy. For him, humans residing in technologically advanced cities live in a bubble of denial. They have been so wealthy and so proud of their technology that they have not considered potential costs. They have bet that technology would allow them to have the benefits of globalization without the risks, especially the risk of emerging disease. They believe they can have conservation of nature and unlimited development, including travel without the risk of contracting a disease.

We have never had a larger human population, population densities have never been greater, and we are hyperconnected. Ian Goldin has made a convincing argument linking many socioeconomic challenges in today's world to the same density and hyperconnectivity that have provided so many benefits.[36] This includes the potential for global pandemics. The conceptual framework of the Stockholm Paradigm, evidence from the real world of pathogen-host associations, and our mathematical modeling all support Goldin's fundamental assertion. Increasing disease risk has been a constant companion of the advance of civilization. All the benefits of human expansion—agriculture and domestication, urbanization and globalization—have been offset by, among other things, increased disease risk.

Our global connectivity allows the dissemination of pathogens from wildlands to cities in a very short time. In a world of global connectivity and increasing population density, protection by isolation is no longer possible. Like climate change and population expansion, globalization creates connections among elements of previously disconnected fitness space. This is always an invitation to explore, and we, and our associated species, have enthusiastically accepted the invitation; climate change, population increase, and hyperconnectivity produce a kind of biotic expansion, and they work synergistically, intensifying each other's effects. Enhanced disease risk is an outcome not only of the Malthusian trap of overpopulation, but also of the density trap of urbanization and the connectivity trap of globalization. This has all the makings of a perfect storm for EIDs.

Summary

Disease has always been, and remains, a cost of human civilization.[37] Urbanization and globalization solve some problems associated with diseases but create new ones. Climate change increases the risks, not only by amplifying EID possibilities but by adding additional threats, which are threat multipliers for disease and for which disease is a threat multiplier.

If we have been successful to this point, you the reader will have passed through what are sometimes called the five stages of grief: denial, anger, bargaining, depression, and acceptance.[38] You will have followed a path all three of us have trod during our professional careers. In the first decade, we did not admit there was a problem. After all, everything we needed for our research was readily available and abundant. The second decade made us face the fact that things were changing, and that made us angry. Not angry enough, however—we knew hidden places still largely untouched where we could still find what fascinated us so much. But that decade passed quickly, and we found ourselves spending the 1990s bargaining with governments, granting agencies, nongovernmental organizations, private foundations, and the United Nations for money and support to do something about the problem. All those initiatives withered, and most of them died. And all that happened at just the time we finally realized how serious the problem was. That led to our decade of depression. We regret that it took us such a long time to comprehend fully what was happening just in our small—but important—area of implications of global climate change and to rouse ourselves into action. That period was marked by the emergence of what we call the *Cassandra Collective*. Many dedicated scientists, all of whom had spent most of their careers trying to warn the public of the danger even though those warnings were largely ignored, trying to garner enough resources to fully understand the scope of the approaching danger, were aging and beginning to lose hope. A growing number of them concluded that it was too late to do anything effective and decided they wanted to enjoy the final 20–30 years of quality life left to humanity. And so many of them gave up, beginning in the first decade of the new century, quietly. We are among the scientists of our age who have made it through that depression—which we felt—into a dogged acceptance. The danger is great, time is short, and we are unprepared.[39] *But we are not dead yet.*

We hope you have arrived with us at the final stage. If so, let us all accept this: if we make it through this with technological humanity intact, we must never make the same mistakes again. What we have discovered is sobering indeed. As senior scientists, we do not panic easily. But we are worried. We live in extraordinary times, so the scientific community needs to engage in extraordinary research. It is very late in the game, but it is never too late to try. Going forward, we must think the unthinkable and expect the unexpected. Rather than pretending that it won't affect us, we need to ask, "What will we do when it happens to us?"

It is a characteristic of wisdom not to do desperate things.[40]

Let's be wise before we get desperate. Will we keep postponing the day of reckoning until events overwhelm us, or will we begin to actively prepare to survive in a complex and uncertain future? Return to the tennis adage: never change a winning game, always change a losing game. When this is not followed, there are two reasons. Either you do not recognize that you are playing a losing game, or you have no ability to change. Business as usual is not enough, no matter how much money you pour into it. If our belief that time is short, the danger is great, and we are largely unprepared is correct, then it would be pointless to write this book simply in order to say we have a problem. The Stockholm Paradigm tells us *why* we have a problem. Fortunately, it also gives us insights into what we might be able to do to help mitigate the effects of the problem, and that is the topic of the next chapter. We leave you with this:

Be afraid. Be very afraid.
But do not panic.
We have a plan.

NINE

Taking Action: Evolutionary Triage

Triage: the sorting of and allocation of treatment according to a system of priorities designed to maximize the number of survivors.

We attempt to mitigate damage caused by diseases only after they have emerged. This is understandable in the socioeconomic stewpot of priorities in which we exist. Reactive management policies are unsustainable in a world of climate change and its threat multipliers. The public is not encouraged by such things as a recent call by the World Health Organization for people to be concerned about "disease X." Many specialists believe public policies amount to no more than rearranging the deck chairs on the *Titanic*.

The Intergovernmental Panel on Climate Change (IPCC) is known to the public as the organization that develops comprehensive global assessments of the impacts of climate change. The IPCC also hosts meetings to consolidate observations and develop options for mitigation and adaptation. These are summarized in Assessment Reports, an early one of which reinforced our habit of reaction by stating there was insufficient evidence of adaptive anticipatory processes on Earth.[1] There is thus no way we can anticipate particular outcomes of impacts of climate change, so we explore ways to respond ever more rapidly to events, but always *after* they occur. Fortunately, the IPCC assessment is wrong—Darwinian evolution is an adaptive process with considerable anticipatory capacities. We can anticipate disease occurrences and mitigate the damage by learning the lessons of evolutionary history. Knowing the specific and evolutionarily conservative traits associated with pathogen survival, we can anticipate the ways in which pathogens may be introduced to a new area, how they will be transmitted within that new area, and how they might be exported to other areas.

Essence of an Action Plan: "Anticipate to Mitigate"

To be forewarned is to be forearmed.[2]

We have had limited success coping with the global EID minefield. Smallpox has been eradicated and polio nearly so. Vaccination campaigns banished rinderpest from Africa and foot and mouth disease is now controlled through complex programs involving surveillance, quarantine, and eradication of infected stock. Texas cattle fever was pushed out of the United States near the end of the nineteenth century as a result of widespread control of tick vectors for the protistan parasite that causes the disease. The tick vectors remain, however, and the pathogen continues to exist in wild ungulates such as deer, which share range with cattle. Medications have been used effectively to nearly eliminate the river blindness and guinea worm infections afflicting people in many parts of Africa. The widespread use of such a simple device as a bed net has dramatically reduced the scourge of malaria throughout Sub-Saharan Africa. These are all good things, but they overshadow the truth that pathogen eradication is seldom achieved, is often tenuous, temporary, and localized, and is strongly resource dependent. Success depends on a tapestry of infrastructure that is easily frayed or tattered by local and regional conditions and conflict.

Humanity suffers death by a thousand cuts daily from episodes of unanticipated disease emergence and ineffective response. This must end. The Stockholm Paradigm offers guidelines for adding the necessary proactive component to ongoing efforts to cope with the EID crisis. And we *must* begin coping. We think the new approach could lead to less consternation and human suffering. The alternative is business as usual, and that is not a viable option, no matter how much money we throw at the problem. Public resources tend to be spent no matter how decisions are made to spend them, so if we want to be more effective, we need to focus on what we do with the money, not on how much we have. When is the best time to repair and upgrade a bridge—during a drought, or after flooding has destroyed it, killing anyone unlucky enough to have been on it when it collapsed?

We need information about pathogens that might but are not currently causing disease. We could then anticipate at least some of the potential disease risk. Increasing the geographic or host range of a pathogen may not always cause disease. Some host populations are less tolerant than others; there may be some pathogen variants that are inherently more pathogenic than others; and there may be some pathogen variants that are either more or less pathogenic in different hosts. This means there may be a lag between the arrival of a pathogen in a host population or a geographic area and disease outbreak.[3] If we can find them before they find us, we might be able to avert or mitigate disease outbreaks. The few cases in which "anticipate to mitigate" has been successful have been those based on fundamental biology and deep ecological knowledge about *conditions* of transmission, leading to an understanding of how to break the *cycles* of transmission. But the principle has not been applied widely. Presumably this has been because we have failed to appreciate that the EID crisis is fundamentally due to inadequate application of evolutionary principles.

The DAMA Protocol: Finding Them before They Find Us

DAMA (Document—Assess—Monitor—Act) is an integrated proposal to build a proactive capacity to understand, anticipate, and respond to the emerging disease crisis.[4] The acronym is an umbrella term for many proposed initiatives, all of which have merit and many of which embody elements of the protocol. Both the Food and Agricultural Organization (FAO) of the United Nations and the Office International des Epizooties (OIE; World Organisation for Animal Health) maintain and coordinate broad-scale surveillance of animal pathogens and diseases, primarily with a focus on what is known, creating a global infrastructure that has often served us well in limiting the impact of emergent diseases.[5] The USDA's Plant Pathogens and Quarantine (PPQ) program follows a set of risk-based assessment protocols.[6] The aim is to describe, document, detect, identify, and mitigate plant pests and pathogens, either by stopping entry or through identification of potential imports prior to dissemination from source countries. This approach has achieved considerable success contributing to global food security.[7]

Among the most ambitious inventory projects under way is PREDICT, a collaboration between the United States Agency for International Development (USAID) and an array of international participants and partners.[8] PREDICT is a project within the Emerging Pandemic Threats program, founded in 2009, recognizing that our business-as-usual approaches are ineffective and that the threat of viral disease is urgent. PREDICT identified its fundamental problem thusly:

Despite intensive, high-quality research efforts globally, we are still not able to predict which viruses will become pathogenic to people; which will cause new epidemics in animals; nor where and under what circumstances disease will emerge.[9]

The goal of the project is understanding the identity of viral pathogens, their host associations, and their potential for host range expansion from mammalian reservoir hosts to humans. PREDICT has focused surveillance efforts concerning zoonotic viruses in hotspots across 21 countries in Africa and Asia targeting wild free-ranging mammals, especially rodents, domestic ungulates and primates (including humans) that live in natural and managed ecosystems.[10]

PREDICT is the most comprehensive project for viral discovery yet undertaken. And yet, it restricts itself mostly to identification of known viral pathogens (such as Ebola virus, Nipah virus, Marburg virus, related coronarvirus agents of SARS and MERS, and, tangentially, avian influenza viruses), with peripheral interest in what seem to be previously unknown "orphan viruses" (viruses without a known host). The first five-year plan for PREDICT was from 2009 to 2014. Once global hot spots were identified—places where the occurrence of emergent disease was predicted to increase over time and where targeted surveillance would be optimal[11]—56,000 people, livestock, and wildlife animals were sampled yielding more than 250,000 specimens.[12] From those samples 984 potentially pathogenic viruses were identified, only 169 of which (21 percent) were previously known. The potential for zoonotic spread was assessed according to the phylogenetic relatedness of previously unknown viruses to relatives known to cause disease.[13] This represents the first widespread effort to apply principles from phylogenetics to the study of viral pathogens, following earlier suggestions for eukaryotic and metazoan pathogens.[14]

These data were used to make initial predictions about the overall numbers of zoonotic viruses in mammals.[15] Concurrently, insights about the abiotic and biotic factors of disease emergence were developed combining field observations with mathematical modeling. The results were not surprising: landscape change, mobility encompassing transportation and trade and climate, with variation in seasonal cycles and long-term change, led the list of phenomena associated with disease emergence. Pathogen diversity was directly related to the richness of host species and to climate, and transmission risk was correlated with population density, bush meat consumption, and the proximity of livestock.[16] DAMA is designed for a world in flux owing to climate change. Fully implemented, it complements other approaches. The protocol focuses on the high-probability/low-impact pathogens that are not the target of larger-scale efforts but are the primary components of pathogen pollution. It encompasses all kinds of pathogens, not just targeted groups. Finally, DAMA is envisioned to be a bottom-up approach, based on the activities of field scientists working in collaboration with citizen scientists and local health and community education personnel. This should allow more rapid and focused action than a typical top-down "big science" project. Nonetheless, DAMA would use the same technology and the same high-level laboratory, analytic, and archival infrastructure as other projects, emphasizing its fundamentally cooperative and integrative nature.

Document

No name, no information, wrong name, wrong information.[17]

Proper biodiversity assessment begins with, and is dependent on, information derived from systematics. Systematics is the branch of biology charged with making certain that every biologist who uses a particular name is in fact referring to the same thing. Assigning a specific epithet to a group of organisms is proposing that they are members of a diagnosable inclusive and exclusive hereditary information system. When we name a species on this planet, we provide a key to valuable information about it, about its origins and history, about its location, and about how it interacts with its surroundings, including other species. Some of that information is embodied in the species itself, but most of it is embodied in the species' histories.

We cannot anticipate anything about the chances of disease, or about the means of mitigating it, caused by unknown pathogens. We live in a veritable sea of pathogens and have been blissfully unaware of most of them because they cause no more than transient problems for a small minority of hosts, often ones we rarely encounter. We only notice them when their explorations of new fitness space produce disease outbreaks. The best estimates are that we have documented less than 10 percent of the pathogens on this planet. In the 1990s, generating a comprehensive inventory of species of pathogens was a daunting challenge, incurring tremendous costs in terms of time, money, and personnel, and for the most part the challenge was not taken up. This is no longer an acceptable excuse. Technological advances made in the years since have allowed us to generate, for relatively little money and in a short time, large amounts of data about who is present in any given site. This has led to a kind of basic field biology on steroids, as demonstrated by projects such as PREDICT and the Beringian Coevolution Project.[18]

The Taxonomic Impediment

Know your enemy.[19]

Systematists are the biologists best trained to expect the unexpected.[20] In fact, they are obsessed with finding species no one has ever seen before. And yet, in 1992 the International Union of Biological Sciences/*Diversitas* recognized a global shortage of professional taxonomists. This lack of capacity was termed the *taxonomic impediment* as early as 1996 (leading to the Global Taxonomy Initiative). It has been raised repeatedly in the House of Lords Reports—in 1992, 2002 (including the Darwin Declaration), and 2008. In the United States, the dire situation was reflected in the reports from the President's Council of Advisors on Science and Technology (PCAST) and other national discussions about invasive species.[21] The authors of this book have been involved in numerous conferences, workshops, planning sessions, even pilot projects designed to highlight the essential role of taxonomic expertise in all areas of biology associated with climate change.[22]

Since then the need for systematics expertise only has increased, and the taxonomic impediment has grown annually. University faculty positions for systematists, government research in systematics, and support for natural history museum collections have declined, as has public funding for taxonomic research. Today no country is self-sufficient when it comes to taxonomic expertise about the species that live within its borders, much less those who arrive from elsewhere.

The significance of good taxonomy was underscored by the recent discovery that Lyme disease in North America is actually caused by more than one species of bacteria in the genus *Borrelia* and is a cluster of pathogens and diseases globally. This previously unrecognized taxonomic diversity is likely responsible for the confusion attendant upon diagnosing and treating Lyme disease.[23] Similarly, great strides in developing effective antiretroviral drugs for treating AIDS were accelerated by the taxonomic discovery that HIV in humans was associated with two different—and not that closely related—species of the virus, HIV-1 and HIV-2. This emphasizes the need for the continual development of species recognition capacities that can only come from trained and observant taxonomists.

Modern systematics is more than just identifying things. We expect to see monkeypox virus in Central Africa, but not in pet prairie dogs and people in the midwestern states, where it emerged in 2003.[24] A natural reservoir is unknown, although it can circulate in primates, rodents, and rabbits and can have nearly a 10 percent mortality rate in humans. Molecular taxonomy led us to understand that imported Gambian pouched rats from Central Africa brought this gift to American shores, and that knowledge led to breaking an emerging cycle of transmission that would have added monkeypox to our growing burden of pathogen pollution. A generation before that particular close call, molecular taxonomists demonstrated that monkeypox disease in humans resulted from ecological fitting by original strains of monkeypox exploring new hosts rather than novel mutations.[25] The world long ago decreased its vaccination program in recognition of the eradication of smallpox. Systematists know that monkeypox is related to smallpox, which is the reason smallpox vaccinations provide significant protection against monkeypox. Recently, however, civil war in Congo has led people to undertake deeper incursions into the forests searching for food. Closer and more frequent contact with mammalian reservoirs soon produced monkeypox infections in people who lack smallpox vaccinations.

Documentation Needs

Personnel

The need is comprehensive. We must have more taxonomists and associated field and laboratory staff, especially parataxonomists—local people with a high degree of knowledge about the biodiversity in their neighborhoods.[26] Given that no country is self-sufficient with respect to taxonomic expertise, proper documentation activities, no matter where they occur, will almost always involve people from "other countries." Parataxonomists have been shown to be essential support staff for professional taxonomists working outside their own countries; an inventory of parasites of vertebrates in the Área de Conservación Guanacaste, in Costa Rica, showed the feasibility of training parataxonomists and their value as colleagues. This kind of collaboration with residents of local communities where inventory projects are being carried out also creates community goodwill and support for scientific activities.

Despite technological assistance, much of this phase of DAMA involves what is variously called "boots on the ground" or "sweat equity." Commitment to human capacity–building of this scope is rare, owing mostly to the expense in time, expertise, and money associated with training activities. PREDICT trained nearly 1000 parataxonomists capable of assisting all field collections and documentation of potential viral pathogens.[27] Current estimates suggest that at least 320,000 species of viruses in nine families circulate among mammals.[28] Discovering just these would cost more than $6 billion. Programs implementing DAMA require levels of funding comparable to the budgets of scientists tracking millions of asteroids that are no threat to find one that could be, or military analysts producing contingency plans for wars they hope they never have to fight. There is considerable potential and reason for optimism if we link field activities to molecular and computer laboratories. Identification is a gateway, not an end.[29]

Infrastructure

Proper documentation requires attention to detail. Significant forethought, as well as significant funding, could produce designs for adapting existing data-collection equipment to find new pathogens before they emerge, rather than looking for a small number of—sometimes only one—species already known to be causing problems. Similarly, there must be laboratories, natural history collections, and databases to receive the reference specimens and data collected in the field.[30] Fortunately, an international initiative, called the Global Biodiversity Information Facility (GBIF),[31] has been established to help provide conduits for sharing biodiversity information freely throughout the world. In this way, all inventory projects working in all parts of the world with all kinds of pathogens can share information readily, increasing the chances that someone will see a problem arising before it blows up in our faces, as has been happening too often.

Activities

Many previous taxonomic inventory projects have been driven by a desire to learn more about species already known and considered significant. A smaller number have been the opposite—let's look at everything and hope that something important will show up. The documentation phase envisioned by the DAMA protocol falls between those extremes; the search for pathogens is certainly focused, but the search for unknown pathogens requires flexibility. The most effective way to document pathogens that are waiting to find us is to investigate *ecological interfaces*—the wildlands and the agroscape, the urbanscape and the agroscape, the urbanscape and the wildlands. Such interfaces provide maximum evolutionary opportunity for pathogens and hosts because they are crossover places, in which pathogens may reside in one area without making themselves known but then break out into adjacent areas where there are hosts that are important to us (including ourselves) but are not usually infected (and therefore have experienced little or no selection for resistance). Climate change and anthropogenic activities alter existing habitats and the interfaces between them, so a focus on dynamic interfaces rather than static hot spots is essential if we are to understand emerging disease in today's context. Disease hot spots are not going to be stationary, and we need to anticipate where they will go and how they will change. The host focus of DAMA within those interfaces is the reservoirs of pathogens. We want to know where pathogens live when they are not infecting us or our crops or livestock. DAMA does geographically focused and site-intensive studies capable of finding rare pathogens or rare variants of pathogens, rather than site-extensive sampling associated with global inventory projects such as PREDICT. Those hosts most diseased are often not the reservoirs, that is, the hosts in which long-term persistence of a pathogen occurs. It is essential to focus on the reservoirs if we are to understand disease recurrences and the host context of changes in host and geographic range for pathogens.[32]

The speed with which the EID crisis is happening and its global context make methods for rapidly sampling large numbers of potential hosts a priority. Historically, quests to find pathogens involved laborious and detailed examinations of single hosts. This is changing dramatically as we fuse classical methods with emergent molecular technologies for identification. This process, which has come to be known as *genetic barcoding*—in other words, having a definitive molecular marker for the identification of each species—has begun to amplify our capacities. Importantly, barcodes always need to be grounded in what we call authoritative identification based on an actual specimen (thus the critical relevance of museums and taxonomists). Advanced technology changes the game by creating possibilities for seeking pathogens in ways that do not injure or kill the host. For example, if we examine feces left by large, mobile animals that were not in the vicinity when we arrived to collect them, preexisting nucleotide-sequence databases allow us to identify the host species as well as the pathogens and link those data to information about geographic ranges and host population structure.[33] We can do this with blood, urine, and tissues as well, greatly enhancing the possibility of microbial discovery. And as we race into the future, we are at the metagenomics frontier, with the possibility of near-simultaneous identification of microparasites (viruses, bacteria, fungi, and protozoa) and macroparasites (worms and arthropods) through this kind of noninvasive sampling.

In short, the more taxonomic inventories humanity can fund, the better. They should be highly focused, however, to provide information for each potential pathogen species: (1) what the species is, (2) what the species infected, (3) where the species occurred geographically, (4) a link to phylogenetic history, and (5) as much natural history and epidemiological information as possible. These efforts must use the most advanced technologies, but also require a fusion of modern technology with old-fashioned sweat equity–based natural history, taxonomy, ecology, and behavior. As a result of technological advances in the past 25 years, inventories for documenting diversity have changed from clawing a manageable number of data points out of the wildlands to being overwhelmed by mountains of data obtained from large-scale field collections and data streams from next-generation sequencing. Our world in the twenty-first century is exploding with complexity. If society supports long-term commitments and programs for documenting diversity, we can manage information from streams of data that could only be imagined 30 years ago. We are once again talking about big science, this time what is called big data, and our capacity for handling large amounts of hyperconnected information.[34]

Although documenting what is out there is necessary for effective response to the EID crisis, the focus of the DAMA protocol is not simply on accumulating a lot of information. Rather, it is on assessing that information in the context of disease risk and converting such assessments into effective action.

Assess (Triage)

Our inventory efforts will document many species that are not a concern, at least not at present. Given the immediacy and magnitude of the EID threat, we cannot to stop and smell the fungi, documenting every single species and painstakingly studying each one to determine just how it might affect humanity. We must make effective decisions in a timely fashion, and that calls for rapid and focused assessments of the pathogen information we encounter. When policy recommendations about documented biodiversity involve known and potential pathogens, assessment becomes risk assessment. When they involve life-and-death decisions, risk assessments become triage. The EID crisis warrants triage.[35] Triage decisions stemming from pathogen inventories should be based on the answers to three linked questions: (1) have we encountered any species that are known pathogens? (2) have we encountered any species that are closely related to known pathogens? and (3) for pathogens of interest, what are their reservoir hosts and where do they live? The emergence of sophisticated methods of molecular taxonomy, in conjunction with ever-faster and more powerful computers, allows the first question to be answered rather easily, by matching data collected during the inventory with archived data from known pathogens. The hope is to find such pathogens in an area where they were not known to occur *before* they produce a disease outbreak, rather than waiting for them to announce themselves by making our livestock, crops, or children sick. A good example of this kind of initiative is the assessment of viral species in mammalian hosts living in biodiversity hot spots in Africa and Asia.[36]

Answering the second question is more time-consuming and requires particular skills on the part of those charged with providing answers. As a result of the phylogenetics revolution, systematics has developed into a multifaceted science capable of providing a wide range of essential biological information in an explicitly evolutionary framework. First and foremost, comparing data for any particular species within a phylogenetic framework will tell us who its closest relatives are. If a given species is closely related to a known pathogen, historical ecology represents a powerful tool for risk assessment (fig. 9.1).[37]

Similarly, we can use intraspecific phylogenies to determine whether

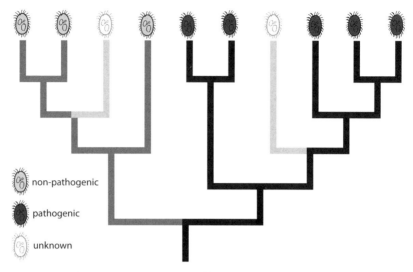

FIGURE 9.1 Diagrammatic representation of how we use phylogenetic relationships to assess the threat that an unknown, newly discovered pathogen might cause disease. In this case, we have a group of ciliated microbes of some sort, two of which are "unknowns." The one on the left is closely related to a group of three other species, none of which cause disease. The one on the right is also a member of a group of three other species, but in this case, all of them are disease-causing pathogens. All other things being equal, we would assess the threat level for the unknowns on the left as "low" and for the one on the right as "high."

we have documented a variant of a given pathogen that is known to be associated with disease or a variant closely related to a pathogen known to be associated with disease (see, e.g., fig. 5.9). In a complementary manner, our evidence indicates that critical traits associated with transmission dynamics, host range, and microhabitat preferences (which affect signs and symptoms of disease) are both specific and evolutionarily conservative. Many of the properties of an unknown species are likely to be shared with its closest relatives. Faced with an unknown species whose closest relatives are pathogens, making it a high-level threat, we can use what we know about the natural history of those closest relatives to make informed inferences about ways in which the new species might be transmitted, where it might live in its host, and even a number of the signs and symptoms of the disease it might provoke.[38]

The cophylogeny tanglegram for the nematode genus *Haemonchus* (fig. 9.2) depicts many cases in which the pathogen-host associations do not reflect cospeciation. If we remove *H. contortus*, the species that has been moved around the globe with domestic livestock most extensively, the total host range for the remaining members of *Haemon-*

chus covers the known host range for *H. contortus*. If, as we suggested in chapter 4, cophylogeny studies tell us something about fundamental fitness space (which, among other things, means potential host range), then the broad host range of *H. contortus* should be no surprise. Furthermore, we should anticipate that other species in the genus have a high potential for host range expansion, given the opportunity (*H. bedfordi* comes immediately to mind). And with global climate change accelerating, that opportunity is approaching rapidly.

The conceptual, modeling, and empirical data supporting the Stockholm Paradigm allow us to assert with confidence that whenever we find a pathogen causing significant disease in one host, there is at least one other host—the reservoir—likely living in an ecological area adja-

FIGURE 9.2 Cophylogeny tanglegram comparing the phylogeny of the barber-pole nematodes *Haemonchus* spp. (*left*) and a simplified phylogeny of their hosts. Numbers in parentheses denote numbers of host genera inhabited by each species. Modified from Hoberg et al. (2004); Hoberg and Zarlenga (2016).

cent to the one in which the diseased host resides. For that reason, we cannot overemphasize the importance of determining which hosts are the reservoirs of pathogens and in which portions of habitat interfaces they reside. And in this case, phylogenetic studies provide help. For example, there is a concentration of many virus families, with known zoonotic members, occurring in a relatively restricted number of chiropteran (bat) families.[39] Many people are already aware of the potential of contracting rabies by coming into close contact with bats. These data indicate that the rabies threat is only the tip of the iceberg when it comes to the potential for viral infections' being transmitted from bats to humans.[40] Another example is the hantaviruses, a global group only recently discovered in the Western Hemisphere—and at first thought to occur only in a few places in the Americas—circulating in rodents and capable of infecting humans when offered the opportunity. They are now known from a considerably greater number of small mammals including Eulipotyphla (shrews), Chiroptera (bats), and an increasing number of rodents throughout the Americas and Eurasia.[41] Human exposures occur in the interfaces between rural and urban environments, and any of the hanta group should be considered as having the potential to cause respiratory and hemorrhagic diseases in humans. There is encouraging evidence of a recent focus on finding the reservoirs.[42]

Assessment Needs

Personnel

We must develop and use our increasingly valuable archival collections of hosts and pathogens, along with informatics resources, to describe the pathogen-risk portion of the biosphere. Systematists trained in integrating information on natural history in a phylogenetic context can present these discoveries in a manner that allows triage decisions to be made. All countries that wish to help mitigate disease threats must allocate funds to train and employ more taxonomists and to support museum infrastructure. In the short term, at least, we cannot assume that sufficient additional resources will be allocated for this purpose. As a consequence, it becomes even more imperative that the taxonomic impediment is not linked to a shortage of other biodiversity specialists. Effective assessment of what we document requires an array of specialists, from scientists who maintain archival collections, to those who perform molecular and bioinformatics analyses, to those who place documented information in both real-time (e.g., specialists working on population dynamics of hosts and pathogens) and deep-history (phy-logenetic) contexts, and, finally, to those who not only maintain these complex information databases but are capable of updating them rap-idly and providing as close to real-time access as possible.

Infrastructure

We need facilities to house all these essential people: molecular tax-onomy laboratories, phylogenetics laboratories, and natural history repositories for specimens, tissues, sequences, and digitized natural history information. There is some good news. First of all, many sys-tematists maintain both molecular taxonomy and phylogenetics labo-ratory capabilities, and many of them work in natural history muse-ums or government laboratories. Networks must be developed that link field-based collections to deposition and maintenance of specimens, tissues, and nucleotide products in archival repositories. The latter are the baselines for documenting pathogen and host distribution and pro-vide the capacity to recognize (often retrospectively) the occurrence of novel pathogens or new host and geographic distributions. In ad-dition, much of the critical infrastructure necessary for documenting pathogen diversity also supports activities associated with producing information necessary to make informed triage decisions. So, while we need significant allocations of funds to undertake these essential ser-vices, the multiuse nature of much of the infrastructure means con-siderable economy, especially if there is cooperation among groups. That multiuse nature of biological collections forms the backdrop for a biological detective story. A disease was lurking among the pinyon pines and undulating landscapes of the Four Corners, a magical desert landscape linking New Mexico, Arizona, Colorado, and Utah. It wasn't there every year, it wasn't known to the disease ecologists in any co-herent way, but it resided in the extended memories and oral history of Navajo and other indigenous peoples. In the spring of 1993 an out-break of a severe respiratory illness killed 10 people in an eight-week period, eventually leading to a 70 percent mortality rate as the disease progressed.[43] The pathogen turned out to be a new virus in the hanta group, pathogens of rodents in many regions of the world that often occur as serious zoonoses. The virus was eventually called Sin Nombre (Spanish for "without a name," abbreviated SNV).[44] Field collections in the Four Corners ensued, and the host involved was identified as the ubiquitous *Peromyscus maniculatus*, the white-footed deer mouse. But where had the virus come from? Had it been introduced? Hantavirus was at that time rarely known in the Western Hemisphere, and prior to 1993 only the Prospect Hill virus had been documented, all the way across the continent in Maryland.

Museum collections of rodent specimens, tissues, and blood, along with human specimens from undiagnosed disease outbreaks, demonstrated that the virus had been present for a long time. Since 1993 more than 25 members of this virus group have been discovered, and phylogenetic analysis documents a deep history linking Eurasia, North America, and South America. It was literally the stuff of legends, a mysterious disease that had persisted in the oral histories of indigenous people. Archival frozen tissues revealed the history of the virus and were initially the window through which the question of origins could be examined. In the absence of these collections, held and databased in a museum repository, our current knowledge of hanta in the Americas would have been delayed for a considerable period of time.[45]

Activities

Although complex and sophisticated, activities producing data for triage decisions are straightforward. They involve molecular taxonomy to identify known pathogens before they trigger disease outbreaks, phylogenetic analysis to identify relatives of known pathogens, and historical ecological comparisons of those close relatives to assess the risk potential of the unknown relative. If a particular documented pathogen is known to cause disease or is a close relative of a species known to cause disease, it is assessed as high priority. This means we want to know its full geographic range, what interfaces it lives in, what its reservoirs are, and as much about its population genetics as possible, including the potential for rare genotypes that can infect species we care about (including us) by the stepping-stone dynamic shown in the modeling discussed in chapter 5. In a world of climate change, we need to track as much of this information as rapidly as possible. And that is the rationale for the next phase of the DAMA protocol.

Monitor (Surveillance)

As minuscule as is our knowledge of the microcosmos of pathogens, there is an even greater lacuna in our understanding of the biology of even those of special concern. We have insufficient understanding of the geographic distribution and natural history of pathogens during those times when they are not engaging our attention. Nor do we always understand the biotic and abiotic factors that catalyze the shift from quiescence to disease. Sometimes it results from changes in ecological conditions, as turned out to be the case with Sin Nombre. The El Niño Southern Oscillation produces unusually heavy rains that lead to an explosion of growth and seed production in pinyon pines. Many animals, including white-footed deer mice, take advantage of this bounty to convert pinyon pine biomass into rodent offspring, many of them carrying Sin Nombre. The expanding rodent population explores new areas, including human habitations, and contact between humans and hanta is achieved.[46]

When pathogens are given an invitation to explore, they are relentless. The easiest way for them to explore, and create the most havoc, is biotic expansion triggered by climate change, which can drive the mass movement of many pathogens capable of infecting many hosts. Trophic changes within existing communities bring previously unexposed hosts into contact with resident pathogens. Pathogens may spread geographically even without host movements along the margins of two ecosystems, where the parasite might expand geographically because there is a related host in the adjacent area. This exploratory ability of pathogens explains why the distribution of a pathogen is broader than the distribution of disease attributable to it.[47] Environmental perturbations that create new connections in fitness space are associated with amplification of pathogen populations, host range expansion, and disease emergence.[48]

Each documented species that warrants special attention (i.e., a known pathogen or a close relative of a known pathogen) must be monitored to more fully document host and geographic ranges, as well as to gather exhaustive knowledge of genetic variation. That information needs to be incorporated into models of anthropogenic climate forcing, indicated by long-term progressive and short-term fluctuating changes in many variables, particularly temperature and precipitation, to give us some idea of what might happen next.[49] Of special attention will be new pathogen-host associations emerging in the interfaces from urban landscapes to wildlands.[50]

Monitoring Needs

Personnel

This aspect of DAMA is perfect for engaging local citizens, neighbors of the parataxonomists, and other locally based support personnel who are the focus of efforts to mitigate disease impacts. The same boots on the ground that help us document will be needed to monitor the pathogens of special interest identified in our assessments. Don't forget that the Texas cattlemen knew ticks were involved with Texas cattle fever long before the scientists did.

Monitoring efforts, however, also require additional trained specialists. In particular, we will need the expertise of population geneticists, population ecologists, and phylogeographers to interpret the real-time data collected by the inventory specialists. We need those data to make projections about future activities of the pathogens, which will require specialists in mathematical modeling. In other words, in addition to boots on the ground, we need fingers on keyboards. And finally, we will require specialists in surveillance to use the data and analyses to propose ways to implement any necessary changes in documentation activities. We envision a cooperative system in which necessary changes in our activities on the ground can be put into practice at the first sign that a pathogen's distribution or behavior is changing. When such changes are the harbingers of a disease outbreak, the DAMA protocol becomes an early warning system for those who are tasked with responding to outbreaks.

Infrastructure

Successful monitoring requires substantial infrastructure. But, again there is some good news: the same personnel and infrastructure used in basic documentation activities can be used for monitoring. All of the data analysis, modeling, and projections, however, will require high levels of computer technology for the various specialists to use. We will also need to engage the emerging monitoring programs linking citizen-scientists with specialists through cell-phone-based apps as an aid to their efforts. The ultimate goal is to link monitoring technology from a single person with a cell phone to the most comprehensive satellite-based global monitoring systems available—from boots on the ground to eyes in the sky.

Activities

Monitoring networks can combine geographically extensive and site-intensive sampling information across thousands of kilometers into series of snapshots of pathogen distribution. Linking those snapshots into time sequences (videos) would allow us to observe changes in pathogen distribution and behavior, such as those driven by subtle climate shifts, in near real time. The ultimate goals are to recognize previously unoccupied fitness space before the pathogen finds it (anticipation) and to recognize in real time when the pathogen enters it (mitigation). As mentioned earlier, ecological niche modeling should play a critical role in these efforts.[51]

There is a pressing need to coordinate multiple levels of society. We can take advantage of ongoing documentation activities and personnel to begin the essential process of getting buy-in from local citizens. Sharing local or indigenous insights and observations—what is often called *traditional ecological knowledge*—is critical. For example, in boreal hamlets and communities, hunters on the tundra and sea ice hold special knowledge and experience and are often among the few who regularly venture across remote regions.[52] Tapping this kind of human capacity, from the poles to the tropics, will produce regular observations by people with great sensitivity to changes in their local ecosystems that might enhance pathogen transmission.[53]

Their observations, sent to coordinating data centers using specially designed cell-phone apps, can provide the essential information for specialists in population genetics, population ecology, and phylogeography. Their results will also allow the modelers to produce projections for various temporal and spatial scales (how fast is it moving? how far is it going?). And those results will go to the surveillance people, who will make recommendations to the boots on the ground. The goal is to make this flow of observation to data to analysis to projection to decision a loop operating as close to real time as possible, beginning with those most at risk.

Act

Never change a winning game, always change a losing game.[54]

The emerging disease crisis may not be the most important threat from climate change, but it is costly, and diseases are threat multipliers for all other elements of climate change. Disease may never have the first or even the primary cause of previous human catastrophes, but disease may well have been the final straw. We are playing a losing game with respect to climate change and emerging disease, and we must act to change that. The DAMA protocol brings a sense of urgency with it, even while it brings a sense of hope—we can do something proactive and effective about the emerging disease crisis if, once disease threats have been identified, there is prompt and decisive action. Many colleagues with whom we have spoken tell us that the most difficult transition in all areas related to the climate-change threat is from assessment to action or from monitoring to action.

Health practitioners follow a simple, straightforward, ancient and honorable philosophy known as "do no harm." This principle is intended to protect not the patient but the doctor. It originated as a response by practitioners to an aspect of the Babylonian legal system called the Code of Hammurabi. In that code, if a doctor treated someone for, let's say, an eye disease and the person lost the sight of the eye, the doctor's corresponding eye was removed (hence the origin of the phrase "an eye for an eye"). Health practitioners have had a bunker mentality ever since. And in a world where nothing changes, once you know what the diseases are in your world, you can wait to be certain what you have, then treat it with little possibility of legal or corporal repercussions. And with advancing health knowledge, we now have many proactive approaches that save money and maintain health better than waiting for acute disease while not putting the clinician at risk. Changing your diet and getting some exercise is far cheaper and less traumatic than open-heart surgery following a heart attack not only for the patient, but for the clinician.

In a world of climate change and emerging disease, however, "do no harm" has become a prescription for waiting until a pandemic occurs and then trying to contain it. A passive complacency emerges as each outbreak burns out into the pathogen background, and no comprehensive accounting of the socioeconomic impact of the pandemic is ever made. As the world becomes increasingly urbanized, the problem will worsen. High population density in urban settings makes it less likely that you know, much less are related to, your neighbors. And that makes it difficult to know whom you can count on in an emergency. Not knowing whom to cooperate with puts people at risk of not being able to mobilize effectively. This problem of a lack of trust is generally compensated for by government services, assumed to be neutral and objective with respect to all citizens. But even this assumption is regularly doubted, especially in the largest urban centers, where services too often fail to be delivered in an evenly distributed manner.

We suggest that "do no harm" be replaced with what biodiversity specialists call "the precautionary principle," which admonishes researchers and policy makers that incomplete knowledge should not stop us from action when faced with a crisis. In the case of the EID crisis, the precautionary principle gives us a mandate to investigate pathogens that represent a risk of disease before they become the agents of an outbreak. This reverses thousands of years of human fear.

Only wise men rush in where fools fear to tread.

We must track known pathogens and their relatives even when they are quiescent. When that information is combined with what is known about the likely movements of pathogens from other places as a result of climate change, human population changes, and globalization, we can anticipate what pathogens may be coming to a particular place and what hosts are most at risk. Consequently, DAMA action plans involve two things. First, we can understand which pathogens are likely to migrate to a particular area, and we can inform people about measures they can take to reduce the chances that a given pathogen will successfully take up residence, or at least extend the time period before the establishment of that pathogen. Secondly, both for pathogens that are in residence but have not yet caused disease and for those potential new arrivals, we can inform people about ways to recognize outbreaks quickly so that they can be reported and additional measures enacted.

DAMA gives guidelines about what is needed in order to act decisively and effectively. It does not, however, tell any region, country, or even municipality how to act, except that action must be cooperative. This means that action can be tailored to the needs and capabilities of each place.[55] Perhaps the most difficult obstacle to be overcome is the recognition that in many cases your own future safety requires that you help your neighbors, even if you don't like them very much.

Needs

Personnel

Few people know how to act in a decisive manner; such skills, however, can be taught. This requires specialists in public activism, public outreach, education, and training. It also requires that there be permanent jobs for paraprofessionals working in adjunct with specialists to implement DAMA to the benefit of everyone. We think this kind of cooperative behavior might result in an extension of Aldo Leopold's "environmental ethic" to a "biodiversity and health ethic."[56]

Infrastructure

We require two kinds of infrastructure to support effective action. One is educational and job-training programs, as described above. The other is facilities associated with museums, universities, and government laboratories tasked with recommending actions based on documentation, assessment, and monitoring activities. Linking high-level scientific and technical infrastructure to the activities of paraprofessionals and citizen-scientists on the local level also serves as an important symbol to the public that the scientific and academic community and the government support these initiatives and appreciate the efforts of everyone involved.

Activities: Education and Cooperation Are Essential

Do not aim technology at the problem: educate people, arm them with technology, and aim them at the problem.

There are two components of activities associated with the "Act" section of DAMA. The first has to do with public outreach and jobs training. Public health means public education. People need to know what to look for: check your kids for ticks, dump the birdbath, empty the water from old tires; stay inside at dawn and dusk; wear shoes; maintain clean drinking water and unspoiled food. A cost-effective way to do this is to work it into the schools' curricula.

Everyone should be educated to make a contribution, so adult education through government-funded outreach programs is also essential. Not only will this provide critical information, it will reinforce to the average citizen that this is a national security priority. Then there can be a partnership in which the average citizen understands that there is a crisis, and although government-supported scientists deal with existing problems and try to anticipate new threats, actions by the public are essential.

The second has to do with responses whenever DAMA activities produce an "Act" recommendation. Most such recommendations at the local level will involve getting the word out to people about a particular threat and how to minimize their and their family's exposure to it. This requires input from local residents who can place scientific data within the appropriate cultural context. The monitoring aspect of DAMA is perfect for engaging the help of local citizens, neighbors of the parataxonomists and other locally based support personnel who are the focus of efforts to mitigate disease impacts. It is likely they who have reported some of the information leading to an "Act" recommendation in any event. So, not only are they in the best position to implement an "Act" recommendation, but being recognized for doing so provides public approval that represents positive feedback for the people involved, encouraging them to continue and others to participate.[57]

To this point, we have emphasized grassroots or bottom-up activities. The EID crisis is a global concern, however, so "Act" recommendations must also be shared with relevant governmental bodies and NGOs that can spread an alert as widely as is necessary and take appropriate action. In return, those organizations bear the responsibility of supporting local DAMA projects. One of the best proof-of-concept studies that we know of for DAMA has been done by a Hungarian biologist, Gábor Földvári, working at the University of Veterinary Medicine in Budapest. Margaret Island sits in the middle of the Danube River, just upstream from the Hungarian Parliament. It is a place of history, natural beauty, and intensive human recreational use. Among the wildlife species on the island are its adorably cute hedgehogs. They, in turn, host an array of ticks, some of which may, given the opportunity, attach themselves to humans. Földvári and his research group investigated the hedgehogs of the island for three consecutive years. In documenting basic ecological parameters of the animals, they made several important observations. First, these hedgehogs, as one of the most successful urban adapters, reach several times higher densities in the urban park than in nearby rural areas. Secondly, they have an exceptionally high tick infestation. The intensive fieldwork that enabled them to study over 400 individual hedgehogs was possible only with substantial involvement of citizen-scientists. As many as 40 or more enthusiastic volunteers, students, family members, and friends of scientists helped to collect, transport, and release the hedgehogs during the late-night hours. Most of the nearly 10,000 ticks removed and identified were of a single species, *Ixodes ricinus*, the sheep tick, for which the most important host seemed to be the hedgehog itself. However, an exotic species, *Hyalomma marginatum* (vector of the Crimean-Congo hemorrhagic fever virus) was also identified, highlighting the importance of monitoring these ectoparasites even within our closest neighborhood. They also discovered that although they had no symptoms, hedgehogs carry an array of bacteria, including *Borrelia burgdorferi* sensu lato, *Anaplasma phagocytophilum*, *Neoehrlichia mikurensis*, and *Rickettsia* species, which are potentially pathogenic in humans. They also found that very few humans actu-

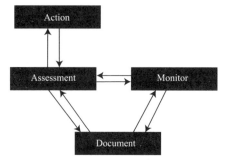

FIGURE 9.3 Summary diagram of DAMA activities. *Document* what is where, focus on interfaces. *Assess* who the real reservoirs are and the potential for changes in the interfaces, leading to changes in host range, possibly to changes in who the reservoirs are, and to the arrival of new pathogens. Determine monitoring priorities, make action recommendations, request additional documentation; *Monitor* the critical interfaces and reservoirs in anticipation of changes or emergences; *Act* on changes in interfaces, reservoirs, pathogens in residence, ways to limit the arrival of anticipated new pathogens, and early warning signs of pathogens anticipated to arrive at some point.

ally found ticks on themselves when they returned from a day at the island. And they discovered two additional pieces of information. First, no one who spent the day lying on grass that was regularly mowed by city personnel ever reported finding a tick on themselves: evidently the mown grass is too short to maintain high humidity, the limiting environmental factor for the ticks. Secondly, a significantly higher density of ticks occurred in the shrubberies and "hidden" parts of the island where the dense vegetation favored tick survival. As a consequence, all those people who reported finding ticks on themselves had one thing in common: finding that the island had only two public toilets, which were not always open, they had availed themselves of the underbrush beside the areas of mowed grass. Because the people using the bushes as latrines are at highest risk, a relatively cheap and effective way of significantly reducing disease risk is installing additional public toilets on the island.[58]

Policy as Lifestyle Change

There is no more neutrality in the world. You either have to be part of the solution, or you're going to be part of the problem.[59]

We believe climate change poses a great danger to humanity, that there is little time to act, and that we are not prepared to cope with it. And no one is acting with a sufficient sense of urgency. The DAMA protocol (fig. 9.3) does not compel anyone to act when confronted with news of an impending disease threat. But some things are clearly indicated. Let's update and paraphrase Elton:[60]

We must make no mistake, we are facing a set of circumstances that could eliminate technological humanity from this planet.

The terrible irony is that we know what to do to stave off that possibility. So, more good news and bad. The good news is that we can do this, and it will be economically feasible, because it will reduce the need for crisis response, which is unsustainably expensive. On the other hand, coping with the EID crisis—or any other aspect of climate change—will not be cheap, and things will never be the same. This may be the last chance technological humanity has to preserve itself. If our civilization emerged during a unique period of climate stability, we may not get a second chance to rebuild. And finally, always remember your children—nearly 65 percent of humans will live in cities by 2030, and very few urbanized children can survive in a nontechnological world. The bad news is that society doesn't seem to want to act in its own behalf. Everyone thinks that this is someone else's responsibility and that it won't happen to them. This is especially true for those who live in affluent and technologically advanced societies, like the ones in which there are many people who have the finances to buy this book and the education to understand it. But everyone is at risk.

It is time to implement what biodiversity scientists have called the "Jane Austen principle," replacing *Pride and Prejudice* with *Sense and Sensibility*. Why should those who feel that there is no true existential threat from EIDs, or that, if there is a threat, technological advances will save the day, nonetheless support DAMA initiatives? Pathogens are more than agents of disease; they are also compact units of information about biocomplexity, thanks to their highly specialized and phylogenetically conservative transmission dynamics and microhabitat preferences. Our focus in the first decade of the twenty-first century was on understanding how knowledge about pathogens in their environments offers cost-effective information essential for coping with the biodiversity of the future, including a better understanding of ecosystem services. The benefits we identified then still apply.[61]

If we are wrong, and the accumulation of high-probability/low-impact EIDs does not represent an existential threat to technological humanity, they nonetheless add to the overall costs of coping with climate change and take time away from efforts to develop technological solutions. Implementing DAMA will thus always buy time and save money. At worst, we will find a small number of high-probability/low-impact pathogens, and an enormous number of nonpathogenic species, some of which may turn out to have commercial uses. In addition, DAMA activities will expand our database of the biodiversity—and potential ecosystems services—of pathogens that are not associated with disease. If we are correct, however, as we believe the evidence suggests, the costs of inaction will be unacceptable. Humanity spends a lot of money attempting to anticipate and mitigate asteroid collisions and wars. Should we not do the same for emerging diseases that cost the world a trillion dollars each year?

There must be a sense of urgency and purpose. But our protocols must be evolvable; there is no static solution for problems involving an evolving Earth and an evolving biosphere. And we, as field biologists with well over a century of combined experience in boreal, temperate and tropical regions of the world, are well aware that

No plan of operations extends with any certainty beyond the first contact with the main hostile force.[62]

In this case that hostile force is not so much the pathogens exploring sloppy fitness space, but ourselves. Perhaps the greatest challenge facing any attempt to cope with the EID crisis is our general inability to cooperate effectively. On an individual basis, humans are very good at recognizing circumstances in which cooperation is the best option, and we cooperate quite well. Collectively, we have not been so good at it. But history is not destiny, so it is possible that a large number of humans can recognize that climate change and the EID crisis, intensified by population growth, urbanization, and globalization, is precisely a circumstance in which cooperation on a scale rarely if ever seen before is nonetheless the best option. The DAMA protocol will certainly fail without cooperation. More significant, if human civilization emerged from a unique set of environmental opportunities in the past 10,000 years, we may not get a second chance to mold ourselves into a technological species.

As veterinary and human medicine entered the late twentieth century there was a growing and explicit understanding of the connectivity between animal and human disease, facilitated by environmental opportunities. This led veterinary scientists and clinicians to embrace what is called One Health, more recently Planetary Health.[63] In the

1940s it brought together a coalition of veterinarians, physicians, and a broader range of organismal biologists, with foundations in the Centers for Disease Control and the U.S. Public Health Service. In some ways, this was a natural evolution from the vaccine world of the nineteenth and twentieth centuries.

One Health originated and grew out of the early discussions of Sir William Osler and Rudolf Virchow and was espoused by Charles Schwabe in *Veterinary Medicine and Human Health* in 1984.[64] Planetary Health is a complementary and more explicit approach that transcends pathogens to understand the connection between accelerating climate change, unprecedented environmental perturbation, and human health, with origins in the Rockefeller Foundation–*Lancet* Commission on Planetary Health in 2014.[65]

What is common to all these approaches is the recognition that transboundary and interdisciplinary cooperation is needed, bringing ecology, biodiversity, environmental health, and medicine to the forefront in addressing increasingly complex challenges for humanity and the human landscape.[66] In contrast with earlier independent and limited explorations emphasizing single organisms or regions, our understanding that rapidly changing environments from which viral, bacterial, fungal, arthropod, and helminth pathogens emerge demonstrates the urgency for new ways to coexist and live with our world.[67]

We endorse this perspective wholeheartedly. In order to cope with the onrushing multiple threats associated with climate change—of which EID is only one—we must integrate diverse human activities on multiple scales, and everyone must contribute. This calls for truly long-term planning—beyond the event horizon of most politicians. We have to assume these changes will be permanent. Coping with changes of that magnitude requires the cooperation of many people within and among countries, and on an unprecedented scale.

In the final chapter we encourage humanity to take individual and collective responsibility for the future of civilization and act accordingly. Our focus is on emerging disease, but all threats from climate change influence disease emergence and disease emergence affects all other threats, so we hope our call is broadly applicable.

If you want something you never had, you must do something you have never done.[68]

TEN

Time to Own It:
It's Nobody's Fault but
Everyone's to Blame

Everyone talks about the weather, but no one does anything about it.
—ATTRIBUTED TO MARK TWAIN

The last time our species coped with global climate change, we were able only to resent what was happening; we could do nothing about it. Now it is happening again. This time, however, we are not limited to resenting it; instead, thanks to our science and technology, we can do something about it. If we wish to. Let's begin with focusing on what is most at risk. Some fear the demise of the biosphere, but we do not. Ecological fitting provides a means of spreading your bets—diversifying your portfolio—when conditions change. Species with minimal capacity to oscillate between generalizing and specializing fall victim to the Gambler's Ruin and go extinct. Those that survive have indefinite built-in evolutionary capabilities, so unless humans decide to blow up the planet, enough of the biosphere will survive to enable new biodiversity to evolve. All we know of the history of life on this planet supports our optimistic view. That does not mean we advocate accelerating losses, only that we think the persistence of the biosphere is not in doubt. But that evidence brings with it a sobering insight; the biosphere is beginning to cope with climate change without asking our permission and without waiting to see whether we will participate. Whatever massive evolutionary changes occur, the biosphere will be indifferent to the fate of any particular species, no matter how self-important. Some believe climate change may cause the extinction of *Homo sapiens*. We believe that evolution in general does not favor us in particular, but humans have an abundance of the oscillatory capacity that is associated with long-term persistence. There are so many of us, living in so many parts of the world, occupying such an enormous and varied fitness space, that it is unlikely the entire species can be extinguished. We cannot draw too much comfort from that realization, however. There are many ways in which evolution could change our future role in the biosphere, possibly limiting us to a marginal existence. We believe that what is most at risk is technological humanity.[1] This planet has tolerated the modern version of our species for more than 100,000 years, especially in the unusually tranquil last 10,000, when we became sedentary agronomists and began building our vast technological fitness space. The history of technological humanity, beginning with the emergence of agriculture and domestication, which set the stage for urbanization and globalization, occupies only the last 10,000 years. That period includes multiple cases in which the most technologically advanced societies of the day catastrophically dissolved at what seemed to be the height of their powers, taken down by environmental fluctuations. Each one of those societies believed strongly that it could never happen to them and probably purged anyone who dared say otherwise. They fell precisely because they felt they were invulnerable, and that left them unprepared to cope with abrupt changes in their conditions of existence. *This is the human default we argue against.* Milanković's insights about climate cycling led us to expect a 15,000- to 20,000-year interval of relative stability and "benign" environmental conditions typical of many interglacial periods. We should be only 60 percent of the way through that period of climate stability, with a lot of time to understand and prepare for future climate change. But we have changed the atmosphere and oceans to such an extent that global system scientists now recognize a new period in Earth history, the Anthropocene, which began as recently as the 1750s. Its primary characteristics are the expanding footprints of human dispersal, burgeoning human population, agriculture, technology, and global connectivity. Environmental thresholds and tipping points we thought we would face in the distant future are being breached with increasing frequency because of accelerating climate warming driven by increased levels of atmospheric carbon dioxide of anthropogenic origin. Anthropocene humanity is living beyond its means.

The Blind Men and the Elephant: An Abundance of Good Intentions

No sense of common purpose has emerged in technological human-ity. One reason is that we have a complex relationship with climate change—the whole is greater than the sum of its parts, and there are no simple answers. Like the parable of the blind men and the elephant, people focus on a small part of the problem while assuming they understand its entirety. Only when they all realize they must share their knowledge do they truly understand what they don't know. The good news is that people are beginning to ask tough questions about our future. This has led to a diversity of opinions about what we need to do, suggesting that we have begun to see the problem in a more pro-ductive light. Let's start by understanding what the groups of people representing the various "blind men" in the parable want.

We Want a Guaranteed Future with No Costs (Politicians, Government Bureaucrats)

Politicians and the bureaucrats who operate governments are a criti-cal part of "understanding the elephant." Biologists can guarantee only two things—we will all die eventually, and there is no free lunch. No one lives outside those boundaries. So when members of governmental agencies publicly demand a guaranteed future with no costs, they do not truly expect it. They want to do as much public good as they can without jeopardizing their job security. In this regard, they are like the majority of humans on this planet. Unlike most people, however, they control substantial aspects of the flow of economic power and have the ability to influence enormous amounts of public opinion; they have the power to say yes.

We Want It Simple and Easy to Understand (Policy Makers)

If there were a simple answer, we would have solved the problem by now. As the philosopher of biology David Hull once said, "All the prob-lems that could be solved by common sense have been solved."[2] This is a particularly thorny problem when trying to explain complex sys-tems, the behaviors of which are not smooth and predictable. Scientists cannot make a complex situation simple; they can, however, explain the complexity as clearly as possible.

Policy makers form an essential link between the scientific community and the politicians and bureaucrats. The best policy makers are exceptionally bright people who often combine training in some area of science with expertise in translating science into public action. It is reasonable for policy makers to ask scientists to make their positions understandable to them, so they can explain the scientific findings to people who lack technical training but control the material means to implement actions based on scientific findings. The risk for policy makers is the desire to please their listeners' desire for simplicity.

We Want It the Way We Want to Believe It Used to Be (Ecologists and Conservationists)

Conservationists understand connections and consequences. Commanding detailed information about local ecosystems, they know a lot about the immediate consequences of environmental perturbations. More important, they understand them as complex systems, and that means they have an appreciation for thresholds and tipping points. As a result, they may see the signs of an impending tipping point before it occurs. They also understand that the outcomes of some tipping points may be more significant than those of others.[3] And if they share that knowledge with people who can act promptly, we may be able to mitigate the outcomes of those tipping points.

Despite those advantages, conservationists too often fall into the trap of believing in a mythical past state of the planet that was better for human existence than today. They believe it is possible to get to that past by reversing climate change. And they believe we actually *can* reverse climate change. They are wrong on all counts. The past 100,000 years was not such a nice place for human existence. Returning contemporaneous biodiversity to a previous set of environmental conditions will not cause evolution to rewind. And we cannot reverse climate change even if we thought it would help.

We Want It the Way We Want It (Engineers)

A broad coalition of humans acts as if the biosphere should be static and change needs to be resisted by all the technological capabilities we can muster. Engineers and preservationists (environmentalists who believe we should freeze existing biodiversity into an unchanging state) often think they are fundamentally in disagreement. Engineers have been quite willing to say, "Tell us what you want, and we will make it that way." Without realizing it, preservationists have an analogous perspective: "Tell us what we need to do to stop further changes, and we will do it." Neither group has been able to engineer sustainable stasis, because the context within which humanity resides is not static. Our species is an evolving system within a larger evolutionary system. Efforts by preservationists and engineers may alter the evolutionary timeline and trajectory but do not stop change from occurring.

Burgeoning technology has helped us ask more questions leading to better answers. But new problems now surface faster than we can find technological fixes for them. We must consider whether there are better ways to use the tools we have. Recently, preservationists have begun to understand that we should be trying to conserve not given areas or species, but rather the evolutionary process: the only way to cope with change is to allow change to occur.[4] This is a good beginning.

We Want to Survive (Evolutionists)

Evolution is intimate, yet impersonal. The conditions of life are completely indifferent to our pain, our suffering, our aspirations, and our hopes. But life includes a built-in drive to survive by oscillating between exploration and exploitation. The biosphere is not a collection of variable yet static entities; it is an evolutionary system with an indefinite ability to survive even massive environmental insult through a sort of rope-a-dope. Life lurches along, always slightly out of synch with the surroundings, but imbued with that relentless drive to survive. Each mass extinction triggered by global climate change has been a mass evolutionary reset. The biosphere has lost a lot of species, but those that remained have been capable of sustaining a transitional biosphere and evolving a new one. A fundamental principle of life is that no matter what happens, there are consequences. This flows directly from the realization that everything in life costs something, and everything is connected. Sometimes it seems wonderful—two species that originally competed for the same resource are able to coexist because one or both of them were flexible enough to switch to different resources. Other times it is not so nice. When the first photosynthetic microbe emerged, it catalyzed a global change in biodiversity that saw an explosion of species made possible by the large quantities of oxygen being released into the oceans and atmosphere. This literally life-changing—for the world as it exists now, life-giving—event was accompanied by a global mass extinction, the deaths of virtually all species for which oxygen was a deadly toxin.

Conflict resolution by trade-offs is a hallmark of evolution. And the most fundamental source of conflict is growth. Humanity embarked on a flight of fancy about unlimited growth when it decided to ignore the warnings of the Reverend Malthus and then of Charles Darwin. This needs to end. We must have the wisdom to understand that our prosperity will not protect us. And technology is taking us to population levels that are increasingly unsustainable. *Limits to Growth* predicted that technological humanity could fail quickly if we do not begin paying close attention and taking action. If the second half century of predictions in *Limits to Growth* are correct,[5] then Arrhenius gave us 175 years of advance warning, 125 of which we have largely wasted. Only recently have discussions of humanity's future seriously considered the possibility of the collapse of civilization. We may not have much time, so we need to keep our eyes on the prize, which is the survival of technological civilization. The evolutionary viewpoint is fundamental in that quest: (1) it gives us a reason to believe in the continuation of life on this planet, including humans in some form, no matter what climate changes occur; (2) it explains the dangers of the growth delusion and warns us about the consequences of ignoring it; and (3) it unifies the useful insights we have briefly categorized above, so we have the opportunity to make maximum use of all suggestions by people of good will.

What Prevents Our Taking Action?

Nothing is so much to be feared as fear.[6]

Our biological roots are part of the reason humanity is not doing a better job of coping with climate change and emerging disease. Anyone who has observed wild primates knows that when faced with uncertainty—such as a leopard walking through the undergrowth beneath the trees where they have been feeding—primates freeze. They assess the situation, and if they think there is a real threat, they scatter as fast as they can, each one hoping someone else is slower. In the great debate about "fight or flight," we inherited a decided bias toward flight.

When people are afraid, there is almost always a reason. We are descended from prey animals, primates, but ones whose cleverness, coupled with caution and suspicion, allowed us to survive myriad fearsome predators long enough to begin making the weapons evolution neglected to provide. We began styling ourselves Man the Hunter after we successfully used those weapons on species that once were fellow prey. Being curious and clever, we have invented additional ways of attempting to cope with our fears, from tool making, to fire, to domestication and agriculture, to urbanization, to the Internet.

The physical fragility of early hominids placed a premium on accurately perceiving and generalizing the complexity of their surroundings. Underestimate it and you are lunch, overestimate it and you starve. This is a case in which evolution clearly worked to our advantage. We, the survivors of multiple evolutionary experiments with human beings, have an excellent ability to perceive complex patterns in nature. Our ability to accurately assess complexity in our surroundings allowed us to generalize many things in our existence.[7] Such successful generalizing gave us a sense of security, so we trusted it; generalizing gave us a sense of control over our lives. And yet, there were many features of the world in which our ancestors lived that could not be generalized; many aspects of life are unpredictable. These basic contingencies of life, the complexities that cannot be generalized, frighten us because that's where the saber-toothed cats, cave bears, and boogeymen live. We came to associate "good" with the sense of security that came with successful generalizing and "evil" with the sense of danger that came with complexity.

Classical Greek philosophers also recognized this dual nature of life. Parmenides thought that the fundamental nature of the world was stasis. Any appearance of change is illusion or accident, an indication that something has gone wrong. Thus, the world is fundamentally "good," and we can hope to eliminate contingencies that plague us. Heraclitus thought the fundamental nature of the world was change. Any appearance of stasis is illusion or accident, based on our limited attention, career, and life spans, a perspective that tends to trap us in small slices of time. Thus, we can never hope to eliminate contingencies that plague us, so the world appears fundamentally "evil." Aristotle refined the Parmenidean view into the principle of simplicity upon which all Western science became predicated: "nature operates in the shortest way possible." As it was in our nature to find comfort in simplicity and to distrust complexity, we wanted to believe Aristotle's message. And we did.

Centuries later, the Newtonian Revolution (named for the seventeenth-century natural philosopher Isaac Newton) codified Aristotle's message into what we call Western science. In a universe structured by a few simple but powerful laws, science is the search for those laws, and the promulgation of explanations about the world based on them. The Newtonian Revolution suggests that the universe *ought* to be simple, and therefore it *is* simple. Any appearance of complexity is our fault—the result of incomplete information, something that science is tasked with correcting.

Immanuel Kant saw that the Newtonian Revolution had not solved the ancient conundrum. Our reason tells us the world ought to be simple. However, our experience tells us the world is complex, and the more we learn about it, the more complex it becomes. When reason and experience conflict, when what ought to be is not what is, which do we choose? There is no easy answer, and that leads to what scientists now recognize as the tension between theory and data. When that tension stimulates us to cooperative efforts, good things happen, because humans evolved in a world of patterned complexity, and we have an excellent ability to accurately assess the complexity of our surroundings, to generalize and to particularize. But we still tend to couch generalizations as good and particularities as bad or even evil.[8] Even today, most scientists conduct their research according to the principle that nature is simple, placing more credence in simple than in complex theories.

Humans are thus naturally wary of change and an unknown future and a bit afraid of the complexity they recognize so easily. No matter what we wish the nature of the world to be, no matter how much we think the world ought to be a certain way, the world will be what it is. And that seems to be both complex and changing in ways our species has not experienced since the last ice age. Humanity thinks it ought to be able to withstand those changes better now, thanks to advanced technology, and yet we seem to be losing the battle. Something is not quite right with our traditional view of how the world ought to be. Life is dynamic and complex, the past influencing the present, the present influencing the future. The present is a mixture of good and evil; the future is uncertain and fearful. Our evolutionary legacy allows us to function quite well in complex situations if we can overcome our fear of them.

Living with Fear

Our ability to accurately assess the complexities of our surroundings seems to have been coupled with an ability to plan for future contingencies, including all the natural evils of the world. But we are afraid of much that confronts us in the world today. And in a twist of evolutionary irony, our reasoning ability has been coupled with a pronounced capacity for psychological denial. It is in our nature to freeze while as-sessing a particular potential danger, then scatter in the face of it. And that is followed by denying that the danger was as bad as we thought at the time, or that it will ever happen again. Many people wish to avoid the implication that they should be doing something. This cre-ates what we call the *denial spectrum syndrome*: (1) there is no climate change, there is no EID crisis, everything is fine (this is often tied to an accusation that scientists are just trying to scare us so they can get more grant money); (2) we can stop and even reverse climate change, so when it gets bad enough, we'll take care of it, and it will be fine (this is often followed by the statement that technology has always saved us in the past, so it will again); (3) we are better off than 100 years ago, so it's all fine; (4) it's not as bad as we thought it would be (or as it could have been), so it's all fine; (5) if we admit the truth publicly, it will create panic and things will be worse (in these cases, "worse" usually means that the public will make increasing demands on their leaders and policy makers to do something), so saying and doing nothing is fine; (6) it's not our fault, someone else needs to take care of it, and they will, so it's all fine; and our favorite, (7) I'll be dead, but my children will figure it out, and they'll be fine. These are all examples of psycho-logical and emotional denial, of running away. They are understand-able responses, but they are not helpful in promoting any unified plan of action. In fact, they give rise to some decidedly unhelpful attitudes.

Stance No. 1: Pretending Nothing Is Happening

The evidence for climate change and anthropogenic influence on it is not only objective and overwhelming, it is terrifying.[9] The basic argu-ment for adopting this form of denial is that if our technology created the problem, how can our technology solve it? And if it cannot solve it, then we are doomed or reduced to praying for divine intervention, so we might as well pretend nothing is happening.

Climate change won't let us pretend much longer. The consequences of anthropogenic forcing clearly separate our current time from prior episodes of climate oscillations—no regions of the world are free of the human footprint.[10] Roads bisect the planet, a high percentage of land-scapes have been transformed, and agricultural land and fisheries in-fluence most corners of the globe.[11] We are in a world of encroachment and perturbation, leading to new contact points for pathogen exchange and transmission.[12] We alter entire biotas and then walk across those transformed landscapes. Pathogens enthusiastically explore newly connected fitness space, including more than 1400 pathogens in humans alone.

No wonder that many religious people, faced with such fears, veer away from such concepts as free will and self-determination, or stewardship of all creation, into nihilism and despair. The space they traditionally occupied ("It is God's will") is now crowded with nonreligious people who hope that "Mother Nature" will save us. We do not know about divine intervention, but evolutionary biology tells us that Mother Nature is short on compassion.

Pretending that climate change is not happening is a recipe for disaster. Keep installing solar panels on your house, recycling your waste, buying electric cars (and demanding national networks of charging stations), and vaccinating your children. These actions will not solve the problem, but they may help us buy enough time to find solutions.

Stance No. 2: Hiding Inside the Castle

The concept of progress acts as a protective mechanism to shield us from the terrors of the future.[13]

This is the twenty-first century's version of *The Masque of the Red Death*.[14] Let's hide inside our big, technologically advanced cities and everything will be fine. The basic argument is that cities are centers of technology, with the best health technology, the best public health, the best food and water security. Emerging diseases lurk in the tropics and in the wildlands, not in the cities. The reality is that cities are shockingly vulnerable to emerging disease. Many large cities, especially port cities, which supply inland ones, sit on the shores of oceans and are especially vulnerable to sea level rises and other natural catastrophes. The globalized hyperconnectivity of cities makes them vulnerable to security issues concerning energy, food, and water. Cities have special concerns with respect to building and housing security and delivering basic services such as water, electricity, and the Internet. And all these security issues for cities are threat multipliers for disease. The crisis of emerging disease is only one part of a complex, connected whole that portends great danger for urbanized and hyperconnected humanity.[15] Disease is a great leveler—no matter how it begins, once a disease enters the castle, no one's wealth will protect them. In today's world, the wealthier you are, the more people you depend on for your existence. For this reason, many major public figures are worried about our unpreparedness for global pandemics, as the May 4, 2017, issue of *Time* indicated. They would be those "black elephants," rare catastrophic events we knew about and should have prepared for but ignored. As bad as such pandemics will be, we are even more worried about the long-term socioeconomic impact of pathogen pollution in cities—what we have called the death-by-a-thousand-cuts disease scenario. Pathogen pollution represents a greater existential threat than one-off pandemics, because once they arrive, they never leave. Most major urban centers of the world experience seasonal viral pandemics that temporarily incapacitate large numbers of people—colds, diarrhea, "flulike symptoms." No one dies, and each episode is finished before severe economic harm is done, so they are not reported as pandemics; but that is what they are.

Stance No. 3: Running Away from Home

Since the Enlightenment, science has been our primary means of dealing with the world. It is sometimes called "the magic that works." For the past 300 years the chief magicians have been physicists. Physics has certainly delivered—we have immense technological capabilities thanks to that discipline. Of course, we also have atomic weapons thanks to physics, but no one is perfect. The development of nuclear weapons stripped physicists of their status as impartial observers, absolved from responsibility for the social applications of their research. Although physics delivers smaller, faster, more powerful, and cheaper electronic gadgets almost daily, it has abdicated its responsibility for humanity. It is offering us a vision that this planet has failed us and therefore we need to colonize another one that will do a better job of providing for us.

Talk of abandoning this planet and all but a handful of the 7.7 billion inhabitants is really a way for physicists to say, "we have no answers." There is no way 7.7 billion people will agree to spend money to build a spaceship for 5000 colonists rather than spending it on security issues for themselves and their children right here on Earth. It is unlikely that any single country could do it. And even if it were built, we would never agree on which 5000 people would be included. We personally believe that the physicists are wrong if they think the spaceship would be filled with them. More likely it would be filled with farmers and ranchers and hunters and warriors—pioneers, explorers. They may carry information about physics, but that will be for distant future generations of colonists to rediscover. This proposal is silly. Let's get over that childish fit of pique and talk to the adults in the room.

The abdication by the physics community leaves it to biologists to be the adults in the room. Climate change and its attendant impacts on biodiversity (at least half of which are pathogens) has likewise stripped biologists of their status as impartial observers, absolved from responsibility for the social applications of their research. Whether we like it or not, contemporary biologists are either part of the solution or they are part of the problem. The technological agenda for coping with climate change must be driven by the recognition that human needs are biological needs. Biology has been delivering the goods for humanity in the twenty-first century. But we can to do better.

There is no escaping the reality of climate change—and the increased risk of emerging disease that goes along with it—and our role in amplifying and accelerating climate change. There are no places on the planet that are guaranteed to be safe from emerging disease, especially not our big, wealthy cities. And finally, there is no other place we can feasibly go; there is no Planet B for us. We must face the world as it is and master our fearful biological legacy.

I must not fear. Fear is the mind-killer. Fear is the little-death that brings total obliteration. I will face my fear. I will permit it to pass over me and through me. And when it has gone past I will turn the inner eye to see its path. Where the fear has gone there will be nothing. Only I will remain.[16]

Plan B

There may not be a Planet B, but we promised you a plan B for coping with the emerging disease crisis, and it flows from the Stockholm Paradigm. Evolution is not the first thing the public thinks about when someone mentions an outbreak of cholera. We hope this book has contributed to changing that perception. In the previous chapter, we presented DAMA as a policy recommendation stemming from the Stockholm Paradigm. In reality, it is a policy *imperative*.

A world of climate change is a world of changing opportunity space for the biosphere. Evolutionary biology tells us that while many species are at risk—and many go extinct—during periods of intense climate change, these are also the times when the power of evolvable life emerges most strikingly. The simple truth is that climate change increases opportunities for all life by increasing connectivity in previously disconnected fitness space. This means climate change will produce increased opportunity space for pathogens. What is the pathogen opportunity space? Any place on the planet where life exists that is home to pathogens. That includes pathogens known to cause, or to be potentially capable of causing, diseases in humans and species of socioeconomic importance. In other words, *the threat space for EIDs is exactly the same as the opportunity space for pathogens.* Every part of this planet houses some level of EID threat, as well as some combination of threat multipliers. There is no place to run away to. The castle will not protect you, and abandoning the castle is not a solution, either. The entire planet is a minefield of evolutionary accidents waiting to happen.[17] Given the state of the planet and our influence on that state, there will be a continuing parade of pathogens, diseases, and unanticipated effects from those diseases for as long as climate change continues. And that will be our reality for the foreseeable future.

It is time for some hard reckoning. Climate change, population expansion, urbanization, and globalization are threat multipliers for emerging disease. Emerging disease, in turn, exacerbates poverty, famine, drought, conflict, and migration and is therefore a threat multiplier for those phenomena. What technological humanity is doing, both in ignoring and in trying to cope with emerging disease, is not sustainable. For evolutionary biologists like ourselves, this means that our species is going to experience a population correction. If we were talking about a population of introduced ants, no one would worry when we said the population would crash. Extending that thinking to humans, however, sends up the warning flags. Any rapid and widespread loss of human life in one place will affect the technological infrastructure of that place, as well as those places with which it is connected. The hyperconnected world of today does not allow any single country to maintain its way of life without multiple inputs from elsewhere. Few, if indeed any, countries in the world are self-sufficient for food and water, and none is independent when it comes to technological goods and services. Beyond that, in every country the heavily urbanized areas with the highest concentrations of advanced technology are the most dependent on support from outside.

We worry that humanity may choose, out of shortsightedness, to allow the future to play out as it will, changing nothing about how it approaches climate change, population expansion, urbanization, globalization, and disease. This is analogous to the business-as-usual scenarios in *Limits to Growth* and would not be, to put it mildly, a good idea. Business as usual will lead to pandemics. Those pandemics may be regional or global. They may involve pathogens infecting humans or those infecting crops or livestock. They may begin in urban areas or in rural areas. No matter the details, they will have some features in common. Pandemics infecting crops and livestock may leave enough food for farmers but will reduce food supplies to urban centers. This will tend to increase migration from those urban centers in search of food, either to rural areas or to other urban centers. In either case, the migrants may create or exacerbate food shortages in the areas to which they move. People remaining in cities will spend increasing amounts of time obtaining food. The combination of technically trained people preoccupied with finding food and those leaving the city will adversely affect the urban technological infrastructure. And any adverse impact on the technological infrastructure of any city will have adverse impacts on other cities and on rural areas connected to the city.

Pandemics infecting humans will likewise be associated with socioeconomic disruption. If the pandemic begins in a rural area, the first thing that will happen is that food production will be disrupted, because fewer people will be able to work. The reduction in food production will impact urban centers in the same way as would a pandemic caused by a pathogen infecting crops or livestock; food supplies will diminish. In addition, such a pandemic would catalyze migration to other places, both urban and other rural. Migrants displaced from rural areas will increase demands on food, water, and social services wherever they go. Some pose a health risk wherever they go. This will create socioeconomic disruptions that will increase the chances of conflict between residents and migrants. Those who bring the pathogen into an urban area will risk creating an urban pandemic that would disrupt the technological infrastructure and stimulate an exodus of people from the city. Pandemics caused by pathogens infecting humans that begin in urban centers will have the same net outcomes as those beginning in rural areas, except that the first impact will be disruption of technological infrastructure. Migration from the city would follow, leading to increased demand for food and services in the rural areas and the potential of spreading the pathogen to those involved in food production. Conflicts due to infrastructure disruption and migration will not remain local, however. Human populations living outside a pandemic zone will wish to quarantine potentially infected people (or crops or livestock) out of health concerns. They will wish to exclude migrants because of fear they might create unsustainable demands on infrastructure (especially food and water). Some outside the pandemic zone might view the crippled infrastructure of their neighbors as a target of opportunity. Business as usual will not end well. No matter where the pandemic begins, and no matter whether we are talking about pathogens infecting humans or crops or livestock, the human population correction will be especially acute in urban centers with high population densities. All pandemics will include at least some major socioeconomic disruption through loss of technological infrastructure and conflict. The hyperconnectivity of today's society means that pandemics will fray the tapestry of global technology. They may destroy it.

Time to Own It

The public and its decision-makers throughout the world are fearful and largely demoralized. Can we energize them, and you, to take action with respect to all climate-change threats, including emerging diseases? We think so, if we capitalize on the same human nature that has led us first to try to run away. Remember that we said that humans will always tend to run away from danger. When escape is not possible, however, humans can be ferocious especially in defense of family and home. We need to harness some of *that* capacity.

Putting Human Nature to Work

We must be masters of our own fate or we will be overmastered by events. Individual responsibility and collective action must be undertaken with a sense of urgency. We must find a middle ground between fatalism and nihilism, between "there is no problem" and "it's too late to do anything about it." We think there is a way. And ironically, that way involves harnessing the power of the same psychological denial that has led us to the brink of disaster. Denial indicates that there is still hope. All climate-change denialists hope the future will turn out okay. Those who do not deny the reality of climate change have the same hopes. In the Greek myth, Hope was left at the bottom of Pandora's box, trampled by all the evils of the world in their haste to escape. Hope is the form of denial that allows us to continue when our optimism fades. If nowhere else, we always think there is hope in the members of the next generation—with them come the opportunities for innovations to emerge. Think of it as a lottery for finding people to solve the problems we face and allow technological humanity to survive, tapping into human resources that transcend traditional political, religious, and socioeconomic boundaries. We cannot predict where they will come from, what they will look like, or how fast and how often they will emerge. But we can be certain that if we buy them time, they will find new opportunities to create new capacities. Therefore, along with the increased threat of emerging disease comes increased hope of finding people who will discover workable solutions.

Think of emerging diseases—and all other climate-change threats—as emanating from a Pandora's box that has been kept closed by climate stability. Now that climate change has opened the box, let's not keep Hope imprisoned inside.[18]

Hope Breeds Science

At some point early in human evolution, one or more of our ancestors noticed something. Every so often, when members of their group went to get water, a leopard ate one of them. When this first happened, the group avoided the watering hole. But that lasted about two days. Cautiously approaching the watering hole, the group encountered no leopard. Most members of the group decided this meant the leopard had moved on and the watering hole was safe. And it was, for a while. Then one day, the leopard attacked again. Great consternation seized our ancestors. Except for one or a few of them. They had been watching and counting. And they had discovered something. The leopard only showed up once every seven days. And they told the rest of the group that the watering hole was now safe for six days. Some believed this; most likely many did not. But the problem began to correct itself—those who believed survived, while the rest continued to be eaten, one every week. At some point, the nonbelievers got tired of seeing their friends being eaten and joined the believers. By this time, though, the believers had progressed in their own thinking. On the day when the leopard attacks, it must already be at the watering hole, hidden, when people arrive. What would happen if we arrived earlier than the leopard, waited for it to arrive, and then used our sharp pointy sticks to kill it? Of course, it could not have happened in such a direct fashion—after all, there is more than one leopard in the world, and leopards do not necessarily keep to a rigid schedule. Often we were wrong, but sometimes we would be right. When we were wrong, we hoped we could figure it out. And often we could—at least often enough to keep us believing in reason. Eventually our ancestors were able to understand enough leopard behavior to anticipate potential danger and avoid it much more often than before. No one was awarded a Nobel Prize, but we had become observers of our world and science had been born nonetheless.

CHAPTER 10

Putting Hope to Work: Linking Science and Cooperation

We are not the only biologists who fear technological society will die. Some, including Jared Diamond, believe we have arrived at the kind of tipping point at which previous human societies have always collapsed catastrophically. Others, like Eörs Szathmáry, take a more galactic view, believing that failures of wisdom and insight similar to our own explain why we have no evidence of intelligent life in the universe: other intelligent species reached tipping points similar to the ones facing us and failed as we may also fail. It's a grim outlook, but we have been warned for more than a century. Nonetheless, in the ponderings of scientists like Diamond and Szathmáry, we find hope. Those who write about the possibility that humanity will self-destruct hope they are wrong, or they would not bother to discuss the issue. They have lost optimism, not hope.

Even if I knew that tomorrow the world would go to pieces, I would still plant my apple tree.[19]

We are going to ask you to face one last demon: the fear of cooperating. This is about the risks of inaction and the difficulties of cooperating. Let's begin with science and scientists. Individual scientists are as closed, parochial, and selfish as any other humans. And yet the scientific community is open, international, and cooperative, a stance that has produced innumerable contributions to humanity. That is the power of combining objective knowledge and cooperation. Humans have always cooperated when they face a common threat and when they have hope. The threat is evident, and we believe science provides the hope. Science cannot save humanity, but it can give humanity the means by which it can choose to save itself.

Even the small step of adding a proactive component to coping with diseases is going to require more cooperation than most people think is possible. There is a widespread belief that humans do not cooperate very much, and that this is an essential part of our evolutionary legacy; not only are we inherently selfish, our very DNA is selfish. Our fearful nature may have made us unusually susceptible to this neo-Darwinian emphasis on evolution as never-ending conflict: we recognize the potential for conflict readily. In reality, humans cooperate a lot, and they often engage in acts of altruism, cooperating outside their kinship groups. The Darwinian perspective, embodied in a contemporary evolutionary framework, recognizes that evolutionary diversification means conflict resolution, and conflict resolution often involves cooperative behavior.[20]

In the long history of humankind (and animal kind, too) those who learned to collaborate and improvise most effectively have prevailed.[21]

The challenges of climate change in general, and emerging disease in particular, are so great that we cannot hope to survive without helping each other. Hope thrives under conditions in which cooperation is affordable. Right now, the cost of not cooperating is far greater than the possible costs of cooperating. The Stockholm Paradigm conceptual framework and the DAMA protocol that emerges from it give hope to the new research voices that have emerged in the past decade. You are not alone, and your efforts matter.

Hope is not a plan, hope is a reason to have a plan.

Global Security and Global Survival

Technological humanity faces an existential crisis. All parties interested in the well-being of humanity—environmentalists, industrialists, politicians, religious leaders—must come to the same negotiating table. Hope that we will manage to survive should be our starting point for cooperation. Every petroleum executive and every member of Greenpeace uses laptop computers and the Internet. If technological humanity is at risk, then everyone has a vested interest in being at the same table solving that problem.

We must agree that technological humanity should survive. Why? The most fundamental reason is that many of our children could not survive outside the largely technological fitness space we have made. Humans born in the last two generations could represent a major transition for humanity, and we need to give them the chance. Who knows what kinds of creative solutions they might find if we only give them that chance? Our children are always our hope for the future, but in this case, it is a mutual interaction. We have to protect them from sliding back into a world with which they cannot cope by preserving as much as possible of the infrastructure of current human civilization that makes our lives better than those of our ancestors. If the emerging disease crisis is a national security issue for every country, then it's a global security issue, and the survival of humanity should supersede national ambitions, at least long enough to ensure that our civilization survives. The very idea of a collaboration between environmentalism and the military on issues of climate-change threats may initially strike many as completely wrongheaded. But that is the little voice of fear in your heads. What better use of combined military programs like the United Nations peacekeepers than defending global security issues that can provoke conflict anywhere, anytime, to the detriment of everyone? No one has to be friends to engage in this mutual lifesaving activity, they simply have to understand that everyone faces a common adversary that is completely indifferent to the existence, much less the aspirations, of any particular group of humans. We don't have to agree with one another on everything, but we should at least agree to cooperate long enough to save technological humanity so we can continue our other disagreements. And maybe, just maybe, at the other end of that exercise, we may realize that resolving disagreements is better than perpetuating them.[22]

Recognizing the magnitude of these threats, the world's militaries spend enormous amounts each year on information related to climate change and the potential for emerging diseases. They understand that the emerging disease crisis is a national security issue. Military leaders understand that big crises begin as small conflicts and they all take place in a cultural and socioeconomic context. In consequence, they overprepare for wars they hope they never have to fight. Likewise, organizations such as NASA and the European Space Agency overprepare for asteroids we hope will never strike the planet. The chances of any given asteroid's hitting this planet are infinitesimal. The odds of an asteroid's hitting this planet during the time between elections for any politician are infinitesimal. And yet, scientists understand that there is a 100 percent chance of an asteroid hitting this planet with globally devastating effects at some point in the future. The potential for global devastation is sufficient for governments to allocate funds to try to anticipate and mitigate such rare events.

So, why do we not overprepare in the same way for emerging diseases? If disease is a national security risk, we ought to overprepare for it. That is what makes the DAMA protocol a policy imperative (fig. 10.1).

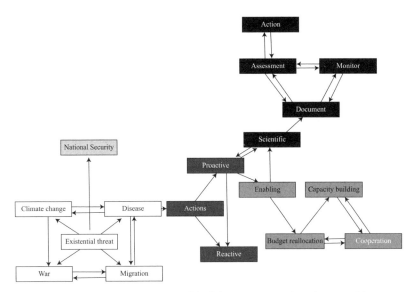

FIGURE 10.1   Diagrammatic depiction of the linkages among various elements of the basis for the DAMA protocol. We identify four existential threats: war, migration, disease, and climate change (which includes such things are famine, drought, and sea level rises). These combine to form the world of national security issues for every country on the planet. Arrows indicate the direction of major impact. Our concern is disease, in this case, the need for action. We can be reactive or proactive. The traditional approach is reactive. The DAMA protocol includes the four elements of scientific proactive action, which feed back into the societal enabling actions. This begins with budget allocations from whatever agencies— national, nonnational, international, extranational—are concerned. Funding is used to build human and technological infrastructure capacity. All of these activities are ultimately driven by cooperation. At the moment, everything connected to the "proactive" box is underfunded.

Summary

We like the idea of humanity's continuing to exist and evolve in ways that allow us to explore the universe and our own potential. We would like to imagine a world of less poverty and hunger, more tolerance of diverse opinions, fewer illnesses, and conflicts resolved without violence. Humanity must grow up and take individual responsibility and collective action for its future together. We must be masters of our own fate, or else we will be overmastered by long-term damage to civilization. We do not know how much time we have, so we must act each day as if what we do will make a difference, acting with a sense of urgency but not of fatalism.

Some facts of life must be acknowledged. Technological humanity is in imminent peril. Crisis response is not cost-effective, and efforts to turn back the clock are doomed to fail; we have to lower the cost of doing business and buy time by being proactive—we have to begin finding them before they find us. Coping with change requires cooperation and evolutionary thinking. We will continue to advocate policies and activities predicated on the notion that we can cope with what is threatening to overwhelm us. Failure could result in a permanent dismantling of technological humanity. Inaction is unacceptable.[23]

Climate change is literally beyond belief. There are no human belief systems—political, economic, social, religious, national, extranational, international, or transnational—that suffer unduly or receive undue respite from the effects of climate change. Climate change thus unites our species like nothing else ever has. And that can be a strength. Let's move forward with humility and determination, rather than recriminations. We are not the fittest; we have never really gotten beyond adequate, and now that's in doubt. But we have a long history of survival based on cleverness.

Let's not dither till we die; let's get to work and do something effective.

We did not write this book simply to leave a record for alien exobiologists confirming that we achieved some level of civilization before we did ourselves in, and that at least some humans saw the calamity approaching. We want our descendants to be there to meet those exobiologists and discuss how we overcame our crisis and compare notes about how they overcame theirs.

Do the right thing. It will gratify some people and astonish the rest.[24]

Notes

1. Compare Morse (1995), Garnett and Holmes (1996), and Sala et al. (2000) with Jones et al. (2008), Hansen et al. (2016), USGRCP (2017), and World Economic Forum (2018).
2. Arrhenius (1896).
3. Elton (1958). Goodell (2018) recently summarized the looming impact of climate change on human movements.
4. IPCC (2013); Trtanj et al. (2016). Water is an ephemeral resource—and availability is determined by different mechanisms relative to geography and history. Water sustainability is a challenge worldwide, for reasons that are not always obvious.
5. McMichael (2015); Hsiang and Sobel (2016).
6. Salkeld and Stapp (2008); Wilder et al. (2008); Salkeld et al. (2010); Brinkerhoff et al. (2010); Cully et al. (2010); Jones and Britten (2010); St. Romain et al. (2013); Makundi et al. (2015); Li et al. (2016); Mize and Britten (2016); Richgels et al. (2016); Mize et al. (2017). See also assessments of global plague diversity: Morelli et al. (2010); Dentovskaya et al. (2016); Respicio-Kingry et al. (2016); and the origins and persistence of plague: Bos et al. (2016); Andrades Valtueña et al. (2016, 2017); Yue et al. (2016).
7. de Kruif (1926); Zinsser (1935).
8. de Kruif (1926); Zinsser (1935).
9. Garrett (1994, 2001); Diamond (1997).
10. Editorial, *New Scientist*, January 30, 2016.
11. Morse (1995); Wolfe et al. (2007); Jones et al. (2008); Woolhouse (2008); Brook and Fordham (2015).

12. Shi et al. (2005); Pfennig et al. (2010); Rolland et al. (2012); Llambí et al. (2013); Roossinck (2013, 2015a, 2015b); Giribet (2015); Mueller et al. (2015); Roche et al. (2015); Roossinck and García-Arenal (2015); Roossinck et al. (2015); Skórka et al. (2015); Springer et al. (2015); Whitmee et al. (2015); Bajracharya et al. (2016); Baneth et al. (2016); El Sawaf et al. (2016); McCoy et al. (2016); Schade et al. (2016); Wu et al. (2016); Zhang et al. (2017); Feldman et al. (2017); Giannelli et al. (2017); Smout et al. (2017); Watts et al. (2017a, 2017b).
13. Scheffers et al. (2016); Pecl et al. (2017).
14. Aguirre-Liguori et al. (2016).
15. Trenberth et al. (2007); Masson-Delmotte et al. (2013); IPCC (2013, 2014a, 2014b, 2014c); Hoberg et al. (2017).
16. Mann et al. (1999); USGCRP (2017).
17. Juarrero (1999) is the definitive source for understanding the significance of "context" in complex systems.
18. Amel et al. (2017); Crist et al. (2017); DeFries and Nagendra (2017); Hall-mann et al. (2017); Johnson et al. (2017); Rothman (2017); Vignieri and Fahrenkamp-Uppenbrink (2017).
19. Whitmee et al. (2015); Stephens et al. (2016); USGCRP (2017); Watts et al. (2017a, 2017b); Williams and Jackson (2007); Buse et al. (2018); Prairie Climate Center (2018); Smith (2018).
20. Henry David Thoreau, journal entry for September 7, 1851 (see Thoreau, 2009).
21. Wiklund (1973); Wanntorp (1983); Brooks (1985); Coddington (1988); Wanntorp et al. (1990); Brooks and McLennan (1991, 1993, 2002); Nylin et al. (2018). By the mid-1980s different research groups held knowledge essential to understanding incompleteness in the others' efforts to explain pathogen-host systems. Who was going to take the lead in cooperating? The answer was Stockholm University. Among the earliest advocates of phylogenetics were two Swedish botanists at Stockholm University, Kåre Bremer and Hans-Erik Wanntorp. Wanntorp provided the foundation for phylogenetic studies of adaptation and functional diversification that other researchers quickly recognized. That time saw the emergence of a tremendously productive and innovative research program, founded by Christer Wiklund, combining field and experimental studies of phy-tophagous insects and their plants, aimed at understanding arms races and resource tracking models of coevolution, that is now approaching a half century and a fifth academic generation. This period also featured a number of collaborative efforts involving scientists at Stockholm University and colleagues from other countries. In a rare instance of cooperation at the time, phylogeneticists, evolutionists, and ecologists talked to each other. A graduate short course sponsored by the Nordic Council of Ecol-ogy, held at the Stockholm University Research Station in Tovetorp in Feb-ruary 1988, produced an article informally called the Tovetorp Manifesto.

One of the authors of this book (Brooks) was an instructor in that course, and his experience catalyzed a 1991 book on the subject. The following year, Brooks and Deborah McLennan began teaching in an undergraduate evolution course at Stockholm University and attending the annual autumn research conference, called the Blod Bad, at Tovetorp. Those annual visits informed two more books, one about parasite evolution and an updated and expanded version of the 1991 book. The annual teaching visits ended after 2000, but the collaborations continued. A major element of that was career development of those who had been students attending the workshop in 1988, and now their students and students' students, along with an extensive international network of collaborators. Given this history, it should be no surprise that the first professional conference bringing together parasite people and insect-plant people to compare theories and observations was organized by Sören Nylin, who had been a Ph.D. student of Christer Wiklund's at the 1988 conference, and Niklas Janz, one of his former students, who had assisted in the course that Brooks and McLennan taught. The Stockholm paradigm is an apt name.

22. Stout (1947).

23. Benjamin Franklin, on the occasion of the signing of the Declaration of Independence of the United States of America in 1776.

24. For example, Anthony et al. (2013); Hornok et al. (2015); Brookes et al. (2014); Rizzoli et al. (2014); Canhos et al. (2015); Mazet et al. (2015); Földvári et al. (2016); Jacobsen et al. (2016); Joly et al. (2016); Ottersen et al. (2016); Silaghi et al. (2015); Stephens et al. (2016); Szekeres et al. (2016); Young and Olival (2016); Clark et al. (2017); Cohen (2017); Doty et al. (2017); Kilpatrick et al. (2017); Millins et al. (2017); Olivero et al. (2017); Snow et al. (2017); Stone et al. (2017); Young et al. (2017); Vandermark et al. (2018).

25. Herbert (1984).

26. Collier (2017).

27. Maynard Smith and Szathmáry (1995).

28. Hull (1988).

29. Watts (2006).

30. Meltzer et al. (1999); Perry and Randolph (1999); Checkley et al. (2000); Githeko et al. (2000); Patz et al. (2000, 2003, 2007, 2008); Epstein (2001, 2005); Rosenzweig et al. (2001); Tong and Hu (2001); Curriero et al. (2002); Sachs and Malaney (2002); Guzman and Kouri (2003); Anderson et al. (2004); Lopman et al. (2004); McMichael (2004, 2013); Frenzen et al. (2005); Pauly et al. (2005); Budke et al. (2006); McMichael et al. (2006); Patz and Olson (2006); Confalonieri et al (2007); Gilbert et al. (2008); Mavalankar et al. (2008); Mas-Coma et al. (2009); Badjeck et al. (2010); Bentz et al. (2010); Brander (2010); Egbendewe-Mondzozo et al. (2011); Kedmi et al. (2010); Shepard (2010); Weaver et al. (2010); Shepard et al. (2011,

2016); Barnosky et al. (2012, 2017); Adelman et al. (2012); Semenza et al.
(2012); Knight-Jones and Rushton (2013); Lee et al. (2013); Singh et al.
(2013); Barnett (2014); Adenowo et al. (2015); Grisi et al. (2014); Shinn
et al. (2014); Staples et al. (2014); Stevens et al. (2015); Wappes (2015);
Whitmee et al. (2015); Beard et al. (2016); Gubler (2012); Ibisch et al.
(2016); Lam et al. (2016); Panzer et al. (2016); Samy and Peterson (2016);
Lee et al. (2017); Trtanj et al. (2016); Richter et al. (2017); Jansen et al.
(2018); Krech et al. (2018).

31. Editorial, *New Scientist*, January 30, 2016; "This Week" editorial, *Nature*,
April 7, 2016; Ottersen et al. (2016).

32. We were introduced to this term by Peter Ho in his presentation opening the conference "Disrupted Balance—Society at Risk," organized by
Nanyang Technological University of Singapore, December 5–7, 2016. A
good example is the recent report by the World Health Organization's expert committee charged with identifying the most threatening infectious
diseases of the upcoming year; it warned of "Disease X," a blanket term
suggesting that some emerging disease will create problems in the coming
year but there is no indication what that might be (Shaikh, 2018).

33. Cunningham et al. (2003).

34. World Economic Forum (2018); Sutherland et al. (2018).

35. Wu et al. (2016).

36. Brooks and Ferrao (2005).

37. Franklin D. Roosevelt, inaugural address, March 4, 1933.

CHAPTER TWO

1. See Cameron (1964).
2. Darwin (1872).
3. Darwin (1872).
4. Robert Koch (1843–1910) was a German microbiologist who helped develop the germ theory proposed by Louis Pasteur. Koch received a Nobel
Prize in 1905 for his research on tuberculosis and is best known for what
are now called Koch's postulates, four criteria designed to establish a causal
relationship between a particular pathogen and a disease. Koch (1876)
began formulating the postulates on the basis of earlier concepts by Jakob
Henle, and applied the postulates to describe the etiology of cholera and
tuberculosis, but they have been generalized to other diseases. Koch's postulates are: (1) the microorganism must be found in abundance in all organisms suffering from the disease but should not be found in healthy organisms; (2) the microorganism must be isolated from a diseased organism
and grown in pure culture; (3) the cultured microorganism should cause
disease when introduced into a healthy organism; and (4) the microorganism must be reisolated from the inoculated, diseased experimental host
and identified as being identical to the original specific causative agent.

Koch himself falsified his first postulate when he discovered asymptomatic carriers of cholera and, later, of typhoid fever. It is now widely understood that hosts may be infected with a pathogen without being diseased ("carriers")—this is a common finding for viruses such as polio, herpes simplex, HIV, and hepatitis C, and all tick-borne pathogens. The second postulate fails for pathogens that cannot (at the present time) be grown in pure culture, such as prions, suspected of causing Creutzfeldt-Jakob disease. In addition, all viruses require host cells to grow and reproduce and therefore cannot be grown in pure cultures. Koch also falsified his own third postulate by finding cases involving tuberculosis and cholera in which not all organisms exposed to an infectious agent became infected. Noninfection may be due to such factors as general health and proper immune functioning; acquired immunity from previous exposure or vaccination; or genetic immunity, as with the resistance to malaria conferred by possessing at least one sickle cell allele. There are other exceptions to Koch's postulates. A single pathogen can cause several disease conditions, while multiple pathogens may cause similar symptoms. Some pathogens cause disease in certain hosts and not others. Koch's postulates were developed in the nineteenth century as general guidelines to identify pathogens that could be isolated with the techniques of the day. Even in Koch's time, it was recognized that some infectious agents were responsible for disease even though they did not fulfill all of the postulates. Attempts to rigidly apply Koch's postulates to the diagnosis of viral diseases in the late nineteenth century, at a time when viruses could not be seen or isolated in culture, may have impeded the early development of virology. Koch's postulates have been recognized as largely obsolete by epidemiologists since the 1950s. Therefore, while Koch's postulates retain historical importance and continue to inform the approach to microbiologic diagnosis, they are not routinely used to demonstrate a causal connection between a pathogen and a disease, having largely been supplanted by criteria such as the Bradford Hill criteria for infectious disease causality in modern public health.

5. Such knowledge predates our origins; when chimpanzees and some other of our relatives are full of intestinal worms, they selectively eat plants they normally avoid, and those plants contain compounds that act as emetics and vermifuges (Huffman, 1997, 2001, 2016; Huffman and Vitázkova, 2007; Huffman et al., 2013; Shurkin, 2014).
6. Complex life cycles are the rule among rust fungi (Roelfs, 1989; Hovmøller et al., 2011).
7. Darwin (1872).
8. What we now call *speciation with gene flow* (sympatric and parapatric speciation); this also includes, along with some forms of peripheral isolates speciation, the newer term *ecological speciation* (Brooks and McLennan, 2002).
9. Wagner (1868).

10. von Ihering (1891).
11. von Ihering (1902). Von Ihering argued that such simultaneous speciation could only be explained by geographic isolation, because one could not envision selective pressures operating equally on such distantly related and ecologically distinct species as hosts and parasites. Further complicating his own ideas, he recognized two categories of hosts: *autochthonous* hosts were assumed to have evolved in the place where they lived at the time when geographic isolation occurred—their parasites were unique to them and found nowhere else. By contrast, *allochthonous* hosts evolved in one area then dispersed, taking their parasites within them into another area—their parasites might thus be found in other hosts in other areas. Only parasites of autochthonous hosts were reliable indicators of lineage age, which was needed to help identify the episodes of geographic isolation that led to the emergence of new species. Von Ihering therefore focused his attention on groups in which each species was associated with only a single host in a single area.
12. Bowler (1983).
13. Although the term *orthogenesis* only appeared in 1893 in a publication by the German biologist Wilhelm Haacke, Albert von Kölliker's theory of heterogenesis proposed 30 years earlier had all the elements of orthogenetic thinking (Wright, 1984; Vucinich, 1988). It was based on three general ideas, all thought to be contrary to Darwin's views: the multiple origin of living forms, the internal causes of particular variants, and "sudden leaps" (heterogenesis) in the evolutionary process.
14. von Nägeli (1884).
15. Eimer (1897, 1898).
16. Spencer (1898–99).
17. Kellogg (1907); see also Brooks (2010) and Brooks and Agosta (2012).
18. For an overview of wheat rust diversity, see Roelfs (1989); Hovmøller et al. (2011); Ali et al. (2014).
19. Houck (1924); Ewing (2016).
20. For an account of Arrhenius, see Baum (2016).
21. By the final year of the nineteenth century, Kellogg had teamed with his fellow Stanford University professor David Starr Jordan to promote these ideas in a more general manner. In 1907 he published *Darwinism Today*, helping set the tone for twentieth-century evolutionary biology. He described Darwinism as a reigning monarch, sitting back complacently as contenders for the throne squabbled, not realizing that one of those contenders would surely replace him. Kellogg wrote that Darwinism was of critical historical importance but little actual scientific value in light of the ultramodernism of the early twentieth century. According to Kellogg, Darwin had two ideas, the tree of life and natural selection, and the former was not original with Darwin while the ability of the latter in producing species remained controversial and might be overrated. Having

quickly dismissed Darwinism, Kellogg discussed a number of aspirants to the soon-to-be-vacated crown, summarizing the different perspectives, rarely taking sides overtly, always careful to alert the reader when he was injecting his personal perspective. He suggested that each of the perspectives he discussed had some merit, though some, he felt, had more merit than others. He did not offer a grand new vision of evolutionary biology, choosing instead to focus on the common unanswered questions that served as flashpoints among the various contenders to replace Darwinism. Like other progressive thinkers of his day, Kellogg anticipated that mathematical and statistical studies of variation and experimental studies of inheritance would eventually produce enough information to resolve ambiguities in inferences drawn from developmental biology, taxonomy (including what he considered affiliated areas of field biology, such as biogeography and ecology), and paleontology. Mendel's work had recently been rediscovered, and Thomas Hunt Morgan was beginning his journey that would lead to a Nobel Prize for his pioneering work in experimental genetics. In other words, when enough data accumulated, the awaited grand synthesis would emerge.

22. Kellogg (1907, 1913a, 1913b, 1914); Kellogg and Kuwana (1902); and collaborations with the ichthyologist David Starr Jordan, e.g., Jordan and Kellogg (1900, 1908) and Jordan et al. (1909).
23. Kellogg (1913a, 1913b).
24. Kellogg (1896).
25. Kellogg (1896).
26. Von Ihering returned to Germany from South America in 1920 and died a decade later. Kellogg left Stanford in 1915–16 and again from 1921 to 1933 to participate in programs of social justice, having become a strong opponent of Social Darwinist ideas espoused by the German high military command just before World War I. He became the head of the U.S. National Research Council and a trustee of what became the Society for Science and the Public. He died in 1937.
27. It was imperative for them to distinguish between cases in which the same species of parasite inhabited two or more host species, one type of which would indicate phylogenetic relationships among hosts while the other, in which the parasite had straggled from its original to a distantly related host, would not. At that time there was no method for reconstructing phylogeny, so the appropriate comparisons of host and parasite phylogenies with each other and with geographic distributions could not be made. Von Ihering was vehemently antiselectionism, while Kellogg accepted the influence of selection in some aspects of biology. Finally Kellogg accepted the Darwinian notion that commonness of association between hosts and parasites was due to phylogenetic history, but he rejected the equally Darwinian notion of phylogenetic conservatism in the evolution of adaptations. Thus, the only means by which these

ideas could be assessed were appeals to logical plausibility directing the interpretation of data, rather than the reverse. This would seem to be an unsatisfactory state of affairs, but other scientists nonetheless found the ideas plausible. These included the Australian parasitologists S. J. Johnston (1912, 1913, 1914a, 1914b, 1916) and Launcelot Harrison (1914, 1915a, 1915b, 1916, 1922, 1924, 1926). Harrison was an early advocate of Alfred Wegener's (1912) theory of continental drift as the geological mechanism underlying geographical isolation. These scientists, like von Ihering and Kellogg, assumed that the phylogeny of host-specific parasite species must parallel the phylogeny of their hosts owing solely to shared geographic history. Like Kellogg, they assumed that the host provided a uniform and stable environment that buffered the parasites from the effects of selection. They and others (e.g., Zschokke 1904, 1933; Ewing 1924, 1928; Kirby 1937; Hopkins 1942) established the circular notion that restricted host range was both the reason for, and a by-product of, common episodes of geographic speciation.

28. Metcalf (1920, 1922, 1923a, 1923b, 1926, 1928a, 1928b, 1929, 1934, 1940); see also Dunn (1925).
29. Metcalf (1923a, 1923b, 1929).
30. Fahrenholz (1909, 1913).
31. Eichler (1940, 1941, 1942, 1948a, 1948b, 1966, 1973, 1982); Baer (1948); Dougherty (1949); Stammer (1955, 1957).
32. von Kéler (1938, 1939).
33. Eichler (1942); Osche (1958, 1960, 1963).
34. Levit and Olsson (2006); Ulett (2014). In North America, major advocates included Coulter (1915), Jordan (1920), Lipman (1922), and Crawley (1923). Even Charles Otis Whitman, an ardent Darwinian in 1899, became a convert to orthogenetic views and wrote a book about orthogenetic evolution in pigeons, which was published in 1919, nine years after his death. Many tried to shine a light on themselves by rebranding orthogenesis. In 1930 the American zoologist Austin Hobart Clark called his version *zoogenesis*. The paleontologist Henry Fairfield Osborn (1934), who had been castigated by the ever combative von Ihering for not paying attention to work about geographical isolation in evolution (Choudhury and Pérez-Ponce de León, 2005), called his view *aristogensis*. In Italy, Daniele Rosa (1918) called it *hologenesis*. To the north and east, Lev Berg (1922, 1926) added a Russian version called *nomogenesis*. Germany was not forgotten in this explosion of names. Edwin Hennig (1922, 1927, 1932) was a German paleontologist with views similar to Osborn's. Not content with using Osborn's term, Hennig's onetime assistant, Karl Beurlen (1932, 1937), invented the term *palingenesis*. Finally, in the 1950s the German paleontologist Otto Schindewolf developed a theory of orthogenesis known as *typostrophism*.
35. Eichler (1948a, 1948b).

36. Verschaffelt (1910).
37. Brues (1920, 1924).
38. A child of immigrants (Belgian father, Irish mother), Dethier finished his Ph.D. in 1939, becoming a biology instructor at John Carroll University in Cleveland, Ohio. When the United States entered World War II in 1941 he joined the Army Air Corps, serving part of his time in Africa and the Middle East. He worked in the Army Chemical Corps as a research physiologist at the Edgewood Arsenal in Maryland (now the Aberdeen Proving Ground) in a long series of experiments analyzing the effects of chemicals on the chemosensors of flies, which was groundbreaking work on biological control. He ended his long and distinguished career as a professor of zoology at the University of Massachusetts, Amherst, where he was the founding director of its neuroscience and behavior program and chaired the Chancellor's Commission on Civility (he published *A University in Search of Civility* in 1984). Dethier also wrote books on natural history for nonspecialists, short stories, essays, and children's books. He reportedly wrote the manuscript for *Chemical Insect Attractants and Repellents* in the bomb bay of a B-25 bomber using an Italian typewriter he characterized as having been "liberated."
39. Dethier (1941, 1953).
40. Dethier (1953).
41. Topley and Wilson (1923); for recent reviews, see Fine et al. (2011) and Anderson (2016).
42. Ono and Hennen (1983) summarized information about the relationships and directionality of host colonization.
43. Bernhard Lauritz Fredrik Bang (1848–1932) was a Danish veterinarian. In 1897, he discovered *Brucella abortus*, the bacterium that causes brucellosis, which can cause abortions in pregnant cattle and undulant fever in humans. He became an adviser to the Danish government concerning issues of the control and eradication of disease in livestock. His advice to slaughter tuberculous cattle rather than engage in mass vaccinations became known as Bang's method. We do not know whether he was related to Peter Bang, who cofounded Bang and Olufsen.
44. Haldane (1949).
45. Dynesius and Jansson (2000); Jansson and Dynesius (2002); Harris (2005); Jansen et al. (2007).
46. deMenocal and Stringer (2016).
47. Ganopolski et al. (2016).
48. Steffen et al. (2011); Barnosky et al. (2012); Brooks and Hoberg (2013); Ruddiman (2013); Capinha et al. (2015); Bell and Tylianakis (2016); Waters et al. (2016); McMichael et al. (2017); Pecl et al. (2017). The Quaternary is a geological bracketing the Pleistocene and Holocene. The Holocene, the current interglacial, began with the termination of the last glacial advances. We anticipated an interval of relative stability and "benign"

environmental conditions typical of many interglacial periods and consistent with climate cycling in the Quaternary; at present we should be about halfway through that. Our influence on the atmosphere and oceans, however, has become so substantial that global system scientists now recognize these disruptions with a specific name in Earth history: the Anthropocene. Having emerged perhaps as recently as the mid-eighteenth century, the Anthropocene is characterized by expanding footprints of human dispersal, burgeoning population and population density, agriculture, technology, and global connectivity. Environmental-state changes to the planet, called *thresholds* and *tipping points*, are rapidly being breached owing to accelerating climate warming, environmental perturbation, and interconnected crises for biodiversity, species invasions, and emerging infectious diseases.

49. Kaufman et al. (2009).

CHAPTER THREE

1. Mode (1958).
2. Manter (1966) summarized his views in three rules; see Brooks and McLennan (1993, 2002).
3. Mayr (1957).
4. Ehrlich and Raven (1964).
5. For a summary of what was known at that time, see Dethier (1975).
6. Harper (1958); Hubble (1958); "The Triumvirate of Heredity" (1958); "Descent of Man" (1959); Comfort (1960).
7. "A Century of Evolution" (1959); Spector (1959).
8. Mode (1961, 1962, 1964).
9. Ono and Hennen (1983).
10. Bergersen and Anderson (1997); Bartholomew and Reno (2002); Gilbert and Granath (2003).
11. Elton (1958); Carson (1962).
12. Wiley (1981).
13. Woese (2000).
14. Mitchell (1901, 1905, 1910); reviewed in Brooks et al. (2007).
15. Hennig (1950).
16. Brooks and McLennan (2002).
17. Brooks (1979).
18. The "California Model," or allopatric cospeciation of Brooks and McLennan (1991); an homage to Kellogg, who was a faculty member at Stanford University.
19. Page (2003).
20. Representative cophylogeny studies. Diverse insect-plant, invertebrate-parasite, and vertebrate-parasite systems are reviewed in part by de Vienne et al. (2013); other selected investigations of macro- and microparasites:

Hafner and Nadler (1988); Perkins (2001); Johnson et al. (2002); Verneau
et al. (2002); Wickström et al. (2003, 2005); Nieberding et al. (2004);
Huyse and Volckaert (2005); Criscione and Blouin (2007); Gómez-Díaz
et al. (2007); Whiteman et al. (2007); Bruyndonckx et al. (2009); Mizukoshi et al. (2012); Bochkov et al. (2011); Nadler et al. (2013); Bellec and
Desdevises (2015); Hu et al. (2016); Lauron et al. (2014); Escudero (2015);
Lima et al. (2015); Fraija-Fernández et al. (2015); Haukisalmi et al. (2016);
Hoyal Cuthill (2016); Doña et al. (2017); Matthews et al. (2018).
21. See chap. 6.
22. Schoonhoven (1996).
23. This terminology arose in Mitter and Brooks (1983; also Brooks and
Mitter, 1984).
24. Jermy (1976, 1984).
25. Taber and Pease (1990); Becerra (1997); Köpf et al. (1998); Berenbaum and
Passoa (1999).
26. Kethley and Johnston (1975).
27. Crespi and Sandoval (2000).
28. Wiklund and Åhrberg (1978); Sperling and Feeny (1995); Armbruster
(1997); Arimura et al. (2000); Halloy (2001); Wahlberg (2001); Gentry
and Dyer (2002); Strauss et al. (2002); Nylin et al. (2005); Peñuelas et al.
(2006); Whiteman and Pierce (2008); Dyer (2011); Whiteman and Mooney
(2012); Gloss et al. (2014); Agrawal and Weber (2015); Cacho et al. (2015);
Fowles et al. (2015); Marinosci et al. (2015); Rahfeld et al. (2015); Richards
et al. (2015); Schäpers et al. (2015); Speed et al. (2015); Stark et al. (2015);
Stevens et al. (2015); Suinyuy et al. (2015); Syed (2015); Vihakas et al.
(2015); Volf et al. (2015a, 2015b); Wiens et al. (2015); Arias et al. (2016);
Cripps et al. (2016); Glassmire et al. (2016); Harrison et al. (2016); Juma
et al. (2016); Lai et al. (2016); Líznarová and Pekár (2016); Moreira et al.
(2016); Orsucci et al. (2016); Smilanich et al. (2016); Warburton et al.
(2016); Züst and Agrawal (2016); Singer (2017).
29. Janzen (1980).
30. Janzen (1985).
31. Janzen and Martin (1982).
32. Maynard Smith and Szathmáry (1995); Szathmáry (2015).
33. Wanntorp (1983); Brooks (1985); Coddington (1988); Wanntorp et al.
(1990); Brooks and McLennan (1991, 1993, 2002); recent updates by Lamsdell et al. (2017) and Cornwell and Nakagawa (2017).
34. Brooks and McLennan (1991, 1993, 2002).
35. Thompson (1988, 1994, 1997, 1999a, 1999b, 1999c, 2005).
36. Ono and Hennen (1983) reviewed historical thoughts about rust fungi
associations with plants, noting that some researchers had previously
suggested that host colonization could proceed from younger hosts to
older hosts (Jackson 1931); Leppik 1953, 1955, 1959, 1967, 1972). Mitter
and Brooks (1983) were the first to note "back-colonization" patterns in

explicit phylogenetic comparisons of phytophagous insects and their plant hosts.

37. Niels Bohr, quoted in Moore (1966).

38. Agosta et al. (2010).

39. In 1893 Henry Baldwin Ward, an American who had studied with Leuckart in Germany, arrived at the University of Nebraska. In 1895 he established the first undergraduate parasitology course in North America. By 1899 he had become professor and chair of zoology, and in 1902 dean of the medical school as well. Ward was a strong advocate of integrating basic biology and clinical biology in the study of parasites and parasitic disease. In 1909 Ward left the University of Nebraska for the University of Illinois, and in 1924 he founded the American Society of Parasitologists, in direct competition with the Society for Tropical Medicine and Hygiene. What happened? It seems that soon after Ward became dean of the medical school there was an academic squabble within medical school faculties across the United States, which resulted in the requirement that all deans of medical schools have MDs. The formation of the American Society of Tropical Medicine and Hygiene more than twenty years before Ward founded the American Society of Parasitologists was an early indication of this professional conflict and splintering of research programs. It is conceivable that if Ward had stayed at Nebraska and remained dean of medicine as well as chair of zoology, we would have been better prepared for climate change and emerging diseases. Instead, research programs in basic biological studies of parasites and studies of parasitic disease lost contact. This was intensified by the emergence of professional specialization in general: scientists no longer called themselves zoologists, they called themselves "entomologists" and "parasitologists," or "systematists" and "ecologists." Each compartmentalization severed contact with other research groups. For example, parasite systematists and insect-plant ecologists diverged so rapidly that there was little contact until recently (Nylin et al., 2018).

40. Historical ecology (Brooks, 1985; Brooks and McLennan, 1991, 1993, 2002); ecological fitting and sloppy fitness space (Agosta, 2006; Agosta and Klemens, 2008); the geographic mosaic theory of coevolution (Thompson, 2005); the oscillation hypothesis (Janz et al., 2006; Janz and Nylin, 2008; Janz, 2011); and the taxon pulse hypothesis (Erwin, 1979, 1981). Early efforts to articulate the Stockholm Paradigm as a synthesis of these research programs included Hoberg and Brooks (2008, 2015).

41. Ono and Hennen (1983).

42. James et al. (2009); Vercruysse et al. (2011); Shalaby (2013); Campbell (2016).

43. Labro (2012); WHO (2005). Although the estimated incidence of malaria globally has been reduced by 17 percent and malaria-specific mortality rates by 26 percent since 2000, approximately 3.3 billion people were

NOTES TO PAGES 67–68

at risk of malaria in 2010, with an estimated 216 million episodes and 655,000 deaths. The public health impact of leishmaniasis has been grossly underestimated, as a substantial number of cases were never recorded. Approximately 1.5–2 million new cases are estimated to occur annually, but only 600,000 are officially declared. Among other parasitic diseases of concern, amebiasis is prevalent throughout the developing nations of the tropics, at times reaching a prevalence of 50 percent of the general population and is estimated to cause more than 100,000 deaths per year. Neglected tropical diseases, a group of infections strongly associated with poverty, impair the lives of an estimated one billion people. Most of the diseases in this group are parasitic diseases, caused by a variety of protozoan and helminth parasites. Schistosomiasis infects more than 200 million people and ranks second only to malaria as the most common parasitic disease, with 20,000 deaths from the consequences of infection (bladder cancer or renal failure [*Schistosoma haematobium*]) or from liver fibrosis and portal hypertension (*S. mansoni*). Trypanosomiasis threatens millions of people in sub-Saharan Africa. As for the soil-transmitted helminths (STHs), more than 807 million people worldwide are infected with ascariasis, resulting in more than 60,000 deaths per year. Hookworm infections afflict an estimated 740 million people worldwide. Among vector-borne helminths, lymphatic filariasis affects more than 120 million people, and trichuriasis affects approximately 604 million people worldwide. Onchocerciasis infects 37 million people living near rivers and is the world's fourth-leading cause of preventable blindness. Additional great concern is the fact that polyparasitism is often present where most neglected tropical diseases occur—for instance, there is widespread overlap between HIV-AIDS and schistosomiasis, TB and intestinal helminth infections, and a high rate of hookworm and malaria coinfections.

44. Proposed by Van Valen (1973), the Red Queen hypothesis embodies a Darwinian duality: on one hand, it suggests that the faster environments change, the faster evolution will occur, while on the other hand it notes that because changes in inheritance systems are more conservative than changes in environments, evolution will always lag behind. Natural selection, as Darwin put it, is a perfecting mechanism, but perfection is never achieved. For discussion of the relationship between coevolutionary arms races and the Red Queen, see Råberg et al. (2014).
45. James et al. (2009); Shalaby (2013); Höglund et al. (2015); Vittecoq et al. (2016); Pichler et al. (2018).
46. See discussion in Brooks and McLennan (2002).
47. Maldonado et al. (2017).
48. Domestic ruminants: annually, $1 billion in Australia and tens of billions of dollars globally (Roeber et al., 2013); $14 billion in Brazil (Grisi et al., 2014). Fishes in aquaculture: Minimally hundreds of millions to more than five billion dollars (Shinn et al., 2014).

49. Kappagoda et al. (2011).
50. E.g., Pardi et al. (2018).
51. Meadows et al. (1972).
52. Hoberg (1997); Brooks (1998, 2000); Brooks and Hoberg (2000, 2001).

CHAPTER FOUR

1. Kellogg (1907).
2. Also called the *new synthesis*, the *modern synthesis*, and the *extended synthesis*.
3. Darwin (1872).
4. Brooks and Hoberg (2007).
5. Darwin (1872).
6. Darwin (1872).
7. Darwin (1872).
8. Darwin (1872).
9. Darwin (1872).
10. Darwin (1872); Maynard Smith and Szathmáry (1995, 1999); Szathmáry (2015).
11. Kellogg (1907).
12. Cope (1887).
13. Page (2011).
14. Janzen (1985).
15. Agosta et al. (2010).
16. See the comprehensive and readable analysis by McLennan (2008).
17. Cuénot (1911, 1925, 1941, 1951).
18. Dobzhansky (1955).
19. Goldschmidt (1940).
20. Simpson (1944).
21. Simpson (1953).
22. Gould and Vrba (1982).
23. Arnold (1994); Armbruster (1996, 1997); Kardon (1998).
24. Eggleton and Belshaw (1993).
25. Armbruster (1997).
26. Wanntorp (1983).
27. Trouvé et al. (1998); Solà et al. (2015).
28. Ross (1972a).
29. Brooks and McLennan (2002); Boucot (1982, 1983, 1990).
30. The neo-Darwinian framework assumes that all traits "cost" something, and selection will eliminate those that do not provide "benefits" as great as, or greater than, their costs. This sounds reasonable at first, but there is a problem. Just as new traits don't magically appear on command, old traits don't magically disappear if they are no longer in the limelight of selection. If a trait does not interfere with any function that is the

focus of current selection, there is no reason to think it will disappear. As well, inheritance systems are complicated, and virtually all traits are inter-connected in some way during development. The maintenance of a not-so-special trait may be a necessary precursor for a different trait that is of critical importance at any given point in time. The evolutionary "cost" of some old traits, like the thorns of those Latin American trees, may not be high enough to provoke selection against them—they may cost more to eliminate than to maintain. This illustrates one form of "survival of the adequate;" traits that once were critical for coping with the environment are now silent partners, but could once again function in the selection spotlight. Anachronisms thus complement chance innovations that are only marginally fit with their surroundings at the time they arise, but may become more fit later on if the spotlight of selection shines on them.

31. West-Eberhard (2003).
32. Hegner (1927).
33. Aleksandr Aleksandrovich Filipchenko had a tragic fate. He introduced the concepts of ecological parasitology into the Soviet Union while teaching in the invertebrate zoology department of Leningrad (now St. Petersburg). State University in the 1930s. Together with V. A. Dogiel he prepared a textbook titled *General Parasitology*. However, in the course of mass repressions at the end of the 1930s, he was arrested and executed in 1938. Filipchenko was fully exonerated posthumously after Stalin's death in the late 1950s. From prison he sent a note to Dogiel asking him to publish the textbook under his (Dogiel's) name alone (it was impossible to publish anything together with an "enemy of the people"). Thus, the first edition of the textbook was published in 1941, and the edition published in 1962 was translated into English (Dogiel, 1962). It remains one of the foundational texts of ecological parasitology.
34. Dougherty (1951); Price (1980); Sprent (1982); Holmes (1983).
35. Anderson (1957, 1958, 1982, 1984); Chabaud (1954, 1955, 1965b, 1982); Durette-Desset (1985); Price (1980); Holmes (1983).
36. Schad (1963); Williams (1960, 1966, 1968); Williams et al. (1970); Suydam (1971); Crompton (1973); McVicar (1979); Rohde (1979, 1981, 1984); Price (1980); Rohde and Hobbs (1986); Holmes (1983); Brooks and McLennan (1993); Whittington and Ernst (2002); ter Hofstede et al. (2004); Chilton et al. (2006); Randhawa and Burt (2008); Matisz et al. (2010); Höfle et al. (2012).
37. For example, *Schistosoma, Fasciola, Clonorchis,* and *Paragonimus* in humans. Researchers who saw this pattern prior to the phylogenetics revolution included McIntosh (1935); McMullen (1938); Buttner (1951); Stunkard (1959); and Grabda-Kazubska (1976).
38. Brooks et al. (1985a); Shoop (1988); Brooks et al. (1989); Carney and Brooks (1991); Brooks and McLennan (1993, 2002); Smythe and Font (2001); Kasl et al. (2015); Galaktionov (2016).

39. Rohde (1979, 1981, 1984); Price (1980); Holmes (1983); Brooks and McLennan (1993).
40. Suydam (1971); Holmes (1973); Poulin (1992); Brooks and McLennan (1993, 2002); Adamson and Caira (1994); Combes et al. (1994); Rohde (1994); Sukhdeo (1995); Buchmann and Uldal (1997); Nutting (1968); Reiczigel and Rózsa (1998); Sasal et al. (1999); Janovy (2002); Whittington and Ernst (2002); ter Hofstede et al. (2004); Alarcos et al. (2006); Chilton et al. (2006); Randhawa and Burt (2008); Matisz et al. (2010); Mele et al. (2012); Fredensborg (2014); Ding et al. (2015).
41. Taber and Pease (1990); Woolhouse and Gowtage-Sequeria (2005); Stacy et al. (2017).
42. Brooks and McLennan (2002) called this the Jagger principle: "You can't always get what you want, but if you try sometimes, you just might find you get what you need."
43. Agosta (2006); Agosta and Klemens (2008).
44. Stenseth and Maynard Smith (1984).
45. Agosta and Klemens (2008).
46. Maynard Smith (1976).
47. Soberón and Arroyo-Peña (2017).
48. Thompson (2005).
49. Peterson and Vieglais (2001); Peterson et al. (2002, 2005); Peterson (2006); Waltari et al. (2007b); Racloz et al. (2008); González et al. (2010); Ogden et al. (2006, 2014); Fischer et al. (2011); Hope et al. (2013a, 2013b); Khormi and Kumar (2014); Samy and Peterson (2016); Escobar and Craft (2016).
50. Watertor (1967); Blankespoor (1974); Ryss et al. (2018).
51. See chap. 3, n. 25.
52. Kurtén and Anderson (1980); Thorington and Hoffmann (2005).
53. Bell et al. (2016).
54. Agosta (2006); Soberón and Arroyo-Peña (2017).
55. Demboski and Sullivan (2003); Sullivan et al. (2014).
56. Darwin (1872).
57. Thompson (2005).
58. Malcicka et al. (2015); Cipollini and Peterson (2018).
59. León-Règagnon et al. (2005); for an example involving a North American digenean inhabiting snakes introduced to Europe, see Santoro et al. (2011); for one involving the introduction of a rust fungus infecting plants, see Beenken et al. (2017).
60. Brooks et al. (2006a).
61. Brooks et al. (2006b); for a similar perspective, see Doolittle and Booth (2016).
62. Hamann et al. (2013); Delatorre et al. (2015); da Graça et al. (2017).
63. Maynard Smith (1982); Maynard Smith and Szathmáry (1995); Szathmáry (2015); Nylin et al. (2018).

CHAPTER FIVE

1. The oscillation hypothesis was proposed by scientists at Stockholm University (Janz et al., 2006; Nylin and Janz, 2009; Wang et al., 2017).
2. Filipchenko (1937).
3. Brooks (2011); Brooks and Agosta (2012).
4. Bascompte et al. (2003); Bascompte and Jordano (2013); Rooney et al. (2006).
5. Araujo et al. (2015); Braga et al. (2018).
6. Araujo et al. (2015).
7. Braga et al. (2018).
8. Altizer et al. (2003); Li and Roossinck (2004); Parrish et al. (2008); Garcia et al. (2011); Raeymaekers et al. (2017); Braidwood et al. (2018).
9. Anderson and May (1982); May and Anderson (1985).
10. Adel et al. (2017); Moens and Pérez-Tris (2016); Rios et al. (2017); Pohlmann et al. (2018).
11. For details see review in Kindhauser et al. (2017).
12. Braga et al. (2015); Kuiken et al. (2017).
13. Modified from Ziętara and Lumme (2002).
14. Filovirus-like particles have been reported in species of insects (e.g., Lundsgaard, 1996).
15. Kupferschmidt (2017).
16. Kroll and Moxon (1990).
17. Modified from synthesis presented by the program *Frontline* (PBS) at http://www.pbs.org/wgbh/pages/frontline/aids/virus/tree.html.
18. Redrawn and modified from Brooks et al. (2015b).
19. Hafner and Nadler (1988).
20. Russell (1968); Spradling et al. (2004).
21. Spradling et al. (2004).
22. Pathogen-host systems showing phylogenetic oscillations in host ranges include monogenoideans and fish (Desdevises et al., 2002; Patella et al., 2017); lice and penguins (Banks et al., 2006); spiruroid nematodes and a range of vertebrates (Wijová et al., 2006; Jorge et al., 2018); and ticks and tick-borne pathogens (Estrada-Peña et al., 2016).
23. Monkeypox virus is a cousin of the smallpox virus that can readily colonize people from wild animal sources, especially other primates and rodents such as squirrels and rats. Discovered during disease outbreaks in monkey colonies in the 1950s and 1960s, it has been known in central Africa since the 1970s. A series of outbreaks in the Democratic Republic of the Congo (Zaire) showed direct human-to-human transmission in 1997 (Cohen, 1997). As reported in the Centers for Disease Control's *Morbidity and Mortality Weekly Reports* in April 1997, virus samples genetically analyzed from the 1996 outbreak did not differ from samples collected in 1970 and 1979. Emergence was thus not associated with the origin of a

novel viral strain, but rather with changes in human behavior (increased use of bush meat, including monkeys and rodents) during ongoing civil war and conflicts that were disrupting food security and displacing people. An emergence of human infections in the north-central United States in 2003 was eventually linked to the pet trade and importation of infected Gambian pouched rats from Africa, colonization of pet prairie dogs, and infection of pet owners. This potentially disastrous outbreak seems to have been self-limiting (Enserink, 2003).

24. Strona et al. (2013); Strona and Castellano (2018). Cichy et al. (2016) discovered that an invasive mollusk threatening to extirpate a native species has acquired parasites previously restricted to the native species.

CHAPTER SIX

1. Thompson (2005).
2. Maynard Smith and Szathmáry (1995).
3. Darlington (1943). Darlington was an American who attended Harvard as an undergraduate student shortly after the end of World War I and never left, eventually becoming the Alexander Agassiz Professor of Zoology.
4. Wilson (1959, 1961).
5. MacArthur and Wilson (1967).
6. Jordan and Kellogg (1908).
7. Erwin (1979, 1981, 1985).
8. Endler (1982); Heaney (2000); Simberloff et al. (1980); Simberloff (1987); Whittaker (2000); Wiley (1986, 1988a, 1988b).
9. Liebherr (1988); Liebherr and Hajek (1990).
10. Lieberman (2000, 2003a, 2003b).
11. Halas et al. (2005).
12. Hoberg and Brooks (2008, 2010); Hoberg (2010).
13. Cold and warm cycles in the Northern Hemisphere drive the expansion and retraction of large ice sheets, leading to changing sea levels. Large direct effects, such as glaciations, in one region may have dramatic indirect effects in distant corners of the world.
14. Darwin (1872).
15. Audy (1958); Elton (1958); Thompson (2005); Hoberg and Brooks (2008).
16. Sala et al. (2000); Harvell et al. (2002); Kutz et al. (2005, 2014); Callaghan et al. (2004a, 2004b); Patz et al. (2005, 2008); Parmesan (2006); Trenberth et al. (2007); Mas-Coma et al. (2008); Hoberg et al. (2008, 2017); Hoberg and Brooks (2008, 2013, 2015); Kaufman et al. (2009); Lafferty (2009); Lawler et al. (2009); Barnosky et al. (2012); Altizer et al. (2013); Hope et al. (2013a, 2015, 2016); Meltofte et al. (2013); Wernberg et al. (2016); Pinsky et al. (2013); Post et al. (2013); Sydeman et al. (2015); Whitmee et al. (2015); deMenocal and Stringer (2016); Scheffers et al. (2016); Timmermann and Friedrich (2016); Trtanj et al. (2016); Kintisch (2017); Pecl et al. (2017); Siepielski et al. (2017).

17. Masson-Delmotte et al. (2013); Rosen (2017); IPCC (2013); Hoffmann et al. (2017); Oppenheimer and Alley (2016); Kintisch (2017); Stocker et al. (2013); deMenocal and Stringer (2016); Hoberg and Brooks (2008).
18. Neiman (2002).
19. Araujo et al. (2015); Braga et al. (2015).
20. Schistosomiasis is one among a number of neglected tropical diseases, a group of parasitic and bacterial diseases that cause substantial illness for more than one billion people globally. These pathogens affect the world's poorest people, contributing to deep cycles of poverty and disease burden (Bruun and Aagaard-Hansen, 2008; Sady et al., 2015).
21. Lawton et al. (2011).
22. Snyder and Loker (2000).
23. Vrba (1985, 1995a, 1995b); Hoberg et al. (2004); Hernández Fernández and Vrba (2005); Brooks and Ferrao (2005); Johnson et al. (2006); Folinsbee and Brooks (2007); Hoorn et al. (2010); Woodburne (2010); Chang et al. (2017); Stigall et al. (2017). Environmental changes in Africa and Eurasia from the Miocene through the Pleistocene influenced the distribution and diversity of mammals, including elephants, large cats, hyenas, ungulates, and our hominoid ancestors. Habitat fluctuations determined the connectivity between Africa and Eurasia that served to control the intercontinental gateways for expansion and exchange among ecological assemblages of mammals and a broad array of parasites. Cooling trends from the Pliocene into the early Quaternary resulted in a shift from forest to grassland-savannah habitats coinciding with radiations among ungulates and hominin ancestors of modern humans. Cooling trends of the late Pliocene leading to glacial-interglacial stages in the Quaternary in South America controlled how faunas expanded, became isolated, and diversified far from glacial centers in the Northern Hemisphere.
24. Halas et al. (2005); Brooks and Ferrao (2005); Hoberg and Brooks (2008). Schistosomiasis has been established in Corsica since 2013 (Ramalli et al., 2018).
25. Audy (1958); Thompson (2005); Hoberg and Brooks (2008).
26. Zoonoses are pathogens that circulate between animals and humans. They can be transmitted directly through vectors such as ticks and mosquitoes or through consumption of contaminated foods and cause considerable human suffering (Daszak et al., 2000; Taylor et al., 2001; Diamond, 2002; Woolhouse and Gowtage-Sequeria, 2005; Wolfe et al., 2007; Kuris, 2012; Robertson et al., 2013, 2014).
27. Gibbons (2013a).
28. Brooks et al. (1995); Yellen et al. (1995); Larick and Ciochon (1996); de Heinzelin et al. (1999); Sponheimer and Lee-Thorp (1999).
29. Defleur et al. (1999).
30. Groucutt et al. (2015); Gibbons (2016).
31. deMenocal and Stringer (2016); Timmermann and Friedrich (2016).
32. Culotta and Gibbons (2016); Gibbons (2017); Nielsen et al. (2017).

33. Dynesius and Jansson (2000); Jansson and Dynesius (2002); Harris (2005); Jansen et al. (2007); Renema et al. (2008); Denton et al. (2010); Tzedakis et al. (2012); Masson-Delmotte et al. (2013); Rasmussen et al. (2014); deMenocal and Stringer (2016); Oppenheimer and Alley (2016); Henry et al. (2016). Orbital forcing explains transitions from glacial to interglacial stages and Milanković cycles, which influenced the duration and extent of cooling and warming during the late Pliocene through the Pleistocene. Oscillations result from the position of the Earth relative to the sun, the tilt of the Earth's axis, and wobbles around that axis. These produce many of the changes in the nature of the conditions that allow biotas to expand or which fragment and isolate them. These climate-change effects can occur on regional, continental, and intercontinental geographic scales, affecting terrestrial and marine biotas globally. During the Pleistocene, oscillations initially occurred at 41,000-year intervals, but beginning about 800,000 years ago, cycles of 100,000-year intervals became typical. The reason for this remains unclear. Late Quaternary glacials were about 80,000–120,000 years in duration, recurring at about 100,000-year intervals. A typical cycle encompassed an irregular period of ice sheet expansion (glacial maximum) followed by rapid warming leading to termination, with accelerated retreat and decay of continental glaciers. Between these waves of glaciation, short intervals (interglacials) of 13,000–28,000 years appeared, characterized by mild to warm conditions. The Quaternary experienced at least 20 episodes of this glacial-interglacial cycling, influencing temperatures, precipitation, drought, sea level, and weather systems around the planet. Extreme glacial maxima were associated with reductions in sea level approaching 120 meters, exposing broad expanses of the continental shelf. Most sea level variation was linked to ice dynamics, expansion and termination of continental glaciers of the Northern Hemisphere, with less contribution from processes in Antarctica. Sunlight, or solar insolation, and the energy arriving on Earth during the summer in the Northern Hemisphere is also key to understanding the ice ages. Insolation and three distinct planetary mechanisms explain episodic events of the ice age in the Quaternary. The position of the Earth around the sun relates to eccentricity or the degree to which the orbit is circular or elliptical (100,000-year cycles); the more elliptical, the more variation. A shift in the tilt of the Earth during passage around the sun is obliquity (41,000-year cycles); increasing tilt results in greater seasonal extremes. Wobbling of the planetary axis results in precession (23,000-year cycles), driving seasonal variations in insolation. Each cycle contains a maximum and a minimum, unfolding on a different time frame, and each determines what is called solar forcing. Eccentricity in particular leads to varying levels of summer insolation in the Northern Hemisphere and in part explains the altitudinal distribution of snow. When snow accumulates and is persistent, glaciations are not far behind. At the other

extreme, precession has been linked to monsoonal climate shifts, either amplifying or dampening rainfall in Africa and expanding or isolating ecosystems and potential pathways to Eurasia.

34. Gokhman et al. (2014); Houldcroft and Underdown (2016).

35. Hopkins (1959); Repenning (1980, 2001); Hoberg (1986, 1992, 1995, 2005); Vermeij (1991a, 1991b); Rausch (1994); Sher (1999); Hoberg and Adams (2000); Hoberg et al. (2003, 2012a, 2012b, 2013, 2017); Lister (2004); Cook et al. (2006, 2017); Zarlenga et al. (2006); Waltari et al. (2007a); Goebel et al. (2008); Koehler et al. (2009); Shafer et al. (2010); Galaktionov et al. (2012); Galbreath and Hoberg (2012, 2015); Hope et al. (2013a, 2013b); Makarikov et al. (2013); Raghavan et al. (2013); Haukisalmi et al. (2014, 2016); Verocai (2015); Verocai et al. (2014); Cooper et al. (2015); Lister and Sher (2015); Gibbons (2016); Galaktionov (2017a, 2017b). Beringia is the intermittent land connection linking Eurasia and North America. Through the Tertiary (the past 20–30 million years or so), Beringia was a near-permanent land connection, and biotic expansion between the continents was largely bidirectional. The world was mostly warm and mild, and humans were not yet on the scene. Boreal forests extended to the shores of the Arctic Basin. Global cooling commenced in the late Pliocene, leading to the great glaciations. By the early Quaternary, only a few million years ago, oscillating climates led to fluctuating sea levels and changes in the physical environment that periodically halted biotic exchange across Beringia. The Russian paleontologist Andrei Sher referred to these as "stoplights." From that point, Beringia alternated between being a land bridge or a seaway controlling biological connectivity between continents and oceans. On land, maximum glaciation led to fragmentation and development of refugial zones, in which new species evolved. Interglacials were associated with the retreat of continental glaciers in North America and Eurasia and sea level rise and oceanic communication linking the Pacific and Atlantic through the Arctic Basin. These changes invited species to explore, and the invitation was readily accepted. No fewer than nine orders of mammals, ranging from the tiniest shrews to the mammoths and the largest baleen whales, and even humans, have Beringian connections and dynamics ruled by the taxon pulses created by these climate cycles. Connectivity and subsequent isolation assembled the historical mosaics that now describe much of contemporary terrestrial and marine faunas across the northern latitudes.

36. Dobson and Carper (1996); Hoberg and Brooks (2013); Zarlenga et al. (2014); McMichael et al. (2017).

37. E.g., Göbekli Tepe in Turkey, ca. 9500 years ago.

38. Francis (2015); Orlando (2016); McMichael et al. (2017).

39. Larson et al. (2005, 2007).

40. Hoberg et al. (2004); Hoberg and Zarlenga (2016).

41. Cleaveland et al. (2001); Rosenthal (2008).

42. Grimm (2015); Shannon et al. (2015); Orlando (2016).
43. Hoberg et al. (2000, 2001); Hoberg (2006).
44. Nakao et al. (2002).
45. Terefe et al. (2014).
46. Hoberg et al. (2001); Wang et al. (2016).
47. Nakao et al. (2002).
48. McMichael et al. (2017).
49. Hoberg (2006); Terefe et al. (2014).
50. deMenocal and Stringer (2016); Timmermann and Friedrich (2016); Houldcroft and Underdown (2016).
51. Larson (2005, 2007, 2010); Francis (2015).
52. Loftus et al. (1994); Larson et al. (2005, 2007).
53. Nakao et al. (2002).
54. Nakao et al. (2002).
55. Larson et al. (2005, 2007, 2010).
56. Martinez-Hernandez et al. (2009).
57. Michelet et al. (2010); Yanagida et al. (2014).
58. McNeill (1976); Pearce-Duvet (2006); Hellenthal et al. (2014); Skoglund et al. (2014); Houldcroft and Underdown (2016).
59. Cleaveland et al. (2001); Wolfe et al. (2007); Jenkins et al. (2013).
60. Rosenthal (2009); Parrish et al. (2008).
61. Lefoulon et al. (2017).
62. Crellen et al. (2016).
63. Bos et al. (2014).
64. Bos et al. (2014).
65. Hoberg et al. (2012b).
66. Mas-Coma et al. (2009).
67. Malcicka et al. (2015).
68. Zarlenga et al. (2006); Pozio et al. (2009); Korhonen et al. (2016).
69. Rosenthal et al. (2008).
70. Rosenthal (2008); Rosenthal et al. (2008).
71. Hoberg et al. (2004); Hoberg and Zarlenga (2016).
72. Vrba (1985, 1995a, 1995b); Vrba and Schaller (2000); Hernández Fernández and Vrba (2005).
73. Araujo et al. (2015).
74. Orlando (2016); Almathen et al. (2016).
75. McMichael et al. (2017).
76. Houldcroft and Underdown (2016).
77. Gibbons (2013b); McMichael et al. (2017).
78. Kintisch (2016); McMichael et al. (2017).
79. McMichael et al. (2017).
80. Andrades Valtueña et al. (2016, 2017).
81. Achtman et al. (1999).
82. Gibbons (2013b).

83. Stenseth et al. (2006); Xu et al. (2011).
84. Schmid et al. (2015).
85. Bos et al. (2016).
86. Morelli et al. (2010).
87. Hu et al. (2016).
88. Hu et al. (2016).
89. McMichael et al. (2017).
90. Vezzulli et al. (2012); Trtanj et al. (2016).
91. Schumann and Leonard (2000); Voegele et al. (2009).
92. Ali et al. (2014).
93. Ali et al. (2014).
94. Ono and Hennen (1983).
95. Liu et al. (2014).
96. Ali et al. (2014).
97. Stakman (1915); Stakman and Piemeisel (1917).
98. Hovmøller et al. (2011); Ali et al. (2014); Beenken et al. (2017).
99. Singh et al. (2011); Bhattacharya (2017).
100. Bhattacharya (2017).
101. ITAP (2008); Macedo et al. (2017); Tenzin et al. (2017).
102. Hoberg et al. (2008).
103. Yates et al. (2002); Hurrell et al. (2003); Chavez et al. (2003); Ben Ari et al. (2008); Stocker et al. (2013); Masson-Delmotte et al. (2013); Sydeman et al. (2015). Internal climate variability on relatively short temporal scales are driven by ocean-atmosphere connections and temperature regime shifts. These influence habitats, productivity cycles, and diversity in marine and terrestrial systems in ecological time. Global mean temperature is influenced directly by the variability of the El Niño–Southern Oscillation (ENSO), persistent since at least the Pliocene, which determines the extent of interannual shifts in warming and cooling. Other oceanic and atmospheric regime shifts are represented by the Pacific Decadal Oscillation (PDO), which drives dynamics between cool and warm phases in sea surface temperatures of the eastern Pacific, and the Atlantic Multi-Decadal Oscillation (AMO) and the North Atlantic Oscillation (NAO), which determine fluctuations and climate shifts across North America and Siberia. Shallow climate shifts are also seen in the warming hiatus (1993–2012), which may reflect decadal-level oscillations that can amplify or dampen short-term trends against the broader background of upward incremental warming of the past century. Among these phenomena, El Niño/La Niña is the primary component of internal natural forces interacting with external solar and volcanic mechanisms. They do not, however, alter the expanding signature of anthropogenic warming and multidecadal climate determinants.
104. IPCC (2013); Meltofte et al. (2013); Cook et al. (2017); Hoberg et al. (2017).
105. Kutz et al. (2005, 2013); Hoberg and Brooks (2015).

106. Laaksonen et al. (2015, 2017).
107. Dobson and Carper (1996); Wolfe et al. (2007); Reinhard and Pucu de Araújo (2014); Morand (2015); Raff (2016); Paseka et al. (2018).
108. McMichael et al. (2017).
109. Patz et al. (2005).
110. de Kruif (1926); Zinsser (1935); Diamond (2002); Orlando (2016); McMichael et al. (2017).

CHAPTER SEVEN

1. Osborn (1910).
2. Osborn (1910, 1918, 1932, 1934). This line of thought corroborates Kellogg's notion that by 1907 no one was paying much attention to what Darwin actually said. Neo-Darwinians and neo-Lamarckians of the day all believed there was a very close fit between organisms and their environments. Darwin knew that there must be a misfit between organisms and their environments if natural selection was to be real.
3. See Brooks and McLennan (2002) for a discussion of Darwin's dual use of adaptation.
4. Osborn (1932).
5. Osborn (1918).
6. Simpson (1944).
7. Miller (1949).
8. Huxley (1942); Wright (1949); Mayr (1963).
9. Rosa (1918, 1931).
10. Hennig (1966).
11. Cracraft (1985).
12. Hoffman and Hercus (2000) use the term *environmental stress*.
13. This was first noticed by in 1878 by Alfred Russell Wallace, codiscoverer of the principal of natural selection and noted critic of neo-Darwinism.
14. Also called "net diversity," this refers to the number of new species formed minus the number of species extinctions (Stanley 1979).
15. In addition to the more stochastic effects of genetic drift; see Masters and Rayner (1993) for a thought-provoking discussion of this problem.
16. Mayr (1954); Newman et al. (1985).
17. Mitter et al. (1988).
18. Cope (1885, 1896).
19. Explicit recognition that organisms impose themselves on their surroundings rather than fitting into predetermined fitness space (also known as niche space) emerged in the last quarter of the twentieth century: Maynard Smith (1976); Lewontin (1978, 1983, 1983); Brooks and Wiley (1988); and Odling-Smee et al. (1996), who coined the term *niche construction hypothesis*.
20. Lewontin (1983). In other words, if the "niche" is real in an evolutionary sense, it must be inherited. If that is true, it must be a property of

the nature of the organism, not of the environment. For us, "niche" is synonymous with "fitness space."

21. Brooks and McLennan (2002).
22. Givnish (2015); Simões et al. (2016).
23. Alycia Stigall (Stockholm Paradigm workshop, Tovetorp, Sweden, March 2016).
24. Stigall (2010, 2012a, 2012b, 2014, 2017); Brame and Stigall (2014); Stigall et al. (2017); Lam et al. (2018).
25. Minter at el. (2017).
26. Darwin (1872).
27. Darwin (1872). For direct discussions of Darwinian evolution catalyzed by climate change, see Jansson and Dynesius (2002); Singer (2017); Pálinkás (2018).
28. West-Eberhard (2003).
29. Wanntorp (1983).
30. McLennan (2008).
31. For a summary of ecological fitting by phylogenetic conservatism in functional traits up to 2000, see Brooks and McLennan (2002); for a summary of works from 2000 to 2010, see Wiens et al. (2010). The following are representative of 400 publications we examined, published from 2010 onward, that also provide evidence of phylogenetic conservatism in functional traits: Robles (2010); Ochoa-Ochoa et al. (2014); Lootvoet et al. (2013); du Preez and Kok (1997); Boeger et al. (2015); Wells et al. (2018); Aguilar-Aguilar et al. (2014); Dimitrov et al. (2014); Wang et al. (2016); Bobeva et al. (2015); Bulgarella and Heimpel (2015); Kasl et al. (2015); Harrison et al. (2016); Norman and Christidis (2016).
32. Taylor and Bruns (1999).
33. Darwin (1872); see Sunny et al. (2015) for a recent review.
34. This is how *Haematoloechus floedae* in Costa Rica and *Fascioloides magna* in Europe became established.
35. See references for ecological niche modeling in chap. 4, n. 49.
36. Soberón and Arroyo-Peña (2017).
37. Simpson (1944). This same partial understanding was echoed more than 20 years later by Grant (1977): "When a species succeeds in establishing itself in a new territory or new habitat, it gains an ecological opportunity for expansion and diversification. The original species may respond to this opportunity by giving rise to an array of daughter species adapted to different niches within the territory or habitat. These daughter species become the ancestors of a series of branch lines when they, in turn, produce new daughter species. The group enters its second phase of development, the phase of proliferation. . . . Adaptive radiation is the pattern of evolution in this phase of proliferation. And speciation is the dominant mode of evolution in adaptive radiation. The first generation of speciational events gives rise to the primary branches in the now growing phylogenetic tree. These primary branches correspond to different pri-

mary ecological niches. Second, third, and later generations of speciation in each primary branch then parcel out the available ecological niches in the territory or habitat into a larger number of more highly specialized species."

38. Maynard Smith and Szathmáry (1995); Szathmáry (2015).
39. Knowles et al. (1999).
40. Nylin (1988). Another critical study from the Stockholm University group. This confirms the Darwinian view that innovations arise and take root in isolation not when they are needed, but rather when conditions allow experimentation. If this is true, the secret to evolutionary success is for innovations to emerge when they are not optimal but conditions allow them to have nonzero fitness, and then persist at low levels until there are changes in the nature of the conditions. This, ironically, is the aspect of Darwinism that Kellogg assigned to irrelevance in 1907.
41. For a fuller discussion of this example, see Brooks and McLennan (2002).
42. Elton (1958).
43. Halas et al (2005); Eckstut et al. (2011).
44. Ross (1972a, 1972b).
45. Brooks and McLennan (2001); Green et al. (2002); Brooks and McLennan (2002); Spironello and Brooks (2003); Bouchard et al. (2004); Halas et al. (2005); Folinsbee and Brooks (2007); Lim (2008); Eckstut et al. (2011).
46. Other studies based on methods that are not highly sensitive to the nuances of taxon pulses, but that have demonstrated histories of episodic expansion and isolation on various spatial and temporal scales include: *Based on an explicit comparative methodology*: van Soest et al. (1997); Winkworth and Donoghue (2005); Swenson et al. (2014); Tagliacollo et al. (2015a, 2015b); Thomaz et al. (2015). *Based on phylogeographic approaches*: Pellegrino et al. (2014); Lebarbenchon et al. (2015); Nakazawa et al. (2015); Papadopoulou and Knowles (2015); Maubecin et al. (2016); Menezes et al. (2016); Prates et al. (2016); Streicker et al. (2016); Samy et al. (2017); Colella et al (2018); Fecchio et al. (2018). *Establishing a regional context with comparative phylogenetic foundations*: Africa: du Toit et al. (2013); Agnarsson et al. (2015); Bohoussou et al. (2015). Asia: Ruedi et al. (2012); Justiniano et al. (2014); Razgour et al. (2015); Zhang et al. (2017). Australia, Antarctica, South America: Gardner and Campbell (1992); Rowe et al. (2008). Beringia, Eurasia, North America: Hoberg (1986, 1992, 1995, 2005); Hoberg and Adams (2000); Repenning (2001); Nakao et al. (2002); Lister (2004); Johnson et al. (2006); Zarlenga et al. (2006); Waltari et al. (2007a, 2007b); Koepfli et al. (2008); Koehler et al. (2009); Shafer et al. (2010); Galaktionov et al. (2012); Galbreath and Hoberg (2012, 2015); Hoberg et al. (2012a, 2012b); Dawson et al. (2014); Verocai et al. (2014, 2015); Kohli et al. (2015); Lister and Sher (2015); Korhonen et al. (2016); Bell et al. (2016); Chang et al. (2017); Cook et al. (2006, 2017). North America, South America: Marshall et al. (1982); Woodburne (2010); Bacon et al.

(2015); Hoorn et al. (2010); Matamoros et al. (2014); McHugh et al. (2014); Pérez-Rodríguez et al. (2015); Stigall et al., 2017). *Insular systems*: Swenson et al. (2014, 2015); Economo et al. (2015); Carvalho et al. (2015); Florencio et al. (2015); Faria et al. (2016); Weigelt et al. (2016); Rutschmann et al. (2017); Jorge et al. (2018). These comprise a representative—but not exhaustive—subset of more than 250 published articles, most published since 2000, that provide evidence supporting the taxon pulse dynamic.
47. Hegner et al. (1938).
48. Price (1980).
49. The epidermis of the members of Neodermata forms a continuously produced protective coating that allows the organism to live in an environment devoted to digesting biological material (a host intestine) without being digested.
50. Brooks et al. (1985b); Brooks (1989); Brooks and McLennan (1993); Zamparo et al. (2001); Galaktionov (2016).
51. Brooks and McLennan (2002).
52. Rogers (1962).
53. Tinsley (1983).
54. Crompton and Joyner (1980).
55. Moore (1981) and references therein.
56. Wilson et al. (1982).
57. Hoberg et al. (1997, 2001).
58. Tinsley (1983); Kearn (1986) and references therein.
59. Tinsley (1983).
60. Maynard Smith and Szathmáry (1995); Szathmary (2015).
61. Hoberg and Brooks (2008, 2015).
62. Ecker et al. (2018) presented an excellent example of a rapid transition from specializing to generalizing to new specializing.
63. Machado (1912).
64. Line spoken by Bette Davis's character in *All About Eve* (dir. Joseph L. Mankiewicz, 1950).

CHAPTER EIGHT

1. Sayers (1935).
2. Brooks and Agosta (2012).
3. Kauffman (1993).
4. Ulanowicz (1997).
5. Brooks and McLennan (2002). Some new voices are beginning to echo these sentiments; see, e.g., Pullin et al. (2013); Bowman et al. (2017); Sato et al. (2017).
6. Hoberg and Brooks (2008, 2013).
7. Representative publications since 2000: Rausher (2001); Christe et al. (2006); Fallahzadeh et al. (2011); Colwell et al. (2012); Young et al. (2013);

Murphy et al. (2013); Dougherty et al. (2015); Cizauskas et al. (2017); Dicks et al. (2016); Firmat et al. (2016); Hafer and Milinksi (2016); Hopper and Mills (2016); Carlson et al. (2017); Morand and Lajaunie (2017); Stone et al. (2017); Frainer et al. (2018).

8. Maynard Smith and Szathmáry (1995); Szathmáry (2015).

9. Attwell and Cotterill (2000); James et al. (2001); Dynesius and Jansson (2014); Gunton et al. (2016).

10. Wilson (1984).

11. Anderson and May (1982); May and Anderson (1985).

12. There is evidence of an emerging consciousness in the health professions surrounding the issues we have raised (Whitmee et al., 2015).

13. Lynch (2016).

14. von Son-de Fernex et al. (2014); Stone (2016); Winemiller et al. (2016); Niles et al. (2017).

15. Sala et al. (2000); Harvell et al. (2002); Barnosky et al. (2012); Altizer et al. (2013); Pecl et al. (2017).

16. Mas-Coma et al. (2008); Jenkins et al. (2013); Robertson et al. (2014); Utaaker and Robertson (2015).

17. Hoberg et al. (2008); Hoberg (2010).

18. Whitmee et al. (2015); USGCRP (2016); MacIntyre et al. (2017); USGCRP (2017); Watts et al. (2017a, 2017b); Rosà et al. (2018).

19. Robertson et al. (2013); Daversa et al. (2018).

20. Hornok et al. (2015); Müller et al. (2018).

21. This is the concern Bill Gates highlighted in the May 7, 2017, issue of *Time*.

22. As an example, the United States of America maintains a substantial network of controls and inspections at border crossings to limit plant, animal, and human pathogens. Agencies coordinating activities include Homeland Security (United States Customs and Border Protection [USCBP]) and the United States Department of Agriculture (USDA—Animal and Plant Health Inspection Service [APHIS], Plant Protection and Quarantine [PPQ], and Veterinary Services [VS]). Other agencies coordinate activities with respect to different priorities for inspection, quarantine, and mitigation. USCPB coordinates daily activities at 328 official ports of entry, involving 2400 agricultural specialists. Every day there are approximately one million passengers entering the United States, representing $6.3 billion in products; 4600 lots of specimens (organisms) are encountered in plants, meat, animal products, and soil products, and 400 pest species are determined. PPQ coordinates with border protection services; it has 5000 inspectors across ports, borders, and farms and involves international survey and inventory in areas of potential origin for real and suspected plant pathogens. PPQ maintains 16 plant-inspection stations at ports of arrival. The goal is risk-based sampling, designed to discover a 6 percent infestation level with 96 percent

confidence. Given the scope and depth of the imports, it is not feasible to sample all shipments. All that can be done is describe, document, detect, identify, and mitigate pests and pathogens, either by stopping entry or through identification of potential imports prior to dissemination from source countries. In 2016 nearly 17,000 import shipments representing 1.64 billion products from seeds to whole plants were stopped. PPQ identified 162,000 specimens of potential pests, including 73,000 regarded as subject to quarantine. Collectively these activities are critical, saving an estimated $90 billion in production losses due to insects and fungal diseases in plants. VS programs focus on surveillance, monitoring, and in some cases control and eradication of infectious diseases, vectors, and vector-borne diseases, many of which are endemic to the United States.

23. Cunningham et al. (2003).
24. Malthus (1798).
25. "Man the Hunter" was the title of a 1966 symposium organized by Richard Lee and Irven DeVore. The symposium attempted to bring together, for the first time, a comprehensive look at ethnographic research on hunter-gatherers (Lee and DeVore, 1968).
26. Challinor et al. (2016).
27. *Green revolution* refers to a set of agricultural development initiatives throughout the developing world that took off in the late 1960s. The term was first used in 1968 by William Gaud, a former director of the United States Agency for International Development (USAID), who noted the spread of the new technologies. The initiatives were led by Norman Borlaug, called the "Father of the Green Revolution," who has been credited with saving over a billion lives, for which he received the Nobel Peace Prize in 1970. Borlaug also presciently warned that if the Green Revolution were not coupled with population control measures, it could backfire. Thus, today we must cope with the "African youth demographic bulge."
28. Agbor et al. (2012).
29. Sokolov et al. (2017).
30. Meadows et al. (1972).
31. Meadows et al. (1972).
32. Meadows et al. (1972).
33. See https://www.edge.org/conversation/daniel_c_dennett-dennett-on -wieseltier-v-pinker-in-the- new-republic.
34. Turner (2014).
35. Jones et al. (2008); Barnosky et al. (2012); Brooks and Hoberg (2013); Capinha et al. (2015); Bell and Tylianakis (2016); Pecl et al. (2017).
36. Goldin and Mariathasan (2014); Goldin and Kutarna (2016); Scheffers et al. (2016); Beddington (2010).
37. See chap. 6, n. 110.
38. Kübler-Ross (1969).

39. Clarke and Eddy (2017).
40. See Thoreau (2009).

CHAPTER NINE

1. Levina and Tirpak (2006); Klein et al. (2007); IPCC (2014a, 2014b); Berthe et al. (2018).
2. It occurs in the shortened form "forewarned, forearmed" in Robert Greene's *A Notable Discovery of Coosnage* (1591) and *The Art of Conny-Catching* (1592), published together in Greene [1923].
3. Audy (1958).
4. Brooks et al. (2014).
5. Portals for access to disease alerts, databases and reports for animal and plant disease and ongoing outbreaks by a number of agencies: OIE, World Organization for Animal Health, www.oie.int; Centers for Disease Control (CDC), www.cdc.gov; United States Department of Agriculture (USDA) and the Animal and Plant Health Inspection Service (APHIS), www.aphis.usda.gov; World Health Organization (WHO), www.who.int.
6. USDA APHIS, www.aphis.usda.gov.
7. Federal Interagency Committee on ITAP (2008).
8. USAID (2014, 2016); Joly et al. (2016).
9. Brookes et al. (2014); Mazet et al. (2015); Jacobsen et al. (2016).
10. Jones et al. (2008).
11. Jones et al. (2008).
12. USAID (2014); Joly et al. (2016).
13. Anthony et al. (2013).
14. Hoberg (1997); Brooks and McLennan (1991, 2002, 2010); Brooks (1998, 2003); Brooks and Hoberg (2000, 2001, 2006, 2013); Hoberg and Brooks (2008, 2013); Hoberg et al. (2015).
15. USAID (2016); Anthony et al. (2013).
16. Dobson (2005); Brierley et al. (2016).
17. Our recollection is that two of us heard this admonition from Meredith Lane, the distinguished botanist and tireless campaigner for public policy and advocacy associated with issues of climate change and biodiversity.
18. Hoberg et al. (2012, 2013); USAID (2014); Cook et al. (2017).
19. Sun Tzu (2017).
20. Consider recent discoveries about diversity and distribution of such pathogens as *Anaplasma*, *Borrelia*, and *Trypanosoma cruzi*; Millins et al. (2017); Stone et al. (2017); Svitálková et al. (2015); Walter et al. (2017); Vandermark et al. (2018).
21. PCAST (1998); ITAP (2008).
22. Cresswell and Bridgewater (2000) summarized ways in which the GTI objectives might be achieved, though none of them ever materialized.
23. Kilpatrick et al. (2017); Millins et al. (2017); Stone et al. (2017).

24. Enserink (2003).
25. Cohen (1997).
26. Dan Janzen is widely credited with formalizing the concept of parataxon-omists in conjunction with notions of large-scale biodiversity inventories (Janzen et al., 1993; Janzen and Hallwachs, 1994, 2011; Schmiedel et al., 2016).
27. USAID (2014).
28. Anthony et al. (2013).
29. Preece et al. (2017); Young et al. (2017); Wells et al. (2018).
30. Yates et al. (2002); DiEuliis et al. (2015); Oliveira et al. (2016); Cook et al. (2017); Dunnum et al. (2017); Lajaunie and Ho (2017); Hope et al. (2018). The Arctos platform is housed at the Museum of Southwestern Biology, University of New Mexico, and links collections at Harvard University, the University of California, Berkeley; the University of Alaska; and the University of Nebraska. The Arctos relational database system, online at http://arctos.database.museum, provides access to global data and specimens, notably the Beringian Coevolution Project.
31. Global Biodiversity Inventory Facility (GBIF), www.gbif.org.
32. Some representative overviews of the importance of interfaces and reservoirs: Hornok et al. (2015); Olivero et al. (2016); Girisgin et al. (2017); Pinsent et al. (2017); Blaizot et al. (2018); Daversa et al. (2018); Clark et al. (2018); Faust et al. (2018); Ragazzo et al. (2018); Wells et al. (2018).
33. E.g., Jenkins et al. (2005); Kutz et al. (2007); Prosser et al. (2013); Gutiérrez-López et al. (2015); Millán et al. (2015); Hemida et al. (2017); Robertson et al. (2017); Makkonen et al. (2018); Ramalli et al. (2018).
34. Scientific Collections International (2015); DiEuliis et al. (2016); Dunnum et al. (2017).
35. Brooks (1998); Brooks and Hoberg (2000).
36. USAID (2014, 2016).
37. Brooks (1986); Brooks and McLennan (1991, 2002); Hassanin and Abdel-Moneim (2013); Osadebe et al. (2014); Gupta et al. (2016); Fountain-Jones et al. (2017); Hemida et al. (2017); Reynolds et al. (2017); Springer et al. (2017); Walter et al. (2017); Yap et al. (2017).
38. Brooks (1992a, 1992b); Brooks and McLennan (2002).
39. Young and Olival (2016); Anthony et al. (2013); USAID (2014).
40. Dobson (2005); Anthony et al. (2013); Young and Olival (2016); Olival et al. (2017).
41. Yates et al. (2002); Kang et al. (2011); Yanagihara et al. (2015).
42. Additional representative studies emphasizing the importance of reservoirs include Hemida et al. (2013); Yabsley and Shock (2013); Hornok et al. (2015); George et al. (2015); Szekeres et al. (2016); Blaizot et al. (2018); Briggs (2017); Cyranoski (2017); González-Salazar et al. (2017); Shrestha and Maharjan (2017); Springer et al. (2017); Stone et al. (2017); Tenzin et al. (2017); Ragazzo et al. (2018); Ramalli et al. (2018); Wells et al. (2018).

43. Yates et al. (2002).
44. Yates et al. (2002).
45. Yates et al. (2002); DiEuliis et al. (2015); Scientific Collections International (2015); Dunnum et al. (2017).
46. Yates et al. (2002).
47. Audy (1958).
48. Hoberg et al. (2008); Kutz et al. (2014); Hoberg and Brooks (2015); USAID (2014).
49. Harvell et al. (2002); Lafferty (2009); Pinsky et al. (2013); Post et al. (2013); IPCC (2013, 2014a, 2014b, 2014c); Hoberg et al. (2008, 2013); Sydeman et al. (2015); Pecl et al. (2017); McMichael et al. (2017).
50. Harvell et al. (2002); Cumming and Van Vuuren (2006); Hoberg and Brooks (2008, 2015); Young et al. (2013); Dornburg et al. (2014); Araujo et al. (2015); Iverson et al. (2016); Wernberg et al. (2016); Daversa et al. (2018); Edler et al. (2017); Navarro et al. (2017); Pecl et al. (2017); Preece et al. (2017); Young et al. (2017); Hayes and Piaggio (2018); Schwab et al. (2017); Strona and Castellano (2018).
51. Kutz et al. (2013); Pickles et al. (2013); Hoberg and Brooks (2015); Ogden and Lindsay (2016); Samy and Peterson (2016).
52. Hoberg et al. (2013); Meltofte et al. (2013).
53. Jenkins et al. (2013); Kutz et al. (2014).
54. Bill Tilden, tennis star.
55. Grant et al. (2017); Canessa et al. (2018).
56. Goldberg and Patz (2015).
57. Pasquini et al. (2014).
58. Földvári et al. (2011, 2014, 2016); Hornok et al. (2015); Rizzoli et al. (2014); Szekeres et al. (2016). Thomaz-Soccol et al. (2018) recently reported a previously unsuspected urban risk of leishmaniasis in the popular international tourist town of Foz do Iguaçu, Brazil.
59. Cleaver (1968).
60. Elton (1958).
61. Brooks (1998, 2000, 2003); Brooks and Hoberg (2000, 2001, 2007b); Hoberg (2002); Hoberg and Brooks (2008, 2013); Kutz et al. (2007); Polley et al. (2010); Preece et al. (2017); Young et al. (2017).
62. Attributed to Helmuth Karl Bernhard Graf von Moltke
63. American Veterinary Medical Association (2008); Horton and Lo (2015); Cunningham et al. (2017). Other programs of note include Dixon et al. (2014); Bloom et al. (2017); Cooke et al. (2017); Grant et al. (2017); and Carroll et al. (2018).
64. Zinsstag et al. (2011).
65. Horton and Lo (2015).
66. Brooks (2014); Brooks et al. (2014); Hoberg and Brooks (2015); Whitmee et al. (2015); Hoberg et al. (2015); Watts et al. (2017a, 2017b).
67. Dobson (2005); Gebreyes et al. (2014); Young et al. (2017).

68. This quotation seems to have appeared in print first in Calderón and Beltrán (2004), who wrote: "It has also been said, 'If you want something you're never had, you must be willing to do something you've never done.'" By 2012 the saying was being widely attributed to Thomas Jefferson, but this seems to be incorrect.

CHAPTER TEN

1. Recent perspectives providing a link between general threats posed by climate change and emerging diseases in particular include WHO (2015); Ottersen et al. (2016); MacIntyre et al. (2017); Rothman (2017); Scheffers et al. (2016); Beddington (2010); Goodell (2018); Prairie Climate Centre (2018).
2. Hull (1988).
3. "The thing the ecologically illiterate don't realize about an ecosystem is that it's a system. A system! A system maintains a certain fluid stability that can be destroyed by a misstep in just one niche. A system has order, a flowing from point to point. If something dams the flow, order collapses. The untrained miss the collapse until too late. That's why the highest function of ecology is the understanding of consequences" (Herbert, 1965).
4. Beginning as early as Brooks and McLennan (2002).
5. Meadows et al. (1972).
6. Henry David Thoreau, journal entry for September 7, 1851 (see Thoreau, 2009).
7. Mark et al. (2010).
8. Neiman (2002).
9. Trenberth et al. (2007); Kaufman et al. (2009); Denton et al. (2010); Barker et al. (2011); IPCC (2013); Stocker et al. (2013); Ganopolski et al. (2016); Hansen et al. (2016); Henry et al. (2016); Oppenheimer and Alley (2016); Stott (2016); Figueres et al. (2017); Hand (2017); Hofer et al. (2017); Hoffman et al. (2017); Kintisch (2017); Pecl et al. (2017); Steffen et al. (2018). Milanković carved out a comfort zone for humanity a century ago. Summer insolation, the all-important value that determines stability and warming, should gradually increase for the next 25,000 years, possibly offset to some degree by orbital processes. Climate would continue in relative stability for an extended period; the next glacial cycle was not anticipated for another 50,000 years. Scientists did not include potential accelerating climate warming in those initial calculations. Yet, human activities (anthropogenic forcing) catalyzed accelerating increase in global atmospheric, oceanic, and terrestrial surface temperatures during the past 250 years. Annual to decadal oscillations, primarily in ocean-atmosphere connections or extreme volcanism, may amplify or dampen long-term trends in global warming, but they will not alter the

trajectory. Temperature and precipitation extremes will occur but will be difficult to predict. Accelerating change has been especially evident at high latitudes of the Northern Hemisphere since the 1970s, where CO_2 and greenhouse gas emissions have reversed the cooling trend of the past 2000 years. Most sea level variation during the Pleistocene has been linked to ice dynamics, and to the expansion and termination of continental glaciers of the Northern Hemisphere, with less contribution from processes in Antarctica. The Pleistocene was a preamble to what will come and in its final stages reveals just how quickly climate can abruptly shift across extremes. Comparisons with the last interglacial optimum about 114,000–131,000 years ago highlight the climate-driven changes on our horizon. Levels of global atmospheric CO_2 we see now (400 parts per million) strongly contrast with those of the last interglacial period (about 280 ppm). Carbon dioxide concentrations have been higher in the past and have always been associated with a warmer world (e.g., the middle Pliocene). And while the absolute values (400 vs. 280 ppm) may seem trivial to nonscientists, the difference is enough to cause the warming trend that we are now experiencing, one that was not predicted in models created a century ago. The level of 400 ppm, only recently attained for the first time in modern human existence, reflects the accumulated change of the past 250 years and most particularly during the last several decades; furthermore, those levels are expected to rise rapidly during the next half century unless they are controlled and limited, a possibility that now appears out of reach. Rising temperatures associated with elevated atmospheric CO_2 are melting major portions of the planet that have been ice-covered or permafrost-entombed since the end of the last Ice Age. And all the water released by that melting must go somewhere. The measurements of atmospheric carbon critical for projecting our future come from an especially remote climate station perched on the appropriately named Cape Grim, at the edge of the Tasman Sea. Recently revised estimates for sea level increases during the coming century predict a two-meter rise, which would pose a serious challenge to coastal land use and occupation. These current projections as extremely conservative, based on assumptions of steady-state loss of ice volume over time; they do not account for tipping points or thresholds, amplifying feedbacks, changing ocean circulation, or CO_2 inertia leading to accelerated melting, especially in Antarctica. Sea surface temperatures during the previous interglacial period were similar to those we see today, although the sea level was six to nine meters higher. This suggests that there may be lag times between maximum atmospheric and sea surface temperatures and rising sea levels; even more troubling, it suggests the possibility, also predicted in *Limits to Growth*, that sea level changes could be abrupt. Unanticipated cascades in the physical environment may be exemplified by the Greenland ice cap, which is experiencing massive and accelerating disintegration. The equivalent of seven meters of sea level rise remains locked in the Greenland ice cap, and perhaps at short centennial scales another three meters or more in Antarctica (complete loss of the Antarctic ice sheet is equivalent to over 60 meters). Current summer conditions in Greenland are being described as completely unpredicted and may indicate an incipient tipping point and a state change with respect to the rate of melting. The current situation leading to predictions of a two-meter rise in sea levels in the coming century may be the calm before the real—and imminent—storm. Not only do more than 50 percent of humans now live in cities, but an increasing number of them live within 150 kilometers of the ocean. Aside from changes in land-bound ice, increasing ocean volume is directly influenced by sea surface temperatures, atmospheric pressure, prevailing winds, and extreme weather events. There is no global infrastructure for preparing effectively for this eventuality, and the effects are already being felt. In the next two generations habitats where humans live and grow food will increasingly be underwater, drought-stricken, and eroding into the sea in a cascade of extreme precipitation fluctuations. In all cases, there will be increased risk of emerging disease. The scale of human displacement will be as Elton warned us in 1958.

10. Patz et al. (2005); Jones et al. (2008); Steffen et al. (2011); Barnosky et al. (2012); Ruddiman (2013); Capinha et al. (2015); Whitmee et al. (2015); Scheffers et al. (2016); Waters et al. (2016); Pecl et al. (2017).

11. Barnosky et al. (2012); Ibisch et al. (2016); Bell and Tylianakis (2016).

12. Jones et al. (2008); Barnosky et al. (2017); Vellend et al. (2017).

13. Soranno et al. (2014).

14. Poe (1842). Many ancient Egyptians prominent enough to have been mummified were infected with schistosomes (Miller et al., 1992, 1994; Kloos and David, 2002; Barakat, 2013; Othman and Soliman, 2015).

15. See Vasbinder (2019).

16. Herbert (1965); for an excellent overview of the ecology of fear in which prey items exist, see Brown et al. (1999); Laundré et al. (2014); Bleicher (2017).

17. Brooks and Ferrao (2005).

18. For an extended discussion of the benefits of hope in this context, see Buzzell (2012).

19. Attributed to Martin Luther.

20. Maynard Smith and Szathmáry (1995); Szathmáry (2015).

21. Darwin (1872).

22. Soranno et al. (2014).

23. Byerly (2013).

24. Attributed to Mark Twain.

Bibliography

Achtman, Mark, Kerstin Zurth, Giovanna Morelli, Gabriela Tor-
rea, Annie Guiyoule, and Elisabeth Carniel. 1999. "*Yersinia
pestis*, the Cause of Plague, Is a Recently Emerged Clone
of *Yersinia pseudotuberculosis.*" *Proceedings of the National
Academy of Sciences of the United States of America* 96, no. 24:
14043–48.

Adamson, M. L., and J. N. Caira. 2004. "Evolutionary Factors
Influencing the Nature of Parasite Specificity." In "Parasites
and Behaviour," edited by M. V. K. Sukhdeo. Supplement,
Parasitology 109, no. S1: S85–95. https://doi.org/10.1017/
S0031182000085103.

Adel, Amany, Abdelsatar Arafa, Hussein A. Hussein, and Ahmed
A. El-Sanousi. 2017. "Molecular and Antigenic Traits on
Hemagglutinin Gene of Avian Influenza H9N2 Viruses: Evi-
dence of a New Escape Mutant in Egypt Adapted in Quails."
Research in Veterinary Science 112: 132–40. https://doi.org/10
.1016/j.rvsc.2017.02.003.

Adelman, Carol, Jeremiah Norris, Yulya Spantchak, and Kacie
Marano. 2012. *Social and Economic Impact Review on Ne-
glected Tropical Diseases*. Economic Policy/Briefing Paper.
Washington, DC: Hudson Institute.

Adenowo, Abiola Fatimah, Babatunji Emmanuel Oyinloye,
Bolajoko Idiat Ogunyinka, and Abidemi Paul Kappo. 2015.
"Impact of Human Schistosomiasis in Sub-Saharan Africa."
Brazilian Journal of Infectious Diseases 19, no. 2: 196–205.
https://doi.org/10.1016/j.bjid.2014.11.004.

Agbor, Julius, Olumide Taiwo, and Jessica Smith. 2012. "Sub-
Saharan Africa's Youth Bulge: A Demographic Dividend
or Disaster?" In *Foresight Africa: Top Priorities for the Continent in 2012*, 9–11. Brookings Institution, Africa Growth
Initiative.

Agnarsson, Ingi, Brian B. Jencik, Giselle M. Veve, Sahondra Hanitriniaina, Diego Agostini, Seok Ping Goh, Jonathan Pruitt, and Matjaž Kuntner. 2015. "Systematics of the Madagascar Anelosimus Spiders: Remarkable Local Richness and Endemism, and Dual Colonization From the Americas." *ZooKeys* 509 (June): 13–52. https://doi.org/10.3897/zookeys.509.8897.

Agosta, Salvatore J. 2006. "On Ecological Fitting, Plant-Insect Associations, Herbivore Host Shifts, and Host Plant Selection." *Oikos* 114: 556–65.

Agosta, Salvatore J., and Jeffrey A. Klemens. 2008. "Ecological Fitting by Phenotypically Flexible Genotypes: Implications for Species Associations, Community Assembly and Evolution." *Ecology Letters* 11: 1123–34.

Agosta, Salvatore J., Niklas Janz, and Daniel R. Brooks. 2010. "How Generalists Can Be Specialists: Resolving The 'Parasite Paradox' and Implications for Emerging Disease." *Zoologia* 27: 151–62.

Agrawal, Anurag A., and Marjorie G. Weber. 2015. "On the Study of Plant Defence and Herbivory Using Comparative Approaches: How Important Are Secondary Plant Compounds." *Ecology Letters* 18, no. 10: 985–91. https://doi.org/10.1111/ele.12482.

Aguilar-Aguilar, Rogelio, Andrés Martínez-Aquino, Héctor Espinosa-Pérez, and Gerardo Pérez-Ponce de León. 2014. "Helminth Parasites of Freshwater Fishes from Cuatro Ciénegas, Coahuila, in the Chihuahuan Desert of Mexico: Inventory and Biogeographical Implications." *Integrative Zoology* 9: 328–39. https://doi.org/10.1111/1749-4877.12038.

Aguirre-Liguori, Jonás A., Erika Aguirre-Planter, and Luis E. Eguiarte. 2016. "Genetics and Ecology of Wild and Cultivated Maize: Domestication and Introgression." In *Ethnobotany of Mexico*, 403–16. New York: Springer. http://doi.org/10.1007/978-1-4614-6669-7_16.

Alarcos, Ana, Verónica Ivanov, and Norma Sardella. 2006. "Distribution Patterns and Interactions of Cestodes in the Spiral Intestine of the Narrownose Smooth-Hound Shark, *Mustelus schmitti* Springer, 1939 (Chondrichthyes, Carcharhiniformes)." *Acta Parasitologica* 51, no. 2: 100–106. https://doi.org/10.2478/s11686-006-0015-7.

Ali, Sajid, Pierre Gladieux, Marc Leconte, Angélique Gautier, Annemarie F. Justesen, Mogens S. Hovmøller, Jérôme Enjalbert, and Claude de Vallavieille-Pope. 2014. "Origin, Migration Routes and Worldwide Population Genetic Structure of the Wheat Yellow Rust Pathogen *Puccinia striiformis* f.sp. *tritici*." *PLoS Pathogens* 10, no. 1: e1003903. https://doi.org/10.1371/journal.ppat.1003903.

Almathen, Faisal, Pauline Charruau, Elmira Mohandesan, Joram M. Mwacharo, Pablo Orozco-terWengel, Daniel Pitt, Abdussamad M. Abdussamad, et al. 2016. "Ancient and Modern DNA Reveal Dynamics of Domestication and Cross-Continental Dispersal of the Dromedary." *Proceedings of the National Academy of Sciences of the United States of America* 113, no. 24: 6707–12.

Altizer, Sonia, Drew Harvell, and Elizabeth Friedle. 2003. "Rapid Evolutionary Dynamics and Disease Threats to Biodiversity." *Trends in Ecology & Evolution* 18, no. 11: 589–96. https://doi.org/10.1016/j.tree.2003.08.013.

Altizer, Sonia, Richard S. Ostfeld, Pieter T. J. Johnson, Susan Kutz, and C. Drew Harvell. 2013. "Climate Change and Infectious Diseases: From Evidence to a Predictive Framework." *Science* 341, no. 6145: 514–19. https://doi.org/10.1126/science.1239401.

Aly, Mahmoud, Mohamed Elrobh, Maha Alzayer, Sameera Aljuhani, and Hanan Balkhy. 2017. "Occurrence of the Middle East Respiratory Syndrome Coronavirus (MERS-CoV) across the Gulf Corporation Council Countries: Four Years Update." *PLoS ONE* 12, no. 10: e0183850. https://doi.org/10.1371/journal.pone.0183850.

Amel, Elise, Christie Manning, Britain Scott, and Susan Koger. 2017. "Beyond the Roots of Human Inaction: Fostering Collective Effort toward Ecosystem Conservation." *Science* 356, no. 6335: 275–79. https://doi.org/10.1126/science.aal1931.

American Veterinary Medical Association. 2008. "One Health Initiative Task Force—One Health: A New Professional Imperative." http://www.avma.org/kb/resources/reports/documents.onehealth_final.pdf.

Anderson, Pamela K., Andrew A. Cunningham, Nikkita G. Patel, Francisco J. Morales, Paul R. Epstein, and Peter Daszak. 2004. "Emerging Infectious Diseases of Plants: Pathogen Pollution, Climate Change and Agrotechnology Drivers." *Trends in Ecology & Evolution* 19, no. 10: 535–44. https://doi.org/10.1016/j.tree.2004.07.021.

Anderson, Roy C. 1957. "The Life Cycles of Dipetalonematid Nematodes (Filarioidea, Dipetalonematidae): The Problem of Their Evolution." *Journal of Helminthology* 31, no. 4: 203–24. https://doi.org/10.1017/s0022149x00004454.

Anderson, Roy C. 1958. "Possible Steps in the Evolution of Filarial Life Cycles." *Proceedings of the Sixth International Congress of Tropical Medicine and Malaria* 2: 444–49.

Anderson, Roy C. 1982. "Host Parasite Relations and the Evolution of the Metastrongyloidea (Nematoda)." *Mémoires du Muséum national histoire naturelle*, n.s., sér. A, *Zoologie* 123: 129–33.

Anderson, Roy C. 1984. "The Origins of Zooparasitic Nematodes." *Canadian Journal of Zoology* 62: 317–28.

Anderson, Roy M. 2016. "The Impact of Vaccination on the Epidemiology of Infectious Diseases." In *The Vaccine Book*, edited by Barry R. Bloom and Paul-Henri Lambert, 3–31. N.p.: Elsevier. https://doi.org/10.1016/B978-0-12-802174-3.00001-1.

Anderson, Roy M., and Robert M. May. 1982. "Coevolution of Hosts and Parasites." *Parasitology* 85: 411–26.

Andrades Valtueña, Aida, Alissa Mittnik, Felix M. Key, Wolfgang Haak, Raili Allmäe, Andrej Belinskij, Mantas Daubaras, et al. 2016. "The Stone Age

Plague: 1000 Years of Persistence in Eurasia." *bioRxiv*: 094243. https://doi
.org/10.1101/094243.

Andrades Valtueña, Aida, Alissa Mittnik, Felix M. Key, Wolfgang Haak, Raili
Allmäe, Andrej Belinskij, Mantas Daubaras, et al. 2017. "The Stone Age
Plague and Its Persistence in Eurasia." *Current Biology* 27, no. 23: 3683–88.
https://doi.org/10.1016/j.cub.2017.10.025.

Anthony, Simon J., Jonathan H. Epstein, Kris A. Murray, Isamara Navarrete-
Macias, Carlos M. Zambrana-Torrelio, Alexander Solovyov, Rafael Ojeda-
Flores, et al. 2013. "A Strategy to Estimate Unknown Viral Diversity in
Mammals." *mBio* 4, no. 5: e00598-13. https://doi.org/10.1128/mBio.00598
-13.

Araujo, Sabrina B. L., Mariana P. Braga, Daniel R. Brooks, Salvatore J. Agosta,
Eric P. Hoberg, Francisco W. von Hartenthal, and Walter A. Boeger. 2015.
"Understanding Host-Switching by Ecological Fitting." *PLoS ONE* 10,
no. 10: e0139225. https://doi.org/10.1371/journal.pone.0139225.

Arias, Mónica, Aimilia Meichanetzoglou, Marianne Elias, Neil Rosser, Donna L.
de-Silva, Bastien Nay, and Violaine Llaurens. 2016. "Variation in Cyano-
genic Compounds Concentration within a Heliconius Butterfly Community: Does Mimicry Explain Everything?" *BMC Evolutionary Biology* 16,
no. 1: 272. https://doi.org/10.1186/s12862-016-0843-5.

Arimura, Gen-ichiro, Rika Ozawa, Takeshi Shimoda, Takaaki Nishioka, Wilhelm Boland, and Junji Takabayashi. 2000. "Herbivory-Induced Volatiles
Elicit Defence Genes in Lima Bean Leaves." *Nature* 406, no. 6795: 512–15.

Armbruster, W. S. 1996. "Exaptation, Adaptation and Homoplasy: Evolution of
Ecological Traits in *Dalechampia* Vines." In *Homoplasy: The Recurrence of
Similarity in Evolution*, edited by Michael J. Sanderson and Larry Hufford,
227–43. New York: Academic Press.

Armbruster, W. Scott. 1997. "Exaptations Link Evolution of Plant-Herbivore and
Plant-Pollinator Interactions: A Phylogenetic Inquiry." *Ecology* 78: 1661–72.

Arnold, Edward N. 1994. "Investigating the Origins of Performance Advantage:
Adaptation, Exaptation and Lineage Effects." In *Phylogenetics and Ecology*,
edited by Paul Eggleton and Richard I. Vane-Wright, 123–68. London:
Academic Press.

Arrhenius, Svante. 1896. "XXXI. On the Influence of Carbonic Acid in the
Air upon the Temperature of the Ground." *London, Edinburgh and Dublin
Philosophical Journal and Magazine of Science* 41: 237–76.

Attwell, C. A. M., and F. P. D. Cotterill. 2000. "Postmodernism and African
Conservation Science." *Biodiversity and Conservation* 9, no. 5: 559–77.
https://doi.org/10.1023/A:1008972612012.

Audy, J. Ralph. 1958. "The Localization of Disease with Special Reference to the
Zoonoses." *Transactions of the Royal Society of Tropical Medicine and Hygiene*
52: 309–28.

Bacon, Christine D., Daniele Silvestro, Carlos Jaramillo, Brian T. Smith,
Prosanta Chakrabarty, and Alexandre Antonelli. 2015. "Biological

Evidence Supports an Early and Complex Emergence of the Isthmus of Panama." *Proceedings of the National Academy of Sciences of the United States of America* 112, no. 19: 6110–15. https://doi.org/10.1073/pnas.1423853112.

Badjeck, Marie-Caroline, Edward H. Allison, Ashley S. Halls, and Nicholas K. Dulvy. 2010. "Impacts of Climate Variability and Change on Fishery-Based Livelihoods." *Marine Policy* 34, no. 3: 375–83. https://doi.org/10.1016/j.marpol.2009.08.007.

Badrane, Hassan, and Noël Tordo. 2001. "Host Switching in *Lyssavirus* History from the Chiroptera to the Carnivora Orders." *Journal of Virology* 75, no. 17: 8096–8104. https://doi.org/10.1128/JVI.75.17.8096–8104.2001.

Baer, J. G. 1948. "Les helminthes, parasites des vertebres: Relations phylogénetique entre leur évolution et celle de leurs hôtes." *Annales scientifiques de l'Université de Franche-Comté Besançon* 2: 99–113.

Bajracharya, A. S. R., R. P. Mainali, B. Bhat, S. Bista, P. R. Shashank, and N. M. Meshram. 2016. "The First Record of South American Tomato Leaf Miner, *Tuta absoluta* (Meyrick, 1917) (Lepidoptera: Gelechiidae) in Nepal." *Journal of Entomology and Zoology Studies* 4, no. 4: 1359–63.

Baneth, G., S. M. Thamsborg, D. Otranto, J. Guillot, R. Blaga, P. Deplazes, and L. Solano-Gallego. 2016. "Major Parasitic Zoonoses Associated with Dogs and Cats in Europe." *Journal of Comparative Pathology* 155, suppl. 1: S54–74. https://doi.org/10.1016/j.jcpa.2015.10.179.

Banks, J. C., R. L. Palma, and A. M. Paterson. 2006. "Cophylogenetic Relationships between Penguins and Their Chewing Lice." *Journal of Evolutionary Biology* 19, no. 1: 156–66. https://doi.org/10.1111/j.1420–9101.2005.00983.x.

Barakat, Rashida M. R. 2013. "Epidemiology of Schistosomiasis in Egypt: Travel through Time: Review." *Journal of Advanced Research* 4, no. 5: 425–32. https://doi.org/10.1016/j.jare.2012.07.003.

Barker, Stephen, Gregor Knorr, R. Lawrence Edwards, Frédéric Parrenin, Aaron E. Putnam, Luke C. Skinner, Eric Wolff, and Martin Ziegler. 2011. "800,000 Years of Abrupt Climate Variability." *Science* 334, no. 6054: 347–51. https://doi.org/10.1126/science.1203580.

Barnett, Alan D. T. 2014. "Economic Burden of West Nile Virus in the United States." *American Journal of Tropical Medicine and Hygiene* 90: 389–90.

Barnosky, Anthony D., Elizabeth A. Hadly, Jordi Bascompte, Eric L. Berlow, James H. Brown, Mikael Fortelius, Wayne M. Getz, et al. 2012. "Approaching a State Shift in Earth's Biosphere." *Nature* 486: 52–58.

Barnosky, Anthony D., Elizabeth A. Hadly, Patrick Gonzalez, Jason Head, P. David Polly, A. Michelle Lawing, Jussi T. Eronen, et al. 2017. "Merging Paleobiology with Conservation Biology to Guide the Future of Terrestrial Ecosystems." *Science* 355, no. 6325: eaah4787.

Bartholomew, Jerri L., and Paul W. Reno. 2002. "The History and Dissemination of Whirling Disease." *American Fisheries Society* 29: 3–24.

Bascompte, Jordi, and Pedro Jordano. 2013. *Mutualistic Networks*. Princeton, NJ: Princeton University Press.

Bascompte, Jordi, Pedro Jordano, Carlos J. Melián, and Jens M. Olesen. 2003. "The Nested Assembly of Plant-Animal Mutualistic Networks." *Proceedings of the National Academy of Sciences of the United States of America* 100, no. 16: 9383–87.

Baum, Rudy M. 2016. "Future Calculations." In *Distillations* (Summer). Science History Institute. https://www.sciencehistory.org/distillations/magazine/future-calculations.

Beard, C. B., R. J. Eisen, C. M. Barker, J. F. Garofalo, M. Hahn, M. Hayden, A. J. Monaghan, et al. 2016. "Vector-Borne Diseases." In *The Impacts of Climate Change on Human Health in the United States; A Scientific Assessment*, edited by A. Crimmins, J. Balbus, J. L. Gamble, C. B. Beard, J. E. Bell, D. Dodgen, R. J. Eisen, et al., 129–56. Washington, DC: U.S. Global Change Research Program. https://dx.doi.org/10.1111/pan.12888.

Becerra, Judith X. 1997. "Insects on Plants: Macroevolutionary Chemical Trends in Host Use." *Science* 276: 253–56.

Beddington, John. 2010. *2030: The "Perfect Storm" Scenario*. Washington, DC: Population Institute. https://www.populationinstitute.org/external/files/reports/The_Perfect_Storm_Scen ario_for_2030.pdf.

Beenken, Ludwig, Matthias Lutz, and Markus Scholler. 2017. "DNA Barcoding and Phylogenetic Analyses of the Genus *Coleosporium* (Pucciniales) Reveal That the North American Goldenrod Rust *C. solidaginis* Is a Neomycete on Introduced and Native *Solidago* Species in Europe." *Mycological Progress* 16, no. 11–12: 1073–85. https://doi.org/10.1007/s11557-017-1357-2.

Bell, Kayce C., Kendall L. Calhoun, Eric P. Hoberg, John R. Demboski, and Joseph A. Cook. 2016. "Temporal and Spatial Mosaics: Deep Host Association and Shallow Geographic Drivers Shape Genetic Structure in a Widespread Pinworm, *Rauschtineria eutamii*." *Biological Journal of the Linnean Society* 119, no. 2: 397–413. https://doi.org/10.1111/bij.12833.

Bell, Thomas, and Jason M. Tylianakis. 2016. "Microbes in the Anthropocene: Spillover of Agriculturally Selected Bacteria and Their Impact on Natural Ecosystems." *Proceedings of the Royal Society B: Biological Sciences* 283, no. 1844: 20160896. https://doi.org/10.1098/rspb.2016.0896.

Bellec, Laure, and Yves Desdevises. 2015. "Quand virus et hôtes évoluent ensemble: La fidélité est-elle la règle?" *Virologie* 19: 140–48. https://doi.org/10.1684/vir.2015.0612

Ben Ari, Tamara, Alexander Gershunov, Kenneth L. Gage, Tord Snäll, Paul Ettestad, Kyrre L. Kausrud, and Nils Chr. Stenseth. 2008. "Human Plague in the USA: The Importance of Regional and Local Climate." *Biology Letters* 4, no. 6: 737–40. https://doi.org/10.1098/rsbl.2008.0363.

Bentz, Barbara J., Jacques Régnière, Christopher J. Fettig, E. Matthew Hansen, Jane L. Hayes, Jeffrey A. Hicke, Rick G. Kelsey, Jose F. Negrón, and Steven J. Seybold. 2010. "Climate Change and Bark Beetles of the Western United States and Canada: Direct and Indirect Effects." *BioScience* 60, no. 7: 602–13. https://doi.org/10.1525/bio.2010.60.8.6.

Berenbaum, May R., and Steven Passoa. 1999. "Generic Phylogeny of North American Depressariinae (Lepidoptera: Elachistidae) and Hypotheses about Coevolution." *Annals of the Entomological Society of America* 92, no. 6: 971–86.

Berg, Lev S. 1922. *Nomogenez*. Petrograd: GIZ [in Russian].

Berg, Lev S. 1926. *Nomogenesis*. Cambridge, MA: MIT Press.

Bergersen, Eric P., and Dennis E. Anderson. 1997. "The Distribution and Spread of *Myxobolus cerebralis* in the United States." *Fisheries* 22: 6–7.

Berthe, Franck Cesar Jean, Timothy Bouley, William B. Karesh, Issa Chabwera Legall, Catherine Christina Machalaba, Caroline Aurelie Plante, and Richard M. Seifman. 2018. biblioWashington, DC: World Bank Group. http://documents.worldbank.org/curated/en/961101524657708673/One -health- operational-framework-for-strengthening-human-animal-and -environmental-public- health-systems-at-their-interface.

Beurlen, Karl. 1930. *Vergleichende Stammesgeschichte*. Berlin: Borntraeger.

Beurlen, Karl. 1937. *Die Stammungsgeschichtlichen Grundlagen der Abstammungs- lehre*. Jena: Gustav Fischer Verlag.

Bhattacharya, Shaoni. 2017. "Wheat Rust Back in Europe." *Nature* 542, no. 7640: 145–46.

Blaizot, Romain, Marie-Catherine Receveur, Pascal Millet, Domenico Otranto, and Denis J. M. Malvy. 2018. "Systemic Infection with *Dirofilaria repens* in Southwestern France." *Annals of Internal Medicine* 168, no. 3: 228–29. https://doi.org/10.7326/L17–0426.

Blakeslee, April M., Amy E. Fowler, and Carolyn L. Keogh. 2013. "Marine Inva- sions and Parasite Escape: Updates and New Perspectives." *Advances in Marine Biology* 66: 87–169. https://doi.org/10.1016/B978-0-12-408096-6.00002-X.

Blankespoor, Harvey D. 1974. "Host-Induced Variation in *Plagiorchis noblei* Park, 1936 (Plagiorchiidae; Trematoda)." *American Midland Naturalist* 91: 415–33.

Bleicher, Sonny S. 2017. "The Landscape of Fear Conceptual Framework: Defi- nition and Review of Current Applications and Misuses." *PeerJ* 5: e3772. https://doi.org/10.7717/peerj.3772.

Bloom, David E., Steven Black, and Rino Rappuoli. 2017. "Emerging Infectious Diseases: A Proactive Approach." *Proceedings of the National Academy of Sci- ence of the United States of America* 114, no. 16: 4055–59. https://doi.org/10 .1073/pnas.1701410114.

Bobeva, A., P. Zehtindjiev, M. Ilieva, D. Dimitrov, A. Mathis, and S. Bensch. 2015. "Host Preferences of Ornithophilic Biting Midges of the Genus *Culi- coides* in the Eastern Balkans." *Medical and Veterinary Entomology* 29, no. 3: 290–96. https://doi.org/10.1111/mve.12108.

Bochkov, Andre V., Pavel B. Klimov, and Georges Wauthy. 2011. "Phylogeny and Coevolutionary Associations of Makialgine mites (Acari, Psoroptidae, Ma- kialginae) Provide Insight into Evolutionary History of Their Hosts, Strep- sirrhine Primates." *Zoological Journal of the Linnean Society* 162, no. 1: 1–4.

Boeger, Walter A., Flávio M. Marteleto, Letícia Zagonel, and Mariana P. Braga. 2015. "Tracking the History of an Invasion: The Freshwater Croakers (Teleostei: Sciaenidae) in South America." *Zoologica Scripta* 44, no. 3: 250–62. https://doi.org/10.1111/zsc.12098.

Bohoussou, Kouakou Hilaire, Raphael Cornette, Bertin Akpatou, Marc Colyn, Julian Kerbis Peterhans, Jan Kennis, et al. 2015. "The Phylogeography of the Rodent Genus *Malacomys* Suggests Multiple Afrotropical Pleistocene Lowland Forest Refugia." *Journal of Biogeography* 42, no. 11: 2049–61. https://doi.org/10.1111/jbi.12570.

Bos, Kirsten I., Kelly M. Harkins, Alexander Herbig, Mireia Coscolla, Nico Weber, Iñaki Comas, Stephen A. Forrest, et al. 2014. "Pre-Columbian Mycobacterial Genomes Reveal Seals as a Source of New World Human Tuberculosis." *Nature* 514, no. 7523: 494–97. https://doi.org/10.1038/nature13591.

Bos, Kirsten I., Alexander Herbig, Jason Sahl, Nicholas Waglechner, Mathieu Fourment, Stephen A. Forrest, Jennifer Klunk, et al. 2016. "Eighteenth Century *Yersinia pestis* Genomes Reveal the Long-Term Persistence of an Historical Plague Focus." *eLife* 5: 17837. https://doi.org/10.7554/eLife.12994.

Bouchard, Patrice, Daniel R. Brooks, and David K. Yeates. 2004. "Mosaic Macroevolution in Australian Wet Tropics Arthropods: Community Assemblage by Taxon Pulses." In *Rainforest: Past, Present, Future*, edited by Craig Moritz and Eldredge Bermingham, 425–69. Chicago: University of Chicago Press.

Boucot, Arthur J. 1982. *Paleobiologic Evidence of Behavioral Evolution and Coevolution*. Corvallis, OR: The author.

Boucot, Arthur J. 1983. "Does Evolution Take Place in an Ecological Vacuum? II." *Journal of Paleontology* 57: 1–30.

Boucot, Arthur J. 1990. "Community Evolution: Its Evolutionary and Biostratigraphic Significance." In *Paleocommunity Temporal Dynamics: The Long-Term Development of Multi-Species Assemblies*, edited by William Miller III, 48–70. Paleontological Society Special Publication 5. N.p.: Department of Geological Sciences, University of Tennessee.

Bower, Deborah S., Karen R. Lips, Lin Schwarzkopf, Arthur Georges, and Simon Clulow. 2017. "Amphibians on the Brink." *Science* 357, no. 6350: 454–55. https://doi.org/10.1126/science.aao0500.

Bowler, P. 1983. *The Eclipse of Darwinism*. Baltimore, MD: Johns Hopkins Press.

Bowman, David M. J. S., Stephen T. Garnett, Snow Barlow, Sarah A. Bekessy, Sean M. Bellairs, Melanie J. Bishop, Ross A. Bradstock, et al. 2017. "Renewal Ecology: Conservation for the Anthropocene." *Restoration Ecology* 25, no. 5: 674–80. https://doi.org/10.1111/rec.12560.

Braga, Mariana P., Sabrina B. L. Araujo, Salvatore Agosta, Daniel Brooks, Eric Hoberg, Sören Nylin, Niklas Janz, and Walter A. Boeger. 2018. "Host Use

Dynamics in a Heterogeneous Fitness Landscape Generates Oscillations in Host Range and Diversification." *Evolution* 72, no. 9: 1773–83. https://doi .org/10.1111/evo.13557.

Braga, Mariana P., Emanuel Razzolini, and Walter A.Boeger. 2015. "Drivers of Parasite Sharing among Neotropical Freshwater Fishes." *Journal of Animal Ecology* 84, no. 2: 487–97. https://doi.org/10.1111/1365-2656.12298/.

Braidwood, Luke, Diego F. Quito-Avila, Darlene Cabanas, Alberto Bressan, Anne Wangai, and David C. Baulcombe. 2018. "Maize Chlorotic Mottle Virus Exhibits Low Divergence between Differentiated Regional Sub-Populations." *Scientific Reports* 8, no. 1: 1173. https://doi.org/10.1038/ s41598-018-19607-4.

Brame, Hannah-Maria R., and Alycia L. Stigall. 2014. "Controls on Niche Stability in Geologic Time: Congruent Responses to Biotic and Abiotic Environmental Changes among Cincinnatian (Late Ordovician) Marine Invertebrates." *Paleobiology* 40, no. 1: 70–90.

Brander, Keith. 2010. "Impacts of Climate Change on Fisheries." *Journal of Marine Systems* 79: 389–402.

Brierley, Liam, Maarten J. Vonhof, Kevin J. Olival, Peter Daszak, and Kate E. Jones. 2016. "Quantifying Global Drivers of Zoonotic Bat Viruses: A Process-Based Perspective." *American Naturalist* 187, no. 2: E53–64. https:// doi.org/10.1086/684391.

Briggs, Helen. 2017. "Gardens under Threat From 'Game Changing' Plant Disease." December 29. BBC News. http://www.bbc.com/news/science -environment-42504931.

Brinkerhoff, Robert Jory, Sharon K. Collinge, Chris Ray, and Ken L. Gage. 2010. "Rodent and Flea Abundance Fail to Predict a Plague Epizootic in Black-Tailed Prairie Dogs." *Vector-Borne and Zoonotic Diseases* 10, no. 1: 47–52. https://doi.org/10.1089/vbz.2009.0044.

Brook, Barry W., and Damien A. Fordham. 2015. "Hot Topics in Biodiversity and Climate Change Research." *F1000Research* 4 (September): 928. https:// doi.org/10.12688/f1000research.6508.1.

Brookes, Victoria, Marta Hernández-Jover, Peter Black, and M. P. Ward. 2015. "Preparedness for Emerging Infectious Diseases: Pathways from Anticipa-tion to Action." *Epidemiology and Infection* 143, no. 10: 2043–58. https://doi .org/10.1017/S095026881400315X.

Brooks, Alison S., David M. Helgren, Jon S. Cramer, Alan Franklin, William Hornyak, Jody M. Keating, Richard G. Klein, et al. 1995. "Dating and Context of Three Middle Stone Age Sites with Bone Points in the Upper Semliki Valley, Zaire." *Science* 268, no. 5210: 548–53. https://doi.org/10 .1126/science.7725099.

Brooks, D. R., and C. Mitter. 1984. "Analytical Basis of Coevolution." In *Fungus-Insect Relationships: Perspectives in Ecology and Evolution*, edited by Quentin Wheeler and Marilyn Blackwell, 42–53. New York: Columbia University Press.

Brooks, Daniel R. 1979. "Testing the Context and Extent of Host-Parasite Coevolution." *Systematic Zoology* 28: 299–307.

Brooks, Daniel R. 1985. "Historical Ecology: A New Approach to Studying
the Evolution of Ecological Associations." *Annals of the Missouri Botanical
Garden* 72: 660–80.

Brooks, Daniel R. 1986. "Analysis of Host-Parasite Coevolution." In
*Parasitology—Quo Vadit? Proceedings of the Sixth International Congress of
Parasitology*, edited by Michael J. Howell, 291–300. Canberra: Australian
Academy of Science.

Brooks, Daniel R. 1992. "Origins, Diversification, and Historical Structure of
the Helminth Fauna Inhabiting Neotropical Freshwater Stingrays (Potamotrygonidae)." *Journal of Parasitology* 78: 588–95.

Brooks, Daniel R. 1998. "Triage for the Biosphere." In *The Brundtland Commission's Report—10 Years*, edited by Guri Bang Søfting, 71–80. Oslo: Scandinavian University Press.

Brooks, Daniel R. 2000. "Parasite Systematics in the 21st Century: Opportunities and Obstacles." *Memorias do Instituto Oswaldo Cruz* 95: 99–109.

Brooks, Daniel R. 2003. "Parasite Systematics in a New Age of Discovery."
Supplement, *Journal of Parasitology* 89: S72–77.

Brooks, Daniel R. 2010. "Sagas of the Children of Time: The Importance of
Phylogenetic Teaching in Biology." *Evolution: Education and Outreach* 3:
495–98.

Brooks, Daniel R. 2011. "The Mastodon in the Room: How Darwinian Is Neo-
Darwinism?" *Studies in the History and Philosophy of Science C* 42: 82–88.

Brooks, Daniel R. 2014. "Solving the Emerging Infectious Disease Crisis." *Comparative Parasitology* 81: 152–54.

Brooks, Daniel R., and Amanda L. Ferrao. 2005. "The Historical Biogeography
of Coevolution: Emerging Infectious Diseases are Evolutionary Accidents
Waiting to Happen." *Journal of Biogeography* 32: 1291–99.

Brooks, Daniel R., and Deborah A. McLennan. 1991. *Phylogeny, Ecology, and
Behavior: A Research Program in Comparative Biology.* Chicago: University of
Chicago Press.

Brooks, Daniel R., and Deborah A. McLennan. 1993. *Parascript: Parasites and the
Language of Evolution.* Washington, DC: Smithsonian Institution Press.

Brooks, Daniel R., and Deborah A. McLennan. 2001. "A Comparison of a
Discovery-Based and an Event-Based Method of Historical Biogeography."
Journal of Biogeography 28: 757–67.

Brooks, Daniel R., and Deborah A. McLennan. 2002. *The Nature of Diversity: An
Evolutionary Voyage of Discovery.* Chicago: University of Chicago Press.

Brooks, Daniel R., and Deborah A. McLennan. 2010. "The Biodiversity Crisis:
Lessons from Phylogenetic Sagas." *Evolution: Education and Outreach* 3:
558–62.

Brooks, Daniel R., and E. O. Wiley. 1988. *Evolution as Entropy: Toward a Unified
Theory of Biology.* 2nd ed. Chicago: University of Chicago Press.

Brooks, Daniel R., and Eric P. Hoberg. 2000. "Triage for the Biosphere: The Need and Rationale for Taxonomic Inventories and Phylogenetic Studies of Parasites." *Comparative Parasitology* 68: 1–25.

Brooks, Daniel R., and Eric P. Hoberg. 2001. "Parasite Systematics in the 21st Century: Opportunities and Obstacles." *Trends in Parasitology* 17: 273–75.

Brooks, Daniel R., and Eric P. Hoberg. 2006. "Systematics and Emerging Infectious Diseases: From Management to Solution." *Journal of Parasitology* 92: 426–29.

Brooks, Daniel R., and Eric P. Hoberg. 2007a. "Darwin's Necessary Misfit and the Sloshing Bucket: The Evolutionary Biology of Emerging Infectious Diseases." *Evolution: Education and Outreach* 1: 2–9.

Brooks, Daniel R., and Eric P. Hoberg. 2007b. "How Will Global Climate Change Affect Parasites?" *Trends in Parasitology* 23: 571–74.

Brooks, Daniel R., and Eric P. Hoberg. 2013. "The Emerging Infectious Disease Crisis and Pathogen Pollution: A Question of Ecology and Evolution." In *The Balance of Nature and Human Impact*, edited by Klaus Rohde, 215–29. Cambridge: Cambridge University Press.

Brooks, Daniel R., and Salvatore J. Agosta. 2012. "Children of Time: The Extended Synthesis and Major Metaphors of Evolution." *Zoologia* 29: 497–514.

Brooks, Daniel R., Deborah A. McLennan, Virginia León Régagnon, and Eric Hoberg. 2006a. "Phylogeny, Ecological Fitting and Lung Flukes: Helping Solve the Problem of Emerging Infectious Diseases." *Revista mexicana de biodiversidad* 77: 225–34.

Brooks, Daniel R., Eric P. Hoberg, and Walter A. Boeger. 2015. "In the Eye of the Cyclops: The Classic Case of Cospeciation and Why Paradigms Are Important." *Comparative Parasitology* 83: 1–8.

Brooks, Daniel R., Eric P. Hoberg, Walter A. Boeger, Scott L. Gardner, Kurt E. Galbreath, Dávid Herczeg, Hugo H. Mejía-Madrid, S. Elizabeth Rácz, and Altangerel Tsogtsaikhan Dursahinhan. 2014. "Finding Them before They Find Us: Informatics, Parasites and Environments in Accelerating Climate Change." *Comparative Parasitology* 81: 155–64.

Brooks, Daniel R., Jaret Bilewitch, Charmaine Condy, David C. Evans, Kaila E. Folinsbee, Jörg Fröbisch, Dominik Halas, et al. 2007. "Quantitative Phylogenetic Analysis in the 21st Century/Análisis filogenéticos cuantitativos en el siglo XXI." *Revista mexicana de biodiversidad* 78: 225–52.

Brooks, Daniel R., Richard L. Mayden, and Deborah A. McLennan. 1992a. "Phylogeny and Biodiversity: Conserving Our Evolutionary Legacy." *Trends in Ecolology and Evolution* 7, no. 2: 55–59.

Brooks, Daniel R., Richard L. Mayden, and Deborah A. McLennan. 1992b. "Letter to the Editor: Phylogeny and Conservation." *Trends in Ecolology and Evolution* 7: 352.

Brooks, Daniel R., Richard T. O'Grady, and David R. Glen. 1985a. "The Phylogeny of the Cercomeria Brooks, 1982 (Platyhelminthes)." *Proceedings of the Helminthological Society of Washington* 52: 1–20.

Brooks, Daniel R., Richard T. O'Grady, and David R. Glen. 1985b. "Phylogenetic Analysis of the Digenea (Platyhelminthes: Cercomeria) with Comments on Their Adaptive Radiation." *Canadian Journal of Zoology* 63: 411–43.

Brooks, Daniel R., Susan M. Bandoni, Cheryl A. Macdonald, and Richard T. O'Grady. 1989. "Aspects of the Phylogeny of the Trematoda Rudolphi, 1808 (Platyhelminthes: Cercomeria)." *Canadian Journal of Zoology* 67: 2609–24.

Brooks, Daniel R., Virginia León-Règagnon, Deborah A. McLennan, and Derek Zelmer. 2006b. "Ecological Fitting as a Determinant of the Community Structure of Platyhelminth Parasites of Anurans." *Ecology* 87 (suppl.): S76–85.

Brown, Joel S., John W. Laundré, and Mahesh Gurung. 1999. "The Ecology of Fear: Optimal Foraging, Game Theory, and Trophic Interactions." *Journal of Mammalogy* 80, no. 2: 385–99. https://doi.org/10.2307/1383287.

Brues, Charles T. 1920. "The selection of Food-Plants by Insects, with Special Reference to Lepidopterous Larvae." *American Naturalist* 54: 313–332.

Brues, Charles T. 1924. "The Specificity of Food-Plants in the Evolution of Phytophagous Insects." *American Naturalist* 58: 127–44.

Bruun, Brigitte, and Jens Aagaard-Hansen. 2008. *The Social Context of Schistosomiasis and Its Control.* Geneva: World Health Organization.

Bruyndonckx, Nadia, Sylvain Dubey, Manuel Ruedi, and Philippe Christe. 2009. "Molecular Cophylogenetic Relationships between European Bats and Their Ectoparasitic Mites (Acari, Spinturnicidae)." *Molecular Phylogenetics and Evolution* 51: 227–37. https://doi.org/10.1016/j.ympev.2009.02 .005.

Buchmann, K., and A. Uldal. 1997. "*Gyrodactylus derjavini* Infections in Four Salmonids: Comparative Host Susceptibility and Site Selection of Parasites." *Diseases of Aquatic Organisms* 28: 201–9. https://doi.org/10.3354/ dao028201.

Budke, Christine M., Peter Deplazes, and Paul R. Torgerson. 2006. "Global Socioeconomic Impact of Cystic Echinococcosis." *Emerging Infectious Diseases* 12, no. 2: 296–303. https://doi.org/10.3201/eid1202.050499.

Bulgarella, Mariana, and George E. Heimpel. 2015. "Host Range and Community Structure of Avian Nest Parasites in the Genus *Philornis* (Diptera: Muscidae) on the Island of Trinidad." *Ecology and Evolution* 5, no. 17: 3695–3703. https://doi.org/10.1002/ece3.1621.

Buse, Chris G., Jordan Sky Oestreicher, Neville R. Ellis, Rebecca Patrick, Ben Brisbois, Aaron P. Jenkins, Kaileah McKellar, et al. 2018. "Public Health Guide to Field Developments Linking Ecosystems, Environments and Health in the Anthropocene." *Journal of Epidemiology and Community Health*, January, jech–2017–210082. https://doi.org/10.1136/ jech- 2017–210082.

Buttner, Alice. 1951. "La progenèse chez les trématodes digénétiques (fin): Étude de quelque métacercaires à évolution inconnue et de certaines formes de développement voisines de la progénèse; Conclusions générales." *Annales de parasitologie humaine et comparée* 26: 279–322.

Buzzell, Linda. 2012. Review of *Active Hope: How to Face the Mess We're In Without Going Crazy* by Joanna Macy and Chris Johnstone. *Ecopsychology* 4, no. 3: 250–51. https://doi.org/10.1089/eco.2012.0048.

Byerly, T. Ryan. 2013. "The Special Value of Epistemic Self-Reliance." *Ratio* 27, no. 1: 53–67. https://doi.org/10.1111/rati.12016.

Cacho, N. Ivalú, Daniel J. Kliebenstein, and Sharon Y. Strauss. 2015. "Macroevolutionary Patterns of Glucosinolate Defense and Tests of Defense-Escalation and Resource Availability Hypotheses." *New Phytologist* 208, no. 3: 915–27. https://doi.org/10.1111/nph.13561.

Calderón, José L., and Robert Beltrán. 2004. "The Phoenix Has Risen but Has Failed to Thrive: Hope on the Horizon for King-Drew Medical Center." *Journal of the National Medical Association* 96: 160–62.

Callaghan, Terry V., Lars Olof Björn, Yuri Chernov, Terry Chapin, Torben R. Christensen, Brian Huntley, Rolf A. Ims, et al. 2004a. "Past Changes in Arctic Terrestrial ecosystems, Climate and UV Radiation." *Ambio* 33: 398–403.

Callaghan, Terry V., Lars Olof Björn, Yuri Chernov, Terry Chapin, Torben R. Christensen, Brian Huntley, Rolf A. Ims, et al. 2004b. "Biodiversity, Distributions and Adaptations of Arctic Species in the Context of Environmental Change." *Ambio* 33: 404–17.

Cameron, Thomas W. M. 1964. "Host Specificity and the Evolution of Helminthic Parasites." *Advances in Parasitology* 2: 1–34.

Campbell, William C. 2016. "Lessons from the History of Ivermectin and Other Antiparasitic Agents." *Annual Review of Animal Biosciences* 4, no. 1: 1–14. https://doi.org/10.1146/annurev-animal-021815-111209.

Canessa, Stefano, Claudio Bozzuto, Evan H. Campbell Grant, Sam S. Cruickshank, Matthew C. Fisher, Jacob C. Koella, Stefan Lötters, et al. 2018. "Decision-Making for Mitigating Wildlife Diseases: From Theory to Practice for an Emerging Fungal Pathogen of Amphibians." *Journal of Applied Ecology* 55, no. 4: 1–10. https://doi.org/10.111/1365-2664.13089.

Canhos, Dora A. L., Mariane S. Sousa-Baena, Sidnei de Souza, Leonor C. Maia, João R. Stehmann, Vanderlei P. Canhos, Renato De Giovanni, et al. 2015. "The Importance of Biodiversity E-Infrastructures for Megadiverse Countries." *PLoS Biology* 13, no. 7: e1002204. https://doi.org/10.1371/journal.pbio.1002204.

Capinha, César, Franz Essl, Hanno Seebens, Dietmar Moser, and Henrique M. Pereira. 2015. "The Dispersal of Alien Species Redefines Biogeography in the Anthropocene." *Science* 348: 1248–51.

Carlson, Colin J., Kevin R. Burgio, Eric R. Dougherty, Anna J. Phillips, Veronica M. Bueno, Christopher F. Clements, Giovanni Castaldo, et al. 2017. "Parasite Biodiversity Faces Extinction and Redistribution in a Changing Climate." *Science Advances* 3, no. 9: e1602422. https://doi.org/10.1126/sciadv.1602422.

Carney, Joseph P., and Daniel R. Brooks. 1991. "Phylogenetic Analysis of *Allo-glossidium* Simer, 1929 (Digenea: Plagiorchiformes: Macroderoididae) with Discussion of the Origin of Truncated Life Cycle Patterns in the Genus." *Journal of Parasitology* 77: 890–900.

Carroll, Dennis, Peter Daszak, Nathan D. Wolfe, George F. Gao, Carlos M. Morel, Subhash Morzaria, Ariel Pablos-Méndez, et al. 2018. "The Global Virome Project." *Science* 359, no. 6378: 872–74. https://doi.org/10.1126/science.aap7463.

Carson, Rachel. 1962. *Silent Spring*. New York: Houghton Mifflin Harcourt.

Carvalho, José C., Pedro Cardoso, François Rigal, Kostas A. Triantis, and Pablo A. V. Borges. 2015. "Modeling Directional Spatio-Temporal Processes in Island Biogeography." *Ecology and Evolution* 5, no. 20: 4671–82. https://doi.org/10.1002/ece3.1632.

Cazelles, Kévin, Miguel B. Araújo, Nicolas Mouquet, and Dominique Gravel. 2016. "A Theory for Species Co-Occurrence in Interaction Networks." *Theoretical Ecology* 9, no. 1: 39–48.

Centers for Disease Control. 2017. *Neglected Tropical Diseases: Global Health—Division of Parasitic Disease and Malaria*. https://www.cdc.gov/globalhealth/ntd.

Centers for Disease Control. 2018. *Cost of the Ebola Epidemic*. https://www.cdc.gov/vhf/ebola/outbreaks/2014-west-africa/cost-of-ebola.html.

"A Century of Evolution." 1959. Editorial, *New England Journal of Medicine* 260 (January 29): 244–45. https://doi.org/10.1056/NEJM195901292600511.

Chabaud, Alain G. 1954. "Sur le cycle évolutif des Spirurides et de Nématodes ayant une biologie comparable: Valeur systématique des caractères biologiques." *Annales de parasitologie humaine et comparée* 29: 42–88.

Chabaud, Alain G. 1955. "Essai d'interprétation phylétique des cycles évolutifs chez les Nématodes parasites de Vertèbres: Conclusions taxonomiques." *Annales de parasitologie humaine et comparée* 30: 83–126.

Chabaud, Alain G. 1965. "Cycles evolutifs des Nematodes parasites de Vertebres." In *Traité de zoologie*, edited by P. P. Grasse, tome 4, fasc. 2: 437–63. Paris: Masson & Cie.

Chabaud, Alain G. 1982. "Évolution et taxonomie des Nématodes: Revue." In *Parasites: Their World and Ours*, edited by David F. Mettrick and Sherwin S. Desser, 216–21. Amsterdam: Elsevier Biomedical Press.

Challinor, Andrew J., A. K. Koehler, Julian Ramirez-Villegas, S. Whitfield, and B. Das. 2016. "Current Warming Will Reduce Yields Unless Maize Breeding and Seed Systems Adapt Immediately." *Nature Climate Change* 6, no. 10: 954–58.

Chang, Dan, Michael Knapp, Jacob Enk, Sebastian Lippold, Martin Kircher, Adrian Lister, Ross de MacPhee, et al. 2017. "The Evolutionary and Phylogeographic History of Woolly Mammoths: A Comprehensive Mitogenomic Analysis." *Scientific Reports* 7: 44585. https://doi.org/10.1038/srep44585.

Chavez, Francisco P., John Ryan, Salvador E. Lluch-Cota, and Miguel Ñiquen C. 2003. "From Anchovies to Sardines and Back: Multidecadal Change in the Pacific Ocean." *Science* 299, no. 5604: 217–21.

Checkley, William, Leonardo D. Epstein, Robert H. Gilman, Dante Figueroa, Rosa I. Cama, Jonathan A. Patz, and Robert E. Black. 2000. "Effects of El Niño and Ambient Temperature on Hospital Admissions for Diarrhoeal Diseases in Peruvian Children." *Lancet* 355: 442–50.

Chilton, Neil B., Florence Huby-Chilton, Robin B. Gasser, and Ian Beveridge. 2006. "The Evolutionary Origins of Nematodes within the Order Strongylida Are Related to Predilection Sites within Hosts." *Molecular Phylogenetics and Evolution* 40, no. 1: 118–28. https://doi.org/10.1016/j.ympev.2006.01.003.

Choudhury, Anindo, and Gerardo Pérez-Ponce de León. 2005. "The Roots of Historical Biogeography in Latin American Parasitology: The Legacy of Hermann von Ihering and Lothar Szidat." In *Regionalización biogeográfica en Iberoamérica y tópicos afines: Primeras Jornadas Biogeográficas de la Red Iberoamericana de Biogeografía y Entomología Sistemática (RIBES XII. I-CYTED)*, edited by Jorge Llorente Bousquets and Juan J. Morrone, 45–53. Mexico City: Las Prensas de Ciencias.

Christe, Philippe, Serge Morand, and Johan Michaux. 2006. "Biological Conservation and Parasitism." In *Micromammals and Macroparasites*, edited by Serge Morand, Boris R. Krasnov, and Robert Poulin, 593–613. Tokyo: Springer Japan. https://doi.org/10.1007/978-4-431-36025-4_27.

Cichy, Anna, Maria Urbańska, Anna Marszewska, Wojciech Andrzejewski, and Elżbieta Żbikowska. 2016. "The Invasive Chinese Pond Mussel *Sinanodonta woodiana* (Lea, 1834) as a Host for Native Symbionts in European Waters." *Journal of Limnology* 75, no. 2: 288–96.

Cipollini, Don, and Donnie L. Peterson. 2018. "The Potential for Host Switching via Ecological Fitting in the Emerald Ash Borer-Host Plant System." *Oecologia* 187, no. 2: 507–19. https://doi.org/10.1007/s00442-018-4089-3.

Cizauskas, Carrie A., Colin J. Carlson, Kevin R. Burgio, Chris F. Clements, Eric R. Dougherty, Nyeema C. Harris, and Anna J. Phillips. 2017. "Parasite Vulnerability to Climate Change: An Evidence-Based Functional Trait Approach." *Royal Society Open Science* 4, no. 1: 160535. https://doi.org/10.1098/rsos.160535.

Clark, Nicholas J., Sonya M. Clegg, Katerina Sam, William Goulding, Bonny Koane, and Konstans Wells. 2017. "Climate, Host Phylogeny and the Connectivity of Host Communities Govern Regional Parasite Assembly." *Diversity and Distributions* 31, no. 1: 13–23. https://doi.org/10.1111/ddi.12661.

Clark, Nicholas J., Jennifer M. Seddon, Jan Šlapeta, and Konstans Wells. 2018. "Parasite Spread at the Domestic Animal–Wildlife Interface: Anthropogenic Habitat Use, Phylogeny and Body Mass Drive Risk of Cat and Dog Flea (*Ctenocephalides* spp.) Infestation in Wild Mammals." *Parasites & Vectors* 11, no. 1: 8. https://doi.org/10.1186/s13071-017-2564-z.

Clarke, Richard A., and R. P. Eddy. 2017. *Warnings: Finding Cassandras to Stop Catastrophes.* New York: Harper Collins.

Cleaveland, S., M. K. Laurenson, and L. H. Taylor. 2001. "Diseases of Humans and Their Domestic Mammals: Pathogen Characteristics, Host Range and the Risk of Emergence." *Philosophical Transactions of the Royal Society B: Biological Sciences* 356, no. 1411: 991–99.

Cleaver, Eldridge. 1968. *Soul on Ice.* New York: Ramparts Press.

Coddington, Jonathan A. 1988. "Cladistic Tests of Adaptational Hypotheses." *Cladistics* 4, no. 1: 3–22.

Cohen, Jon. 1997. "Is an Old Virus Up to New Tricks?" *Science* 277, no. 5324: 312–13. https://doi.org/10.1126/science.277.5324.312.

Cohen, Jon. 2017. "Where Has All the Zika Gone?" *Science* 357, no. 6352: 631–32. https://doi.org/10.1126/science.357.6352.631.

Colella, Jocelyn P, Tianying Lan, Stephan C. Schuster, Sandra L. Talbot, Joseph A. Cook, and Charlotte Lindqvist. 2018. "Whole-Genome Analysis of *Mustela erminea* Finds That Pulsed Hybridization Impacts Evolution at High Latitudes." *Communications Biology* 1: 51. https://doi.org/10.1038/s422003 -018-0058-y.

Collier, John. 2017. "Information Dynamics, Self-Organization and the Implications for Management." In *The Future Information Society: Social and Technological Problems at the Crossroads*, edited by Wolfgang Hofkirchner and Mark Burgin, 35–57. World Scientific Series in Information Studies 8. Vienna: World Scientific.

Colwell, Robert K., Robert R. Dunn, and Nyeema C. Harris. 2012. "Coextinction and Persistence of Dependent Species in a Changing World." *Annual Review of Ecology, Evolution, and Systematics* 43, no. 1: 183–203. https://doi .org/10.1146/annurev-ecolsys-110411-160304.

Combes, C., A. Fournier, H. Moné, and A. Théron. 1994. "Behaviours in Trematode Cercariae That Enhance Parasite Transmission: Patterns and Processes." In "Parasites and Behaviour," edited by M. V. K. Sukhdeo. Supplement, *Parasitology* 109, no. S1: S3–13. https://doi.org/10.1017/ S0031182000085048.

Comfort, Alexander. 1960. "Darwin and Freud." *Lancet* 276, no. 7142 (July 16): 107–11. https://doi.org/10.1016/S0140-6736(60)91261-7.

Confalonieri, Ulisses, Bettina Menne, Rais Akhtar, Kristie L. Ebi, Maria Hauengue, R. Sari Kovats, Boris Revich, et al. 2007. "Human Health." *Climate Change 2007: Impacts, Adaptation and Vulnerability: Contribution of Working Group II to the Fourth Assessment Report of the Intergovernmental Panel on Climate Change.* http://researchspace.auckland.ac.nz/handle/2292/14073.

Conteh, Lesong, Thomas Engels, and David H. Molyneux. 2010. "Socioeconomic Aspects of Neglected Tropical Diseases." *Lancet* 375, no. 9710: 239–47. https://doi.org/10.1016/S0140- 6736(09)61422-7.

Cook, Joseph A., Eric P. Hoberg, Anson Koehler, Heikki Henttonen, Lotta Wickström, Voitto Haukisalmi, Kurt Galbreath, et al. 2006. "Beringia: Intercontinental Exchange and Diversification of High Latitude Mammals and Their Parasites during the Pliocene and Quaternary." *Mammal Study* 30: S33–44. https://doi.org/10.3106/1348-6160(2005)30[33:BIEADO]2.0.CO;2.

Cook, Joseph A., Kurt E. Galbreath, Kayce C. Bell, Mariel L. Campbell, Suzanne Carrière, Jocelyn P. Colella, Natalie G. Dawson, et al. 2017. "The Beringian Coevolution Project: Holistic Collections of Mammals and Associated Parasites Reveal Novel Perspectives on Evolutionary and Environmental Change in the North." *Arctic Science* 3, no. 3: 585–617. https://doi.org/10.1139/as-2016-0042.

Cooper, Alan, Chris Turney, Konrad A. Hughen, Barry W. Brook, H. Gregory McDonald, and Corey J. A. Bradshaw. 2015. "Paleoecology: Abrupt Warming Events Drove Late Pleistocene Holarctic Megafaunal Turnover." *Science* 349, no. 6248: 602–6. https://doi.org/10.1126/science.aac4315.

Cope, Edward D. 1885. "On the Evolution of the Vertebrata, Progressive and Retrogressive." *American Naturalist* 19, no. 2: 140–48.

Cope, Edward D. 1887. *The Origin of the Fittest.* New York: D. Appleton.

Cope, Edward D. 1896. *The Primary Factors of Organic Evolution.* Chicago: Open Court.

Coulter, John M. 1915. "A Suggested Explanation of Orthogenesis in Plants." *Science* 42, no. 1094: 859–63.

Cracraft, Joel. 1985. "Biological Diversification and Its Causes." *Annals of the Missouri Botanical Garden* 72: 794–822.

Crawley, Howard. 1923. "Evolution in the Ciliate Family Ophryoscolecidae." *Proceedings of the Academy of Natural Sciences of Philadelphia* 75: 393–412.

Crellen, Thomas, Fiona Allan, Sophia David, Caroline Durrant, Thomas Huckvale, Nancy Holroyd, Aidan M. Emery, et al. 2016. "Whole Genome Resequencing of the Human Parasite *Schistosoma mansoni* Reveals Population History and Effects of Selection." *Scientific Reports* 6: 20954.

Crespi, B. J., and C. P. Sandoval. 2000. "Phylogenetic Evidence for the Evolution of Ecological Specialization in *Timema* Walking-Sticks." *Journal of Evolutionary Biology* 13, no. 2: 249–62.

Cresswell, I. D., and P. Bridgewater. 2000. "The Global Taxonomy initiative—Quo Vadis." *Biology International* 38: 12–16.

Cripps, Michael G., Sarah D. Jackman, Cristina Roquet, Chikako van Koten, Michael Rostás, Graeme W. Bourdôt, and Alfonso Susanna. 2016. "Evolution of Specialization of *Cassida rubiginosa* on *Cirsium arvense* (Compositae, Cardueae)." *Frontiers in Plant Science* 7: 1261. https://doi.org/10.3389/fpls.2016.01261.

Criscione, Charles D., and Michael S. Blouin. 2007. "Parasite Phylogeographical Congruence with Salmon Host Evolutionarily Significant Units: Implications for Salmon Conservation." *Molecular Ecology* 16, no. 5: 993–1005.

Crist, Eileen, Camilo Mora, and Robert Engelman. 2017. "The Interaction of Human Population, Food Production, and Biodiversity Protection." *Science* 356, no. 6335: 260–64. https://doi.org/10.1126/science.aal2011.

Crompton, David W. T. 1973. "The Sites Occupied by Some Parasitic Helminths in the Alimentary Tract of Vertebrates." *Biological Reviews* 48, no. 1: 27–83.

Crompton, David W. T., and S. M. Joyner. 1980. *Parasitic Worms*. London: Wykeham Publications.

Cuénot, Lucien. 1911. *La genèse des espèces animales*. Paris: Ancienne Librairie Germer Baillière & Cie.

Cuénot, Lucien. 1925. *L'adaptation*. Paris: G. Dion.

Cuénot, Lucien. 1941. *Invention et finalité en biologie*. Paris: Flammarion.

Cuénot, Lucien. 1951. *L'évolution biologique, les faits, les incertitudes*. Published posthumously with the collaboration of A. Tétry. Paris: Mason & Cie.

Culotta, Elizabeth, and Ann Gibbons. 2016. "Aborigines and Eurasians Rode One Migration Wave." *Science* 353, no. 6306: 1352–53.

Cumming, Graeme S., and Detlef P. Van Vuuren. 2006. "Will Climate Change Affect Ectoparasite Species Ranges?" *Global Ecology and Biogeography* 15, no. 5: 486–97. https://doi.org/10.1111/j.1466-822X.2006.00241.x.

Cunningham, Andrew A., Peter Daszak, and J. P. Rodriguez. 2003. "Pathogen Pollution: Defining a Parasitological Threat to Biodiversity Conservation." Supplement, *Journal of Parasitology* 89: S78–83.

Cunningham, Andrew A., Peter Daszak, and James L. N. Wood. 2017. "One Health, Emerging Infectious Diseases and Wildlife: Two Decades of Progress?" *Philosophical Transactions of the Royal Society B: Biological Sciences* 372, no. 1725: 20160167. https://doi.org/10.1098/rstb.2016.0167.

Curriero, Frank C., Karlyn S. Heiner, Jonathan M. Samet, Scott L. Zeger, Lisa Strug, and Jonathan A. Patz. 2002. "Temperature and Mortality in 11 Cities of the Eastern United States." *American Journal of Epidemiology* 155, no. 1: 80–87.

Cyranoski, David. 2017. "Bat Cave Solves Mystery of Deadly SARS Virus—and Suggests New Outbreak Could Occur." *Nature* 552, no. 7683: 15–16. https://doi.org/10.1038/d41586-017-07766-9.

da Graça, Rodrigo J., Fabricio H. Oda, Flávia S. Lima, Vinicius Guerra, Priscilla G. Gambale, and Ricardo M. Takemoto. 2017. "Metazoan Endoparasites of 18 Anuran Species from the Mesophytic Semideciduous Atlantic Forest in Southern Brazil." *Journal of Natural History* 51, no. 13–14: 705–29. https://doi.org/10.1080/00222933.2017.1296197.

Darlington, Philip J. 1943. "Carabidae of Mountains and Islands: Data on the Evolution of Isolated Faunas, and on Atrophy of Wings." *Ecological Monographs* 13, no. 1: 37–61.

Darwin, Charles. 1872. *The Origin of Species*. 6th ed. London: Murray.

Daszak, Peter, Andrew A. Cunningham, and Alex D. Hyatt. 2000. "Emerging Infectious Diseases of Wildlife—Threats to Biodiversity and Human Health." *Science* 287, no. 5452: 443–49.

Daversa, David R., Andrea Manica, Jaime Bosch, Jolle W. Jolles, and Trenton W. J. Garner. 2018. "Routine Habitat Switching Alters the Likelihood and

Persistence of Infection with a Pathogenic Parasite." *Functional Ecology* 32, no. 5: 1262–70. https://doi.org/10.1111/1365-2435.13038.

Dawson, Natalie G., Andrew G. Hope, Sandra L. Talbot, and Joseph A Cook. 2014. "A Multi-Locus Evaluation of Ermine (*Mustela erminea*) across the Holarctic, Testing Hypotheses of Pleistocene Diversification in Response to Climate Change." *Journal of Biogeography* 41, no. 3: 464–75. https://doi.org/10.1111/jbi.12221.

de Heinzelin, Jean, J. Desmond Clark, Tim White, William Hart, Paul Renne, Giday WoldeGabriel, Yonas Beyene, et al. 1999. "Environment and Behavior of 2.5-Million-Year-Old Bouri Hominids." *Science* 284, no. 5414: 625–29.

de Kruif, Paul. 1926. *Microbe Hunters.* San Diego: Harcourt.

Defleur, Alban, Tim White, Patricia Valensi, Ludovic Slimak, and Évelyne Crégut-Bonnoure. 1999. "Neanderthal Cannibalism at Moula-Guercy, Ardèche, France." *Science* 286, no. 5437: 128–31.

DeFries, Ruth, and Harini Nagendra. 2017. "Ecosystem Management as a Wicked Problem." *Science* 356, no. 6335: 265–70. https://doi.org/10.1126/science.aal1950.

Delatorre, Milena, Nicolay Cunha, Josué Raizer, and Vanda Lúcia Ferreira. 2015. "Evidence of Stochasticity Driving Anuran Metacommunity Structure in the Pantanal Wetlands." *Freshwater Biology* 60, no. 11: 2197–2207.

Demboski, John R., and Jack Sullivan. 2003. "Extensive mtDNA Variation within the Yellow-Pine Chipmunk, *Tamias amoenus* (Rodentia: Sciuridae), and Phylogeographic Inferences for Northwest North America." *Molecular Phylogenetics and Evolution* 26, no. 3: 389–408.

deMenocal, Peter B., and Chris Stringer. 2016. "Human Migration: Climate and the Peopling of the World." *Nature* 53, no. 7623: 49–50.

Denton, George H., Robert F. Anderson, J. R. Toggweiler, R. L. Edwards, J. M. Schaefer, and A. E. Putnam. 2010. "The Last Glacial Termination." *Science* 328, no. 5986: 1652–56.

Dentovskaya, Svetlana V., Mikhail E. Platonov, Tat'yana E. Svetoch, Pavel Kh. Kopylov, Tat'yana I. Kombarova, Sergey A. Ivanov, Rima Z. Shaikhutdinova, et al. 2016. "Two Isoforms of *Yersinia pestis* Plasminogen Activator Pla: Intraspecies Distribution, Intrinsic Disorder Propensity, and Contribution to Virulence." *PLoS ONE* 11, no. 12: e0168089. https://doi.org/10.1371/journal.pone.0168089.

"Descent of Man." 1959. *Lancet* 274, no. 7106 (November 7): 779–80. https://doi.org/10.1016/S0140-6736(59)90870-0.

Desdevises, Yves, Serge Morand, Olivier Jousson, and Pierre Legendre. 2002. "Coevolution between *Lamellodiscus* (Monogenea: Diplectanidae) and Sparidae (Teleostei): The Study of a Complex Host-Parasite System." *Evolution* 56, no. 12: 2459–71.

Dethier, Vincent G. 1941. "Chemical Factors Determining the Choice of Food Plants by Papilio Larvae." *American Naturalist* 75, no. 756: 61–73.

Dethier, Vincent G. 1953. "Host Plant Perception in Phytophagous Insects." *Transactions of the IXth International Congress of Entomology* 2: 81–88.

Dethier, V. G. 1975. "The Monarch Revisited." *Journal of the Kansas Entomological Society* 48, no. 2: 129–40.

de Vienne, Damien M., Guislaine Refrégier, Manuela Lopez Villavicencio, Aurelien Tellier, Michael E. Hood, and Tatiana Giraud. 2013. "Cospeciation Versus Host-Shift Speciation: Methods for Testing, Evidence from Natural Associations and Relation to Coevolution." *New Phytologist* 198, no. 2: 347–85. https://doi.org/10.1111/nph.12150.

Diamond, Jared. 1997. *Guns, Germs, and Steel: The Fates of Human Societies.* W. W. Norton.

Diamond, Jared. 2002. "Evolution, Consequences and Future of Plant and Animal Domestication." *Nature* 418, no. 6898: 700–707. https://doi.org/10.1038/nature01019.

Dicks, Lynn V., Blandina Viana, Riccardo Bommarco, Berry Brosi, María del Coro Arizmendi, Saul A. Cunningham, Leonardo Galetto, et al. 2016. "Ten Policies for Pollinators." *Science* 354, no. 6315: 975–76. https://doi.org/10.1126/science.aai9226.

Dietrich, Muriel, Teresa Kearney, Ernest C. J. Seamark, and Wanda Markotter. 2017. "The Excreted Microbiota of Bats: Evidence of Niche Specialization Based on Multiple Body Habitats." *Federation of European Microbiological Societies Microbiology Letters* 364, no. 1: 7. https://doi.org/10.1093/femsle/fnw284.

diEuliis, Diane, Richard Jaffe, Manuela da Silva, Richard P. Lane, Tim Littlewood, Eileen Graham, and David E. Schindel. 2015. "Scientific Collections and Emerging Infectious Diseases: Report of an Interdisciplinary Workshop." Assisted by Adele Crane and Franco Ciammachilli. *Scientific Collections International Report* 1 (March). https://doi.org/10.5479/si.SciColl.

diEuliis, Diane, Kirk R. Johnson, Stephen S. Morse, and David E. Schindel. 2016. "Specimen Collections Should Have a Much Bigger Role in Infectious Disease Research and Response." *Proceedings of the National Academy of Sciences of the United States of America* 113: 4–7.

Dimitrov, Dimitar, Vaidas Palinauskas, Tatjana A. Iezhova, Rasa Bernotiené, Mikas Ilgünas, Dovilé Bukauskaité, Pavel Zehtindjiev, et al. 2014. "*Plasmodium* spp.: An Experimental Study on Vertebrate Host Susceptibility to Avian Malaria." *Experimental Parasitology* 148: 1–6. http://dx.doi.org/10.1016/j.exppara.2014.11.005.

Ding, Qiaoling, Conrad C. Labandeira, Qingmin Meng, and Dong Ren. 2015. "Insect Herbivory, Plant-Host Specialization and Tissue Partitioning on Mid-Mesozoic Broadleaved Conifers of Northeastern China." *Palaeogeography, Palaeoclimatology, Palaeoecology* 440: 259–73. https://doi.org/10.1016/j.palaeo.2015.09.007.

Dixon, B. R., M. Ndao, J. A. Tetro, Rasha Maal-Bared, Sabah Bidawid, and Jeffrey M. Farber. 2014. "Food and Environmental Parasitology in Canada:

A Network for the Facilitation of Collaborative Research." *Food Protection Trends* 34, no. 6: 376–85.

Dobson, Andrew P. 2005. "What Links Bats to Emerging Infectious Diseases?" *Science* 310, no. 5748: 628–29.

Dobson, Andrew P., and E. Robin Carper. 1996. "Infectious Diseases and Human Population History." *Bioscience* 46, no. 2: 115–26.

Dobzhansky, Theodosius. 1955. *Evolution, Genetics, and Man.* New York: John Wiley & Sons.

Dogiel, Valentin Alexandrovitch. 1962. *General Parasitology.* 3rd ed. London: Oliver & Boyd.

Doña, Jorge, Andrew D. Sweet, Kevin P. Johnson, David Serrano, Sergey Mironov, and Roger Jovani. 2017. "Cophylogenetic Analyses Reveal Extensive Host-Shift Speciation in a Highly Specialized and Host-Specific Symbiont System." *Molecular Phylogenetics and Evolution* 115: 190–96.

Doolittle, W. Ford, and Austin Booth. 2016. "It's the Song, Not the Singer: An Exploration of Holobiosis and Evolutionary Theory." *Biology and Philosophy* 32, no. 1: 5–24. https://doi.org/10.1007/s10539-016-9542-2.

Dornburg, Alex, Jon Moore, Jeremy M. Beaulieu, Ron I. Eytan, and Thomas J. Near. 2014. "The Impact of Shifts in Marine Biodiversity Hotspots on Patterns of Range Evolution: Evidence from the Holocentridae (Squirrelfishes and Soldierfishes)." *Evolution* 69, no. 1: 146–61. https://doi.org/10.1111/evo.12562.

Doty, Jeffrey B., Jean M. Malekani, Lem's N. Kalemba, William T. Stanley, Benjamin P. Monroe, Yoshinori U. Nakazawa, Matthew R. Mauldin, et al. 2017. "Assessing Monkeypox Virus Prevalence in Small Mammals at the Human-Animal Interface in the Democratic Republic of the Congo." *Viruses* 9, no. 10: 283. https://doi.org/10.3390/v9100283.

Dougherty, Ellsworth C. 1949. "The Phylogeny of the Nematode Family Metastrongylidae Leiper, 1909: A Correlation of Host and Symbiote Evolution." *Parasitology* 39, no. 3–4: 222–34.

Dougherty, Ellsworth C. 1951. "Evolution of Zoöparasitic Groups in the Phylum Nematoda, with Special Reference to Host-Distribution." *Journal of Parasitology* 37: 353–78.

Dougherty, Eric R., Colin J. Carlson, Veronica M. Bueno, Kevin R. Burgio, Carrie A. Cizauskas, Christopher F. Clements, Dana P. Seidel, and Nyeema C. Harris. 2015. "Paradigms for Parasite Conservation." *Conservation Biology* 30, no. 4: 724–33. https://doi.org/10.1111/cobi.12634.

Dragoo, Jerry W., J. Alden Lackey, Kathryn E. Moore, Enrique P. Lessa, Joseph A. Cook, and Terry L. Yates. 2006. "Phylogeography of the Deer Mouse (*Peromyscus maniculatus*) Provides a Predictive Framework for Research on Hantaviruses." *Journal of General Virology* 87, no. 7: 1997–2003.

Dunn, Emmett R. 1925. "The Host-Parasite Method and the Distribution of Frogs." *American Naturalist* 59, no. 663: 370–75.

Dunnum, Jonathan L., Richard Yanagihara, Karl M. Johnson, Blas Armien, Nyamsuren Batsaikhan, Laura Morgan, and Joseph A. Cook. 2017. "Biospecimen Repositories and Integrated Databases as Critical Infrastructure for Pathogen Discovery and Pathobiology Research." *PLoS Neglected Tropical Diseases* 11, no. 1: e0005133.

du Preez, Louis H., and Dawid J. Kok. 1997. "Supporting Experimental Evidence of Host Specificity among Southern African Polystomes." *Parasitology Research* 83: 558–62.

Durette-Desset, Marie-Claude. 1985. "Trichostrongyloid Nematodes and Their Vertebrate Hosts: Reconstruction of the Phylogeny of a Parasitic Group." *Advances in Parasitology* 24: 239–306.

du Toit, Nina, Bettine J. van Vuuren, Sonja Matthee, and Conrad A. Matthee. 2013. "Biogeography and Host-Related Factors Trump Parasite Life History: Limited Congruence Among the Genetic Structures of Specific Ectoparasitic Lice and their Rodent Hosts." *Molecular Ecology* 22, no. 20: 5185–5204. https://doi.org/10.1111/mec.12459.

Dyer, Lee A. 2011. "New Synthesis—Back to the Future: New Approaches and Directions in Chemical Studies of Coevolution." *Journal of Chemical Ecology* 37, no. 7: 669.

Dynesius, Mats, and Roland Jansson. 2000. "Evolutionary Consequences of Changes in Species' Geographical Distributions Driven by Milankovitch Climate Oscillations." *Proceedings of the National Academy of Sciences of the United States of America* 97, no. 16: 9115–20.

Dynesius, Mats, and Roland Jansson. 2014. "Persistence of Within-Species Lineages: a Neglected Control of Speciation Rates." *Evolution* 68, no. 4: 923–34. https://doi.org/10.1111/evo.12316.

Ecker, Michaela, James Brink, Liora Kolska Horwitz, Louis Scott, and Julia A. Lee-Thorp. 2018. "A 12,000 Year Record of Changes in Herbivore Niche Separation and Palaeoclimate (Wonderwerk Cave, South Africa)." *Quaternary Science Reviews* 180: 132–44. https://doi.org/10.1016/j.quascirev.2017.11.025.

Eckstut, Mallory E., Caleb D. McMahan, Brian I. Crother, Justin M. Ancheta, Deborah A. McLennan, and Daniel R. Brooks. 2011. "PACT in Practice: Comparative Historical Biogeographic Patterns and Species-Area Relationships of the Greater Antillean and Hawaiian Island Terrestrial Biotas." *Global Ecology and Biogeography* 20, no. 4: 545–57.

Economo, Evan P., Eli M. Sarnat, Milan Janda, Ronald Clouse, Pavel B. Klimov, Georg Fischer, et al. 2015. "Breaking Out of Biogeographical Modules: Range Expansion and Taxon Cycles in the Hyperdiverse Ant Genus *Pheidole*." *Journal of Biogeography* 42, no. 12: 2289–2301. https://doi.org/10.1111/jbi.12592.

Edler, Daniel, Thaís Guedes, Alexander Zizka, Martin Rosvall, and Alexandre Antonelli. 2017. "Infomap Bioregions: Interactive Mapping of Biogeographical Regions from Species Distributions." *Systematic Biology* 66, no. 2: 197–204. https://doi.org/10.1093/sysbio/syw087.

Egbendewe-Mondzozo, Aklesso, Mark Musumba, Bruce A. McCarl, and Ximing Wu. 2011. "Climate Change and Vector-Borne Diseases: An Economic Impact Analysis of Malaria in Africa." *International Journal of Environmental Research and Public Health* 8, no. 12: 913–30. https://doi.org/10.3390/ijerph8030913.

Eggleton, Paul, and Robert Belshaw. 1993. "Comparisons of Dipteran, Hymenopteran and Coleopteran Parasitoids: Provisional Phylogenetic Explanations." *Biological Journal of the Linnean Society* 48, no. 3: 213–26.

Ehrlich, Paul R., and Peter H. Raven. 1964. "Butterflies and Plants: A Study in Coevolution." *Evolution* 18, no. 4: 586–608.

Eichler, Wolfdietrich D. 1941a. "Wirtsspezifität und Stammesgeschichtliche Gleichläufigkeit (Fahrenholzsche Regel) bei Parasiten im Allgemeinen und bei Mallophagen im Besonderen." *Zoologische Anzeiger* 132: 254–62.

Eichler, Wolfdietrich D. 1941b. "Korrelation in der Stammesentwicklung von Wirten und Parasiten." *Zeitschrift fur Parasitenkunde* 12: 94.

Eichler, Wolfdietrich D. 1942. "Die Entfaltungsregel und andere Gesetzmäßigkeiten in den parasitogenetischen Beziehungen der Mallphagen und anderer ständiger Parasiten zu ihrer Wirten." *Zoologische Anzeiger* 137: 77–83.

Eichler, Wolfdietrich D. 1948a. "Evolutionsfragen der Wirtsspezifität." *Biologische Zentralblatt* 67: 373–406.

Eichler, Wolfdietrich D. 1948b. "Some Rules in Ectoparasitism." *Annals and Magazine Natural History* 1: 588–98.

Eichler, Wolfdietrich D. 1966. "Two New Evolutionary Terms for Speciation in Parasitic Animals." *Sysematic Zoology* 15: 216–18.

Eichler, Wolfdietrich D. 1973. "Neuere Überlegungen zu den parasite-phyletischen Regeln." *Helminthologia* 14: 1–4.

Eichler, Wolfdietrich D. 1982. "Les règles parasitophylétiques comme phénomène de la théorie de l'évolution." 2nd Symposium on Host Specificity among Parasite of Vertebrates. *Mémoires du Muséum d'histoire naturelle*, n.s. 123: 69–71.

Eimer, Theodor. 1897. *Orthogenesis der Schmetterlinge: Ein Beweis bestimmt gerichteter Entwickelung und Ohnmacht der natürlichen Zuchtwahl bei der Artbildung; Zugleich eine Erwiderung an August Weismann.* Vol. 2. Leipzig: Wilhelm Engelmann.

Eimer, Theodor. 1898. *On Orthogenesis: And the Impotence of Natural Selection in Species-Formation.* Chicago: Open Court.

El Sawaf, Bahira M., Hala A. Kassem, Nabil M. Mogalli, Shabaan S. El Hossary, and Nadia F. Ramadan. 2016. "Current Knowledge of Sand Fly Fauna (Diptera: Psychodidae) of Northwestern Yemen and How It Relates to Leishmaniasis Transmission." *Acta Tropica* 162: 11–19.

Elton, Charles S. 1958. *The Ecology of Invasions by Animals and Plants.* London: Methuen.

Endler, John A. 1982. "Problems in Distinguishing Historical from Ecological Factors in Biogeography." *American Zoologist* 22, no. 2: 441–52.

Enserink, Martin. 2003. "U.S. Monkeypox Outbreak Traced to Wisconsin Pet Dealer." *Science* 300, no. 5626: 1639–39.

Epstein, Paul R. 2001. "Climate, Ecology, and Health." In *Plagues and Politics*, edited by A. T. Price-Smith, 27–54. Basingstoke: Palgrave.

Epstein, Paul R. 2005. "Climate Change and Human Health." *New England Journal of Medicine* 353: 1433–36.

Erwin, Terry L. 1979. "Thoughts on the Evolutionary History of Ground Beetles: Hypotheses Generated from Comparative Faunal Analyses of Lowland Forest Sites in Temperate and Tropical Regions." In *Carabid Beetles*, edited by Terry L Erwin, George E. Ball, Donald R. Whitehead, and Anne L Halpern, 539–92. Dordrecht: Springer.

Erwin, Terry L. 1981. "Taxon Pulses, Vicariance, and Dispersal: An Evolutionary Synthesis Illustrated by Carabid Beetles." In *Vicariance Biogeography: A Critique*, edited by Gareth Nelson and Donn E. Rosen, 159–96. New York: Columbia University Press.

Erwin, Terry L. 1985. "The Taxon Pulse: A General Pattern of Lineage Radiation and Extinction among Carabid Beetles." In *Taxonomy, Phylogeny, and Zoogeography of Beetles and Ants*, edited by George E. Ball, 437–72. Dordrecht: W. Junk.

Escobar, Luis E., and Meggan E. Craft. 2016. "Advances and Limitations of Disease Biogeography Using Ecological Niche Modeling." *Frontiers in Microbiology* 7: 1174. https://www.ncbi.nlm.nih.gov/pmc/articles/PMC4974947/.

Escudero, M. 2015. "Phylogenetic Congruence of Parasitic Smut Fungi (*Anthracoidea*, Anthracoideaceae) and Their Host Plants (*Carex*, Cyperaceae): Cospeciation or Host- Shift Speciation?" *American Journal of Botany* 102, no. 7: 1108–14. https://doi.org/10.3732/ajb.1500130.

Estrada-Peña, Agustín, Hein Sprong, Alejandro Cabezas-Cruz, José de la Fuente, Ana Ramo, and Elena Claudia Coipan. 2016. "Nested Coevolutionary Networks Shape the Ecological Relationships of Ticks, Hosts, and the Lyme Disease Bacteria of the *Borrelia burgdorferi* (*s.l.*) Complex." *Parasites & Vectors* 9: 517. https://doi.org/10.1186/s13071-016-1803-z

Ewing, Henry E. 1924. "On the Taxonomy, Biology and Distribution of the Biting Lice of the Family Gyropidae." *Proceedings of the U.S. National Museum* 63: 1–42.

Ewing, Henry E. 1928. "A Revision of the American Lice of the Genus *Pediculus*, Together with Consideration of the Significance of their Geographical and Host Distribution." *Proceedings of the U. S. National Museum* 68: 1–38.

Ewing, S. A. 2016. "Emergence of Veterinary Parasitology in the United States: Maurice Hall and the Bureau of Animal Industry." *Veterinary Heritage* 39: 33–44.

Fahrenholz, Heinrich. 1909. "Aus dem Myobien-Nachlass des Herrn Poppe." *Abhandlungen herausgegeben vom Naturwissenschaften Verein zu Bremen* 19: 359–70.

Fahrenholz, Heinrich. 1913. "Ectoparasiten und Abstammungslehre." *Zoologischer Anzeiger* 41: 371–74.

Fallahzadeh, Majid, George Japoshvili, Nazila Saghaei, and Kent M. Daane. 2011. "Natural Enemies of *Planococcus ficus* (Hemiptera: Pseudococcidae) in Fars Province Vineyards, Iran." *Biocontrol Science and Technology* 21, no. 4: 427–33. https://doi.org/10.1080/09583157.2011.554801.

Faria, Christiana M. A., Antonio Machado, Isabel R. Amorim, Matthew J. G. Gage, Paulo A. V. Borges, and Brent C. Emerson. 2016. "Evidence for Multiple Founding Lineages and Genetic Admixture in the Evolution of Species within an Oceanic Island Weevil (Coleoptera, Curculionidae) Super-Radiation." *Journal of Biogeography* 43, no. 1: 178–91. https://doi.org/10.1111/jbi.12606.

Faust, Christina L., Hamish I. McCallum, Laura S. P. Bloomfield, Nicole L. Gottdenker, Thomas R. Gillespie, Colin J. Torney, Andrew P. Dobson, et al. 2018. "Pathogen Spillover during Land Conversion." *Ecology Letters* 21, no. 4: 471–83. https://doi.org/10.1111/ele.12904.

Fecchio, Alan, Jeffrey Andrew Bell, Michael David Collins, Izeni Pires Farias, Christopher Harry Trisos, Joseph Andrew Tobias, Vasyl Volodymyr Tkach, et al. 2018. "Diversification by Host Switching and Dispersal Shaped the Diversity and Distribution of Avian Malaria Parasites in Amazonia." *Oikos*, February. https://doi.org/10.1111/oik.05115.

Federal Interagency Committee for Invasive Terrestrial Animals and Pathogens [ITAP] [David J. Chitwood, Hilda Diaz-Soltero, Eric P. Hoberg, Scott Miller, Robert Reynolds, Benjamin M. Rosenthal, Amy Rossman, Marsha Sitnik, and Alma M. Solis]. 2008. "Situation Report on U.S. Systematic Biology." Protecting America's Economy, Environment, Health and Security against Invasive Species Requires a Strong Federal Program in Systematics Biology. Systematics Subcommittee, Interagency Committee for Invasive Terrestrial Animals and Pathogens (ITAP). Washington, DC.

Feldman, Richard E., Michael J. L. Peers, Rob S. A. Pickles, Daniel Thornton, and Dennis L. Murray. 2017. "Climate Driven Range Divergence among Host Species Affects Range-Wide Patterns of Parasitism." *Global Ecology and Conservation* 9: 1–10.

Figueres, C., H. J. Schellnhuber, G. Whitelam, J. Rockstöm, A. Hobley, and S. Rahmstorf. 2017. "Three Years to Safeguard Our Climate." *Nature* 546: 593–95.

Filipchenko, A. A. 1937. "The Ecological Concept of Parasitism and Parasitology as a Distinct Scientific Discipline." *Uchenie Zapiski Leningradskogo Gosuniversitet Seria Biologicheski* 4, no. 13: 4–14.

Fine, P., K. Eames, and D. L. Heymann. 2011. "'Herd Immunity': A Rough Guide." *Clinical Infectious Diseases* 52, no. 7: 911–16. https://doi.org/10.1093/cid/cir007.

Firmat, Cyril, Paul Alibert, Guillaume Mutin, Michèle Losseau, Antoine Pariselle, and Pierre Sasal. 2016. "A Case of Complete Loss of Gill Parasites in the Invasive Cichlid *Oreochromis mossambicus.*" *Parasitology Research* 115, no. 9: 3657–61. https://doi.org/10.1007/s00436-016-5168-1.

Fischer, Dominik, Stephanine Margarete Thomas, Franziska Niemitz, Björn Reineking, and Carl Beierkuhnlein. 2011. "Projection of Climatic Suitability for *Aedes albopictus* Skuse (Culicidae) in Europe under Climate Change Conditions." *Global and Planetary Change* 78, no. 1–2: 54–64.

Fofana, Abdulai, Luiza Toma, Dominic Moran, George J. Gunn, and Alistair W. Stott. 2009. "Measuring the Economic Benefits and Costs of Bluetongue Virus Outbreak and Control Strategies in Scotland." Eighty-Third Annual Conference, March 30–April 1, 2009, Dublin, Ireland, Agricultural Economics Society, 1–19. https://ideas.repec.org/p/ags/aesc09/51052.html.

Florencio, Margarita, Jorge M. Lobo, Pedro Cardoso, Mário Almeida-Neto, and Paulo A. V. Borges. 2015. "The Colonisation of Exotic Species Does Not Have to Trigger Faunal Homogenisation: Lessons from the Assembly Patterns of Arthropods on Oceanic Islands." *PLoS ONE* 10, no. 5: e0128276. https://doi.org/10.1371/journal.pone.0128276.

Földvári, G., S. Jahfari, K. Rigó, Mónika Jablonszky, Sándor Szekeres, Gábor Majoros, Mária Tóth, Viktor Molnár, Elena C. Colpan, and Hein Sprong. 2014. "*Candidatus neoehrlichia mikurensis* and *Anaplasma phagocytophilum* in Urban Hedgehogs." *Emerging Infectious Diseases* 20, no. 3: 496–97. https://doi.org/10.3201/eid2003.130708.

Földvári, Gábor, Krisztina Rigó, Mónika Jablonszky, Nóra Biró, Gábor Majoros, Viktor Molnár, and Mária Tóth. 2011. "Ticks and the City: Ectoparasites of the Northern White-Breasted Hedgehog (*Erinaceus roumanicus*) in an Urban Park." *Ticks and Tick-Borne Diseases* 2, no. 4: 231–34. https://doi.org/10.1016/j.ttbdis.2011.09.001.

Földvári, Gábor, Pavel Široký, Sándor Szekeres, Gábor Majoros, and Hein Sprong. 2016. "*Dermacentor reticulatus*: A Vector on the Rise." *Parasites & Vectors* 9, no. 1: 655. https://doi.org/10.1186/s13071-016-1599-x.

Folinsbee, Kaila E., and Daniel R. Brooks. 2007. "Miocene Hominoid Biogeography: Pulses of Dispersal and Differentiation." *Journal of Biogeography* 34, no. 3: 383–97.

Fountain-Jones, Nicholas M., William D. Pearse, Luis E. Escobar, Ana Alba-Casals, Scott Carver, T. Jonathan Davies, Simona Kraberger, et al. 2017. "Towards an Eco-Phylogenetic Framework for Infectious Disease Ecology." *Biological Reviews* 341 (suppl.): 514–22. https://doi.org/10.1111/brv.12380.

Fowles, Trevor M., Maria del C. Coscarón, Antônio R. Panizzi, and Scott P. Carroll. 2015. "Scentless Plant Bugs (*Rhopalidae*)." In *True Bugs (Heteroptera) of the Neotropics*, edited by Antônio R. Panizzi and Jocélia Grazia, 2:607–37. New York: Springer.

Fraija-Fernández, Natalia, Peter D. Olson, Enrique A. Crespo, Juan A. Raga, Francisco Javier Aznar, and Mercedes Fernández. 2015. "Independent Host Switching Events by Digenean Parasites of Cetaceans Inferred from

Ribosomal DNA." *International Journal for Parasitology* 45, no. 2–3: 167–73. https://doi.org/10.1016/j.ijpara.2014.10.004.

Frainer, André, Brendan G. McKie, Per-Arne Amundsen, Rune Knudsen, and Kevin D. Lafferty. 2018. "Parasitism and the Biodiversity-Functioning Relationship." *Trends in Ecology & Evolution* 33, no. 4: 260–62. https://doi .org/10.1016/j.tree.2018.01.011.

Francis, Richard C. 2015. *Domesticated: Evolution in a Man-Made World*. New York: W. W. Norton.

Fredensborg, B. L. 2014. "Predictors of Host Specificity among Behavior-Manipulating Parasites." *Integrative and Comparative Biology* 54, no. 2: 149–58. https://doi.org/10.1093/icb/icu051.

French, Brett. 2016. "Fish Kill Closes 183 Miles of Yellowstone River, Tributaries to All Recreation." *Missoulian*, August 19.

Frenzen, Paul D., Alison Drake, and Frederick J. Angulo. 2005. "Economic Cost of Illness due to *Escherichia coli* O157 Infections in the United States." *Journal of Food Protection* 68, no. 12: 2623–30.

Galaktionov, Kirill V. 2016. "Evolution and Biological Radiation of Trematodes: A Synopsis of Ideas and Opinions." In *Coevolution of Parasites and Hosts*, edited by Kirill V. Galaktionov, 74–126. Moscow: Zoological Institute, Russian Academy of Sciences.

Galaktionov, Kirill V. 2017a. "Patterns and Processes Influencing Helminth Parasites of Arctic Coastal Communities during Climate Change." *Journal of Helminthology* 91, no. 4: 387–408.

Galaktionov, Kirill V. 2017b. "Transmission of Parasites in the Coastal Waters of the Arctic Seas and Possible Effect of Climate Change." *Biology Bulletin* 9, no. 43: 1129–47.

Galaktionov, Kirill V., Isabel Blasco-Costa, and Peter D. Olson. 2012. "Life Cycles, Molecular Phylogeny and Historical Biogeography of the 'pygmaeus' Microphallids (Digenea: Microphallidae): Widespread Parasites of Marine and Coastal Birds in the Holarctic." *Parasitology* 139, no. 10: 1346–60.

Galbreath, Kurt E., and Eric P. Hoberg. 2012. "Return to Beringia: Parasites Reveal Cryptic Biogeographic History of North American Pikas." *Proceedings of the Royal Society B: Biological Sciences*, 279: 371–78.

Galbreath, Kurt E., and Eric P. Hoberg. 2015. "Host Responses to Cycles of Climate Change Shape Parasite Diversity across North America's Intermountain West." *Folia Zoologica* 64, no. 3: 218–33.

Ganopolski, Andrey, Ricarda Winkelmann, and Hans Joachim Schellnhuber. 2016. "Critical Insolation-CO2 Relation for Diagnosing Past and Future Glacial Inception." *Nature* 529, no. 7585: 200–203.

Garcia, Herakles A., Adriana C. Rodrigues, Franjo Martinkovic, Antonio H. H. Minervino, Marta Campaner, Vania L. B. Nunes, Fernando Paiva, et al. 2011. "Multilocus Phylogeographical Analysis of *Trypanosoma* (*Megatrypanum*) Genotypes from Sympatric Cattle and Water Buffalo Populations Supports Evolutionary Host Constraint and Close Phylogenetic Relationships with Genotypes Found in Other Ruminants." *International Journal for Parasitology* 41, no. 13–14: 1385–96. https://doi.org/10.1016/j.ijpara.2011.09.001.

Gardner, Scott L., and Mariel L. Campbell. 1992. "A New Species of *Linstowia* (Cestoda: Anoplocephalidae) from Marsupials in Bolivia." *Journal of Parasitology* 78, no. 5: 795–99.

Garnett, Geoffrey P., and Edward C. Holmes. 1996. "The Ecology of Emergent Infectious Disease." *BioScience* 46, no. 2: 127–35. https://doi.org/10.2307/1312815.

Garrett, Laurie. 1994. *The Coming Plague: Newly Emerging Diseases in a World Out of Balance.* New York: Farrar Straus Giroux.

Garrett, Laurie. 2001. *Betrayal of Trust: The Collapse of Global Public Health.* Oxford: Oxford University Press.

Gasbarre, Louis C., Larry L. Smith, Eric Hoberg, and Patricia A. Pilitt. 2009. "Further Characterization of a Cattle Nematode Population with Demonstrated Resistance to Current Anthelmintics." *Veterinary Parasitology* 166, no. 3–4: 275–80. https://doi.org/10.1016/j.vetpar.2009.08.019.

Gebreyes, Wondwossen A., Jean Dupouy-Camet, Melanie J. Newport, Celso J. B. Oliveira, Larry S. Schlesinger, Yehia M. Saif, Samuel Kariuki, et al. 2014. "The Global One Health Paradigm: Challenges and Opportunities for Tackling Infectious Diseases at the Human, Animal, and Environment Interface in Low-Resource Settings." *PLoS Neglected Tropical Diseases* 8, no. 11: e3257. https://doi.org/10.1371/journal.pntd.0003257.

Gentry, Grant L., and Lee A. Dyer. 2002. "On the Conditional Nature of Neotropical Caterpillar Defenses against Their Natural Enemies." *Ecology* 83, no. 11: 3108–3119.

George, David R., Robert D. Finn, Kirsty M. Graham, Monique F. Mul, Veronika Maurer, Claire Valiente Moro, and Olivier A. E. Sparagano. 2015. "Should the Poultry Red Mite *Dermanyssus gallinae* Be of Wider Concern for Veterinary and Medical Science?" *Parasites & Vectors* 8, no. 1: 151. https://doi.org/10.1186/s13071-015-0768-7.

Giannelli, Alessio, Gioia Capelli, Anja Joachim, Barbara Hinney, Bertrand Losson, Zvezdelina Kirkova, Magalie René-Martellet, et al. 2017. "Lungworms and Gastrointestinal Parasites of Domestic Cats: A European Perspective." *International Journal for Parasitology* 47, no. 9: 517–28.

Gibbons, Ann. 2013a. "How a Fickle Climate Made Us Human." *Science* 341: 474–79.

Gibbons, Ann. 2013b. "The Thousand Year Graveyard." *Science* 342: 1306–10.

Gibbons, Ann. 2016. *The Wanderers.* http://science.sciencemag.org/content/354/6315/958.summary.

Gibbons, Ann. 2017. *The First Australians Arrived Early.* http://www.sciencemag.org/news/2017/07/find-australia-hints-very-early-human-exit-africa.

Gilbert, M., J. Slingenbergh, and X. Xiao. 2008. "Climate Change and Avian Influenza." *Revue scientifique et technique* 27, no. 2: 459–66.

Gilbert, Michael A., and Willard O. Granath Jr. 2003. "Whirling Disease of Salmonid Fish: Life Cycle, Biology, and Disease." *Journal of Parasitology* 89, no. 4: 658–67.

Giribet, Gonzalo. 2015. "Morphology Should Not Be Forgotten in the Era of Genomics: A Phylogenetic Perspective." *Zoologischer Anzeiger: A Journal of Comparative Zoology* 256: 96–103. https://doi.org/10.1016/j.jcz.2015.01.003.

Girisgin, Ahmet Onur, Sezen Birlik, Bayram Senlik, and Hikmet Sami Yildirimhan. 2017. "Intestinal Helminths of the White Stork (*Ciconia ciconia* Linnaeus 1758) from an Inter-Route Site in Turkey." *Acta Veterinaria Hungarica* 65, no. 2: 221–33. https://doi.org/10.1556/004.2017.022.

Githeko, Andrew K., Steve W. Lindsay, Ulisses E. Confalonieri, and Jonathan A. Patz. 2000. "Climate Change and Vector-Borne Diseases: A Regional Analysis." *Bulletin of the World Health Organization* 78, no. 9: 1136–47.

Givnish, Thomas J. 2015. "Adaptive Radiation versus 'Radiation' and 'Explosive Diversification': Why Conceptual Distinctions Are Fundamental to Understanding Evolution." *New Phytologist* 207, no. 2: 297–303.

Glassmire, Andrea E., Christopher S. Jeffrey, Matthew L. Forister, Thomas L. Parchman, Chris C. Nice, Joshua P. Jahner, Joseph S. Wilson, et al. 2016. "Intraspecific Phytochemical Variation Shapes Community and Population Structure for Specialist Caterpillars." *New Phytologist* 212 , no. 1: 208–19. https://doi.org/10.1111/nph.14038.

"Global Warming—A Rising Cost to Australia's Livestock Industries." 2017. *Vet Practice Magazine*, June 20. https://vetpracticemag.com.au/global-warming-rising-cost-australias-livestock-industries.

Gloss, Andrew D., Daniel G. Vassão, Alexander L. Hailey, Anna C. Nelson Dittrich, Katharina Schramm, Michael Reichelt, Timothy J. Rast, et al. 2014. "Evolution in an Ancient Detoxification Pathway Is Coupled with a Transition to Herbivory in the Drosophilidae." *Molecular Biology and Evolution* 31, no. 9: 2441–56. https://doi.org/10.1093/molbev/msu201.

Goebel, Ted, Michael R. Waters, and Dennis H. O'Rourke. 2008. "The Late Pleistocene Dispersal of Modern Humans in the Americas." *Science* 319, no. 5869: 1497–1502.

Gokhman, David, Eitan Lavi, Kay Prüfer, Mario F. Fraga, José A. Riancho, Janet Kelso, Svante Pääbo, et al. 2014. "Reconstructing the DNA Methylation Maps of the Neandertal and the Denisovan." *Science* 344, no. 6183: 523–27.

Goldberg, Tony L., and Jonathan A. Patz. 2015. "The Need for a Global Health Ethic." *Lancet* 386, no. 10007: 37–39.

Goldin, Ian, and Chris Kutarna. 2016. *Age of Discovery: Navigating the Risks and Rewards of Our New Renaissance*. London: Bloomsbury.

Goldin, Ian, and Mike Mariathasan. 2014. *The Butterfly Defect: How Globalization Creates Systemic Risks, and What to Do about It*. Princeton, NJ: Princeton University Press.

Goldschmidt, Richard. 1940. *The Material Basis of Evolution*. New Haven, CT: Yale University Press.

Gómez-Díaz, Elena, Jacob González-Solís, M. A. Peinado, and Roderic D. M. Page. 2007. "Lack of Host-Dependent Genetic Structure in Ectoparasites of *Calonectris* Shearwaters." *Molecular Ecology* 16, no. 24: 5204–15.

González, Camila, Ophelia Wang, Stavana E. Strutz, Constantino González-Salazar, Víctor Sánchez-Cordero, and Sahotra Sarkar. 2010. "Climate Change and Risk of Leishmaniasis in North America: Predictions from Ecological Niche Models of Vector and Reservoir Species." *PLoS Neglected Tropical Diseases* 4, no. 1: e585.

González-Salazar, Constantino, Christopher R. Stephens, and Víctor Sánchez-Cordero. 2017. "Predicting the Potential Role of Non-Human Hosts in Zika Virus Maintenance." *EcoHealth* 14, no. 1: 171–77. https://doi.org/10.1007/s10393-017-1206-4.

Goodell, Jeff. 2018. "Welcome to the Age of Climate Migration." *Rolling Stone*, February 25.

Gould, Stephen J., and Elisabeth S. Vrba. 1982. "Exaptation: A Missing Term in the Science of Form." *Paleobiology* 8, no. 1: 4–15.

Grabda-Kazubska, B. 1976. "Abbreviation of the Life Cycles in Plagiorchid Trematodes: General Remarks." *Acta Parasitologica Polonica* 24: 125–41.

Grant, Verne. 1977. *Organismic Evolution*. San Francisco: W. H. Freeman.

Grant, Evan H. Campbell, Erin Muths, Rachel A. Katz, Stefano Canessa, Michael J. Adams, Jennifer R. Ballard, Lee Berger, et al. 2017. "Using Decision Analysis to Support Proactive Management of Emerging Infectious Wildlife Diseases." *Frontiers in Ecology and the Environment* 15, no. 4: 214–21. https://doi.org/10.1002/fee.1481.

Green, Marc D., Marco G. P. Veller, and Daniel R. Brooks. 2002. "Assessing Modes of Speciation: Range Asymmetry and Biogeographical Congruence." *Cladistics* 18, no. 1: 112–24.

Greene, Robert. [1923]. *A Notable Discovery of Coosnage, 1591; The Second Part of Conny-Catching, 1592*. Edited by G. B. Harrison. London: John Lane.

Grimm, David. 2015. "Dawn of the Dog." *Science* 348, no. 6232: 274–79.

Grisi, Laerte, Romário C. Leite, João R. de S. Martins, Antonio T. M. de Barros, Renato Andreotti, Paulo H. D. Cançado, Adalberto A. P. de León, et al. 2014. "Reassessment of the Potential Economic Impact of Cattle Parasites in Brazil." *Revista brasileira de parasitologia veterinária* 23, no. 2: 150–56. https://doi.org/10.1590/S1984-29612014042.

Groucutt, Huw S., Michael D. Petraglia, Geoff Bailey, Eleanor M. L. Scerri, Ash Parton, Laine Clark-Balzan, Richard P. Jennings, et al. 2015. "Rethinking the Dispersal of *Homo sapiens* out of Africa." *Evolutionary Anthropology: Issues, News, and Reviews* 24, no. 4: 149–64.

Gubler, Duane J. 2012. "The Economic Burden of Dengue." *American Journal of Tropical Medicine and Hygiene* 86, no. 5: 743–44. https://doi.org/10.4269/ajtmh.2012.12-0157.

Gunton, Richard M., Charles J. Marsh, Sylvain Moulherat, Anne-Kathleen Malchow, Greta Bocedi, Reinhard A. Klenke, and William E. Kunin. 2016.

"Multicriterion Trade-Offs and Synergies for Spatial Conservation Planning." *Journal of Applied Ecology* 54, no. 3: 903–13.

Gupta, Amit Kumar, Karambir Kaur, Akanksha Rajput, Sandeep Kumar Dhanda, Manika Sehgal, Md Shoaib Khan, Isha Monga, et al. 2016. "ZikaVR: An Integrated Zika Virus Resource for Genomics, Proteomics, Phylogenetic and Therapeutic Analysis." *Scientific Reports* 6, no. 1: 1347. https://doi.org/10.1038/srep32713.

Gutiérrez-López, Rafael, Josué Martínez-de la Puente, Laura Gangoso, Ramón C. Soriguer, and Jordi Figuerola. 2015. "Comparison of Manual and Semi-Automatic DNA Extraction Protocols for the Barcoding Characterization of Hematophagous Louse Flies (Diptera: Hippoboscidae)." *Journal of Vector Ecology* 40, no. 1: 11–15. https://doi.org/10.1111/jvec.12127.

Guzman, María G., and Gustavo Kouri. 2003. "Dengue and Dengue Hemorrhagic Fever in the Americas: Lessons and Challenges." *Journal of Clinical Virology* 27, no. 1: 1–13. https://doi.org/10.1016/s1386-6532(03)00010-6.

Hafer, Nina, and Manfred Milinksi. 2016. "Inter- and Intraspecific Conflicts between Parasites over Host Manipulation." *Proceedings of the Royal Society B: Biological Sciences* 283, no. 1824: 20152870. http://dx.doi.org/10.1098/rspb.2015.2870.

Hafner, Mark S., and Steven A. Nadler. 1988. "Phylogenetic Trees Support the Coevolution of Parasites and Their Hosts." *Nature* 332, no. 6161: 258–59.

Halas, Dominik, David Zamparo, and Daniel R. Brooks. 2005. "A Historical Biogeographical Protocol for Studying Biotic Diversification by Taxon Pulses." *Journal of Biogeography* 32, no. 2: 249–60.

Haldane, J. B. S. 1949. "Disease and Evolution." *Rice Science Supplement* 19: 68–76.

Hallmann, Caspar A., Martin Sorg, Eelke Jongejans, Henk Siepel, Nick Hofland, Heinz Schwan, Werner Stenmans, et al. 2017. "More Than 75 Percent Decline over 27 Years in Total Flying Insect Biomass in Protected Areas." *PLoS ONE* 12, no. 10: e0185809. https://doi.org/10.1371/journal.pone.0185809.

Halloy, S. R. P. 2001. "Effects on System Structure, Diversity and Stability of the Distance Exponent in a Resource Attraction Model." *Complexity International* 8: 18.

Hamann, M. I., A. I. Kehr, and C. E. González. 2013. "Biodiversity of Trematodes Associated with Amphibians from a Variety of Habitats in Corrientes Province, Argentina." *Journal of Helminthology* 87, no. 3: 286–300.

Hand, Eric. 2017. "Fossil Leaves Bear Witness to Ancient Carbon Dioxide Levels." *Science* 355, no. 6320: 14–15. https://doi.org/10.1126/science.355.6320.14.

Hansen, James, Makiko Sato, Paul Hearty, Reto Ruedy, Maxwell Kelley, Valerie Masson-Delmotte, Gary Russell, et al. 2016. "Ice Melt, Sea Level Rise and Superstorms: Evidence from Paleoclimate Data, Climate Modeling, and Modern Observations That 2°C Global Warming Could Be Dangerous." *Atmospheric Chemistry and Physics* 16, no. 6: 3761–3812.

Harper, R. M. J. 1958. "Evolution and Illness." *Lancet* 272, no. 7037 (July 12): 92–94. https://doi.org/10.1016/S0140-6736(58)91259-5.

Harris, Stuart A. 2005. "Thermal History of the Arctic Ocean Environs Adjacent to North America during the Last 3.5 Ma and a Possible Mechanism for the Cause of the Cold Events (Major Glaciations and Permafrost Events)." *Progress in Physical Geography* 29, no. 2: 218–37.

Harrison, Joshua G., Zachariah Gompert, James A. Fordyce, C. Alex Buerkle, Rachel Grinstead, Joshua P. Jahner, Scott Mikel, Christopher C. Nice, Aldrin Santamaria, and Matthew L. Forister. 2016. "The Many Dimensions of Diet Breadth: Phytochemical, Genetic, Behavioral, and Physiological Perspectives on the Interaction between a Native Herbivore and an Exotic Host." *PLoS ONE* 11, no. 2: e0147971.

Harrison, Launcelot. 1914. "The Mallophaga as a Possible Clue to Bird Phylogeny." *Australian Zoologist* 1: 7–11.

Harrison, Launcelot. 1915a. "Mallophaga from *Apteryx*, and Their Significance: With a Note on *Rallicola*." *Parasitology* 8: 88–100.

Harrison, Launcelot. 1915b. "The Relationship of the Phylogeny of the Parasite to That of the Host." *Report of the British Association for the Advancement of Science* 85: 476–77.

Harrison, Launcelot. 1916. "Bird-Parasites and Bird-Phylogeny." *Ibis* 10: 254–63.

Harrison, Launcelot. 1922. "On the Mallophagan Family Trimenoponidae, with a Description of a New Genus and Species from an Australian Marsupial." *Australian Zoologist* 2: 154–59.

Harrison, Launcelot. 1924. "The Migration Route of the Australian Marsupial Fauna." *Australian Zoologist* 3: 247–63.

Harrison, Launcelot. 1926. "Crucial Evidence for Antarctic Radiation." *American Naturalist* 60: 374–83.

Harvell, C. Drew, Charles E. Mitchell, Jessica R. Ward, Sonia Altizer, Andrew P. Dobson, Richard S. Ostfeld, and Michael D. Samuel. 2002. "Climate Warming and Disease Risks for Terrestrial and Marine Biota." *Science* 296, no. 5576: 2158–62.

Hassanin, Kamel M. A., and Ahmed S. Abdel-Moneim. 2013. "Evolution of an Avian H5N1 Influenza A Virus Escape Mutant." *World Journal of Virology* 2, no. 4: 160. https://doi.org/10.5501/wjv.v2.i4.160.

Haukisalmi, Voitto, Lotta M. Hardman, Eric P. Hoberg, and Heikki Henttonen. 2014. "Phylogenetic Relationships and Taxonomic Revision of *Paranoplocephala* Lühe, 1910 Sensu Lato (Cestoda, Cyclophyllidea, Anoplocephalidae)." *Zootaxa* 3873, no. 4: 371–415.

Haukisalmi, Voitto, Lotta M. Hardman, Vadim B. Fedorov, Eric P. Hoberg, and Heikki Henttonen. 2016. "Molecular Systematics and Holarctic Phylogeography of Cestodes of the Genus *Anoplocephaloides* Baer, 1923 Ss (Cyclophyllidea, Anoplocephalidae) in Lemmings (Lemmus, Synaptomys)." *Zoologica Scripta* 45, no. 1: 88–102.

Hayes, Mark A., and Antoinette J. Piaggio. 2018. "Assessing the Potential Impacts of a Changing Climate on the Distribution of a Rabies Virus Vector." *PLoS ONE* 13, no. 2: e0192887. https://doi.org/10.1371/journal.pone .0192887.

Heaney, Lawrence R. 2000. "Dynamic Disequilibrium: A Long-Term, Large-Scale Perspective on the Equilibrium Model of Island Biogeography." *Global Ecology and Biogeography* 9, no. 1: 59–74.

Hegner, Robert W. 1927. *Host-Parasite Relations between Man and His Intestinal Protozoa.* New York: Century.

Hegner, Robert W., Francis M. Root, Donald L. Augustine, and Clay G. Huff. 1938. *Parasitology.* New York: D. Appleton-Century.

Hellenthal, Garrett, George B. J. Busby, Gavin Band, James F. Wilson, Cristian Capelli, Daniel Falush, and Simon Myers. 2014. "A Genetic Atlas of Human Admixture History." *Science* 343, no. 6172: 747–51.

Hemida, M., R. Perera, P. Wang, M. Alhammadi, L. Siu, M. Li, L. Poon, L. Saif, A. Alnaeem, and M. Peiris. 2013. "Middle East Respiratory Syndrome (MERS) Coronavirus Seroprevalence in Domestic Livestock in Saudi Arabia, 2010 to 2013." *Eurosurveillance* 18, no. 50: 20659. https://doi.org/10 .2807/1560-7917.ES2013.18.50.20659.

Hemida, Maged Gomaa, Mohammed A. Al-Hammadi, Abdul Hafeed S. Daleb, and Cecilio R. Gonsalves. 2017. "Molecular Characterization and Phylogenetic Analyses of Virulent Infectious Bronchitis Viruses Isolated From Chickens in Eastern Saudi Arabia." *VirusDisease* 28, no. 2: 189–99. https:// doi.org/10.1007/s13337-017-0375-7.

Hennig, Edwin. 1922. *Paläontologische Beiträge zur Entwicklungslehre.* Tübingen: J. C. B. Mohr in Kommission.

Hennig, Edwin. 1927. "Wege und Triebkräfte organischer Entfaltung." *Naturwissenschaften* 18: 260–62.

Hennig, Edwin. 1932. *Wesen und Wege der Paläontologie: Eine Einführung in die Versteinerungslehre als Wissenschaft.* Berlin: Borntraeger.

Hennig, Willi. 1950. *Grundzüge einer Theorie der phylogenetischen Systematik.* Berlin: Deutscher Zentralverlag.

Hennig, Willi. 1966. *Phylogenetic Systematics.* Urbana: University of Illinois Press.

Henry, L. G., Jerry F. McManus, William B. Curry, Natalie L. Roberts, Alexander M. Piotrowski, and L. D. Keigwin. 2016. "North Atlantic Ocean Circulation and Abrupt Climate Change during the Last Glaciation." *Science* 353, no. 6298: 470–74.

Herbert, Brian, and Kevin J. Anderson. 2012. *Hunters of Dune.* London: Hachette.

Herbert, Frank. 1965. *Dune.* New York: Chilton Books.

Herbert, Frank. 1984. *Heretics of Dune.* New York: G. P. Putnam's Sons.

Hernández Fernández, Manuel, and Elisabeth S. Vrba. 2005. "A Complete Estimate of the Phylogenetic Relationships in Ruminantia: A Dated Species-

Level Supertree of the Extant Ruminants." *Biological Reviews* 80, no. 2: 269–302.

Hoberg, Eric P. 1986. "Evolution and Historical Biogeography of a Parasite-Host Assemblage: *Alcataenia* spp. (Cyclophyllidea: Dilepididae) in Alcidae (Charadriiformes)." *Canadian Journal of Zoology* 64, no. 11: 2576–89.

Hoberg, Eric P. 1992. "Congruent and Synchronic Patterns in Biogeography and Speciation among Seabirds, Pinnipeds and Cestodes." *Journal of Parasitology* 78: 601–15.

Hoberg, Eric P. 1995. "Historical Biogeography and Modes of Speciation across High Latitude Seas of the Holarctic: Concepts for Host-Parasite Coevolution among the Phocini (Phocidae) and Tetrabothriidae (Eucestoda)." *Canadian Journal of Zoology* 73: 45–57.

Hoberg, Eric P. 1997. "Phylogeny and Historical Reconstruction: Host-Parasite Systems as Keystones in Biogeography and Ecology." In *Biodiversity II: Understanding and Protecting Our Biological Resources*, edited by Marjorie L. Reaka-Kudla, Don E. Wilson, and Edward Wilson, 243–61. Washington, DC: Joseph Henry Press.

Hoberg, Eric P. 2002. "Foundations for an Integrative Parasitology: Collections, Archives, and Biodiversity Informatics." *Comparative Parasitology* 69, no. 2: 124–31.

Hoberg, Eric P. 2005. "Coevolution and Biogeography among Nematodirinae (Nematoda: Trichostrongylina) Lagomorpha and Artiodactyla (Mammalia): Exploring Determinants of History and Structure for the Northern Fauna across the Holarctic." *Journal of Parasitology* 91, no. 2: 358–69.

Hoberg, Eric P. 2006. "Phylogeny of *Taenia*: Species Definitions and Origins of Human Parasites." *Parasitology International* 55: 23–30.

Hoberg, Eric P. 2010. "Invasive Processes, Mosaics and the Structure of Helminth Parasite Faunas." *Revue scientifique et technique (International Office of Epizootics)* 29: 255–72.

Hoberg, Eric P., and A. M. Adams. 2000. "Phylogeny, History and Biodiversity: Understanding Faunal Structure and Biogeography in the Marine Realm." *Bulletin of the Scandinavian Society for Parasitology* 10, no. 2: 19–37.

Hoberg, Eric P., and Daniel R. Brooks. 2008. "A Macroevolutionary Mosaic: Episodic Host-Switching, Geographical Colonization and Diversification in Complex Host-Parasite Systems." *Journal of Biogeography* 35, no. 9: 1533–50.

Hoberg, Eric P., and Daniel R. Brooks. 2010. "Beyond Vicariance: Integrating Taxon Pulses, Ecological Fitting, and Oscillation in Evolution and Historical Biogeography." In *The Biogeography of Host-Parasite Interactions*, edited by Serge Morand and Boris R. Krasnov, 7–20. New York: Oxford University Press.

Hoberg, Eric P., and Daniel R. Brooks. 2013. "Episodic Processes, Invasion and Faunal Mosaics in Evolutionary and Ecological Time." In *The Balance of Nature and Human Impact*, edited by Klaus Rohde, 199–214. Cambridge: Cambridge University Press.

Hoberg, Eric P., and Daniel R. Brooks. 2015. "Evolution in Action: Climate Change, Biodiversity Dynamics and Emerging Infectious Disease." *Philosophical Transactions of the Royal Society B: Biological Sciences* 370, no. 1665: 20130553.

Hoberg, Eric P., and Dante S. Zarlenga. 2016. "Evolution and Biogeography of *Haemonchus contortus*: Linking Faunal Dynamics in Space and Time." *Advances in Parasitology* 93: 1–30.

Hoberg, Eric P., Anson Koehler, and Joseph Cook. 2012a. "Complex Host-Parasite Systems in *Martes*: Implications for Conservation Biology of Endemic Faunas." In *Biology and Conservation of Martens, Sables and Fishers*, edited by Keith B. Aubry, William J. Zielinski, Martin G. Raphael, Gilbert Proulx, and Steven W. Buskirk, 39–57. Ithaca, NY: Cornell University Press.

Hoberg, Eric P., Arlene Jones, Robert L. Rausch, Keeseon S. Eom, and Scott Lyell Gardner. 2000. "A Phylogenetic Hypothesis for Species of the Genus *Taenia* (Eucestoda: Taeniidae)." *Journal of Parasitology* 86, no. 1: 89–98.

Hoberg, Eric P., Daniel R. Brooks, and Douglas Siegel-Causey. 1997. "Host-parasite cospeciation: History, principles and prospects." In *Host-Parasite Evolution: General Principles and Avian Models*, edited by Dale H. Clayton and Janice Moore, 212–35. New York: Oxford University Press.

Hoberg, Eric P., J. Ralph Lichtenfels, and Lynda Gibbons. 2004. "Phylogeny for Species of *Haemonchus* (Nematoda: Trichostrongyloidea): Considerations of Their Evolutionary History and Global Biogeography among Camelidae and Pecora (Artiodactyla)." *Journal of Parasitology* 90, no. 5: 1085–1102.

Hoberg, Eric P., Joseph A. Cook, Salvatore J. Agosta, Walter Boeger, Kurt E. Galbreath, Sauli Laaksonen, Susan J. Kutz, and Daniel R. Brooks. 2017. "Arctic Systems in the Quaternary: Ecological Collision, Faunal Mosaics and the Consequences of a Wobbling Climate." *Journal of Helminthology* 91, no. 4: 409–21.

Hoberg, Eric P., Kurt E. Galbreath, Joseph A. Cook, Susan J. Kutz, and Lydden Polley. 2012b. "Northern Host-Parasite Assemblages: History and Biogeography on the Borderlands of Episodic Climate and Environmental Transition." *Advances in Parasitology* 79: 1–97.

Hoberg, Eric P., Lydden Polley, E. J. Jenkins, and S. J. Kutz. 2008. "Pathogens of Domestic and Free-Ranging Ungulates: Global Climate Change in Temperate to Boreal Latitudes across North America." *Revue scientifique et technique (International Office of Epizootics)* 27: 511–28.

Hoberg, Eric P., Nancy L. Alkire, A. D. Queiroz, and Arlene Jones. 2001. "Out of Africa: Origins of the *Taenia* Tapeworms in Humans." *Proceedings of the Royal Society B: Biological Sciences* 268, no. 1469: 781–87.

Hoberg, Eric P., Salvatore J. Agosta, Walter A. Boeger, and Daniel R. Brooks. 2015. "An Integrated Parasitology: Revealing the Elephant through Tradition and Invention." *Trends in Parasitology* 31, no. 4: 128–33.

Hoberg, Eric P., Susan J. Kutz, J. A. Cook, Kirill Galaktionov, Voitto Haukisalmi, Heikki Henttonen, Sauli Laaksonen, A. Makarikov, and D. J. Marcogliese.

2013. "Parasites in Terrestrial, Freshwater and Marine Systems." In *Arctic Biodiversity Assessment: Status and Trends in Arctic Biodiversity; Conservation of Arctic Flora and Fauna*, 476–505. Akureyi, Iceland: Arctic Council.

Hoberg, Eric P., Susan J. Kutz, Kurt Galbreath, and Joseph A. Cook. 2003. "Arctic Biodiversity: From Discovery to Faunal Baselines—Revealing the History of a Dynamic Ecosystem." Supplement, *Journal of Parasitology* 89: S84–95.

Hofer, Stefan, Andrew J. Tedstone, Xavier Fettweis, and Jonathan L. Bamber. 2017. "Decreasing Cloud Cover Drives the Recent Mass Loss on the Greenland Ice Sheet." *Science Advances* 3, no. 6: e1700584.

Hoffman, Jeremy S., Peter U. Clark, Andrew C. Parnell, and Feng He. 2017. "Regional and Global Sea-Surface Temperatures during the Last Interglaciation." *Science* 355, no. 6322: 276–79.

Hoffmann, Ary A., and Miriam J. Hercus. 2000. "Environmental Stress as an Evolutionary Force." *BioScience* 50, no. 3: 217–26.

Höfle, Ursula, Marco W. G. van de Bildt, Lonneke M. Leijten, Geert van Amerongen, Josanne H. Verhagen, Ron A. M. Fouchier, Albert D. M. E. Osterhaus, et al. 2012. "Tissue Tropism and Pathology of Natural Influenza Virus Infection in Black-Headed Gulls (Chroicocephalus Ridibundus)." *Avian Pathology* 41, no. 6: 547–53. https://doi.org/10.1080/03079457.2012.744447.

Höglund, Johan, Katarina Gustafsson, Britt-Louise Ljungström, Moa Skarin, Marian Varady, and Fredrik Engström. 2015. "Failure of Ivermectin Treatment in *Haemonchus contortus* Infected-Swedish Sheep Flocks." *Veterinary Parasitology: Regional Studies and Reports* 1–2 (December): 10–15. https://doi.org/10.1016/j.vprsr.2016.02.001.

Holden, Constance. 1997. "Monkeypox Not Mutating." *Science* 278, no. 5345: 1885.

Holmes, John C. 1973. "Site Selection by Parasitic Helminths: Interspecific Interactions, Site Segregation, and Their Importance to the Development of Helminth Communities." *Canadian Journal of Zoology* 51, no. 3: 333–47. https://doi.org/10.1139/z73-047.

Holmes, John C. 1983. "Evolutionary Relationships between Parasitic Helmiths and Their Hosts." In *Coevolution*, edited by Douglas J. Futuyma and Montgomery Slatkin, 161–85. Sunderland, MA: Sinauer Associates.

Hoorn, Carina, F. P. Wesselingh, H. Ter Steege, M. A. Bermudez, A. Mora, J. Sevink, I. Sanmartín, et al. 2010. "Amazonia through Time: Andean Uplift, Climate Change, Landscape Evolution, and Biodiversity." *Science* 330, no. 6006: 927–31.

Hope, Andrew G., Eric Waltari, David C. Payer, Joseph A. Cook, and Sandra L. Talbot. 2013. "Future Distribution of Tundra Refugia in Northern Alaska." *Nature Climate Change* 3 , no. 10: 931–38.

Hope, Andrew G., Brett K. Sandercock, and Jason L. Malaney. 2018. "Collection of Scientific Specimens: Benefits for Biodiversity Sciences and Limited Impacts on Communities of Small Mammals." *BioScience* 68, no. 1: 35–42. https://doi.org/10.1093/biosci/bix141.

Hope, Andrew G., Eric Waltari, Jason L. Malaney, David C. Payer, J. A. Cook, and S. L. Talbot. 2015. "Arctic Biodiversity: Increasing Richness Accompanies Shrinking Refugia for a Cold-Associated Tundra Fauna." *Ecosphere* 6, no. 9: 1–67.

Hope, Andrew G., Naoki Takebayashi, Kurt E. Galbreath, Sandra L. Talbot, and Joseph A. Cook. 2013. "Temporal, Spatial and Ecological Dynamics of Speciation among Amphi-Beringian Small Mammals." *Journal of Biogeography* 40, no. 3: 415–29.

Hope, Andrew G., S. E. Greiman, V. V. Tkach, E. P. Hoberg, and J. A. Cook. 2016. "Shrews and Their Parasites: Small Species Indicate Big Changes?" *Arctic Report Card.* http://www.arctic.noaa.gov/Report-Card/Report-Card -2016.

Hopkins, David M. 1959. "Cenozoic History of the Bering Land Bridge." *Science* 129: 1519–28.

Hopkins, G. H. E. 1942. "The Mallophaga as an Aid to the Classification of Birds." *Ibis* 6: 94–106.

Hopper, Julie V., and Nicholas J. Mills. 2016. "Novel Multitrophic Interactions among an Exotic, Generalist Herbivore, Its Host Plants and Resident Enemies in California." *Oecologia* 182, no. 4: 1117–28. doi 10.1007/ s00442-016-3722-2.

Hornok, Sándor, Gábor Földvári, Krisztina Rigó, Marina L. Meli, Enikő Gönczi, Attila Répási, Róbert Farkas, et al. 2015. "Synanthropic Rodents and Their Ectoparasites as Carriers of a Novel Haemoplasma and Vector-Borne, Zoonotic Pathogens Indoors." *Parasites & Vectors* 8, no. 1: 27. https://doi.org/10 .1186/s13071-014-0630-3.

Horton, Richard, and Selina Lo. 2015. "Planetary Health: A New Science for Exceptional Action." *Lancet* 386, no. 10007: 1921–22.

Houck, Ulysses G. 1924. "History of the Bureau of Animal Industry and Zoological Division." U.S. National Animal Parasite Collection Records. Box 98, Folder 5. Washington, DC: Special Collections, National Agricultural Library.

Houldcroft, Charlotte J., and Simon J. Underdown. 2016. "Neanderthal Genomics Suggests a Pleistocene Time Frame for the First Epidemiologic Transition." *American Journal of Physical Anthropology* 160, no. 3: 379–88.

Hovmøller, Mogens S., Chris K. Sørensen, Stephanie Walter, and Annemarie F. Justesen. 2011. "Diversity of *Puccinia striiformis* on Cereals and Grasses." *Annual Review of Phytopathology* 49, no. 1: 197–217. https://doi.org/10.1146/ annurev-phyto-072910-095230.

Hoyal Cuthill, Jennifer F., Kim B. Sewell, Lester R. G. Cannon, Michael A. Charleston, Susan Lawler, D. Timothy J. Littlewood, Peter D. Olson, and David Blair. 2016. "Australian Spiny Mountain Crayfish and Their Temnocephalan Ectosymbionts: An Ancient Association on the Edge of Coextinction?" *Proceedings of the Royal Society B: Biological Sciences* 283, no. 1831: 2016 0585. http://dx.doi.org/10.1098/rspb.2016.0585.

Hsiang, Solomon M., and Adam H. Sobel. 2016. "Potentially Extreme Population Displacement and Concentration in the Tropics Under Non-Extreme
Warming." *Scientific Reports* 6: 25697.

Hu, Dalong, Bin Liu, Lu Feng, Peng Ding, Xi Guo, Min Wang, Boyang Cao,
Peter R. Reeves et al. 2016. "Origins of the Current Seventh Cholera Pan-
demic." *Proceedings of the National Academy of Sciences of the United States of
America* 113, no. 48: E7730–39. 201608732.

Hubble, Douglas. 1958. "The Autobiography of Charles Darwin." *Lancet* 272,
no. 7036 (July 5): 37–39. https.//doi.org/10.1016/S0140-6376(58)90020-5.

Huffman, Michael A. 1997. "Current Evidence for Self-Medication in Primates: A Multidisciplinary Perspective." In "Yearbook of Physical Anthropology." Supplement, *American Journal of Physical Anthropology*
104, no. 25: 171–200. https://doi.org/10.1002/(SICI)1096-8644(1997)
25+<171::AID-AJPAJ>3.0.CO;2-7.

Huffman, Michael A. 2001. "Self-Medicative Behavior in the African Great
Apes: An Evolutionary Perspective into the Origins of Human Traditional
Medicine: In Addition to Giving Us a Deeper Understanding of Our Clos-
est Living Relatives, the Study of Great Ape Self-Medication Provides a
Window into the Origins of Herbal Medicine Use by Humans and Prom-
ises to Provide New Insights into Ways of Treating Parasite Infections and
Other Serious Diseases." *American Institute of Biological Sciences Bulletin* 51,
no. 7: 651–61.

Huffman, Michael A. 2016. "Primate Self-Medication, Passive Prevention and
Active Treatment—A Brief Review." *International Journal of Multidisciplinary
Studies* 3: 1–10.

Huffman, Michael A., and Sylvia K. Vitazkova. 2007. "Primates, Plants, and Parasites: The Evolution of Animal Self-Medication and Ethnomedicine." In *Eth-
nopharmacology*, vol. 2, *Primates, Plants, and Parasites: The Evolution of Animal
Self-Medication and Ethnomedicine*. Oxford: EOLSS. http://www.eolss.net.

Huffman, Michael A., Naofumi Nakagawa, Yasuhiro Go, Hiroo Imai, and
Masaki Tomonaga. 2013. *Monkeys, Apes, and Humans: Primatology in Japan*.
New York: Springer Science & Business Media.

Hull, David L. 1988. *Science as a Process*. Chicago: University of Chicago Press.

Hurrell, James W., Yochanan Kushnir, Geir Ottersen, and Martin Visbeck.
2003. "An Overview of the North Atlantic Oscillation." In *The North Atlantic Oscillation: Climatic Significance and Environmental Impact*, edited by
James W. Hurrell, Yochanan Kushnir, Geir Ottersen, and Martin Visbeck,
1–35. Geophysical Monograph Series 134. Washington, DC: American
Geophysical Union. https://doi.org/10.1029/134GM01.

Huxley, Julian. 1942. *Evolution: The Modern Synthesis*. London: George Allen &
Unwin. http://krishikosh.egranth.ac.in/bitstream/1/2057456/1/ANAND-52
.pdf.

Huyse, Tine, and Filip A. M. Volckaert. 2005. "Comparing Host and Parasite
Phylogenies: *Gyrodactylus* Flatworms Jumping from Goby to Goby." *Systematic Biology* 54, no. 5: 710–718.

Ibisch, Pierre L., Monika T. Hoffmann, Stefan Kreft, Guy Pe'er, Vassiliki Kati, Lisa Biber-Freudenberger, Dominick A. DellaSala, et al. 2016. "A Global Map of Roadless Areas and Their Conservation Status." *Science* 354, no. 6318: 1423–27.

Intergovernmental Panel on Climate Change [IPCC]. 2013. "Summary for Policymakers." In *Climate Change 2013: The Physical Science Basis; Contribution of Working Group I to the Fifth Assessment Report of the Intergovernmental Panel on Climate Change,* edited by T. F. Stocker, D. Qin, G.-K. Plattner, M. Tignor, S. K. Allen, J. Boschung, A. Nauels, Y. Xia, V. Bex, and P. M. Midgley, 1–27. Cambridge: Cambridge University Press.

Intergovernmental Panel on Climate Change [IPCC]. 2014a. "Summary for Policymakers." In *Climate Change 2014: Impacts, Adaptation, and Vulnerability; Part A: Global and Sectoral aspects; Contribution of Working Group II to the Fifth Assessment Report of the Intergovernmental Panel on Climate Change,* edited by C. B. Field, V. R. Barros, D. J. Dokken, K. J. Mach, M. D. Mastrandrea, T. E. Bilir, M. Chatterjee, et al,, 1–32. Cambridge: Cambridge University Press.

Intergovernmental Panel on Climate Change [IPCC]. 2014b. *Climate Change 2014—Impacts, Adaptation and Vulnerability: Global and Sectoral Aspects.* Edited by Christopher B. Field, Vicente R. Barros, David Jon Dokken, Katharine J. Mach, and Michael D Mastrandrea. Cambridge: Cambridge University Press. https://doi.org/10.1017/CBO9781107415379.

Intergovernmental Panel on Climate Change [IPCC]. 2014c. *Climate Change 2014: Synthesis Report. Contribution of Working Groups I, II and III to the Fifth Assessment Report of the Intergovernmental Panel on Climate Change.* [Core writing team: R. K. Pachauri and L. A. Meyer (eds.)]. Geneva: IPCC.

Iverson, Samuel A., Mark R. Forbes, Manon Simard, Catherine Soos, and H. Grant Gilchrist. 2016. "Avian Cholera Emergence in Arctic-Nesting Northern Common Eiders: Using Community-Based, Participatory Surveillance to Delineate Disease Outbreak Patterns and Predict Transmission Risk." *Ecology and Society* 21, no. 4: 12. https://doi.org/10.5751/ES-08873-210412.

Jacobsen, Kathryn H., A. Alonso Aguirre, Charles L. Bailey, Ancha V. Baranova, Andrew T. Crooks, Arie Croitoru, Paul L. Delamater, et al. 2016. "Lessons from the Ebola Outbreak: Action Items for Emerging Infectious Disease Preparedness and Response." *EcoHealth* 13, no. 1: 200–212. https://doi.org/10.1007/s10393-016-1100-5.

Jackson, H. S. 1931. "Present Evolutionary Tendencies and Origin of the Life Cycles in the Uredinales." *Memoirs of the Torrey Botanical Club* 18: 1–108.

Jacquiet, P., J. F. Humbert, A. M. Comes, J. Cabaret, A. Thiam, and D. Cheikh. 1995. "Ecological, Morphological and Genetic Characterization of Sympatric *Haemonchus* spp. Parasites of Domestic Ruminants in Mauritania." *Parasitology* 110, no. 4: 483–92.

James, Alexander, Kevin J. Gaston, and Andrew Balmford. 2001. "Can We Afford to Conserve Biodiversity?" *Bioscience* 51, no. 1: 43–52. https://doi.org/10.1641/0006-3568(2001)051[0043:CWATCB]2.0.CO;2.

James, Catherine E., Amanda L. Hudson, and Mary W. Davey. 2009. "Drug Resistance Mechanisms in Helminths: Is It Survival of the Fittest?" *Trends in Parasitology* 25, no. 7: 328–35.

Janovy, John, Jr. 2002. "Concurrent Infections and the Community Ecology of Helminth Parasites." *Journal of Parasitology* 88, no. 3: 440–45. https://doi .org/10.2307/3285429.

Jansen, E., J. Overpeck, K. R. Briffa, J. C. Duplessy, F. Joos, V. Masson-Delmotte, D. Olago, et al. 2007. "Palaeoclimate." In *Climate Change 2007: The Physical Science Basis: Contribution of Working Group I to the Fourth Assessment Report of the Intergovernmental Panel on Climate Change*, edited by S. Solomon, D. Qin, M. Manning, Z. Chen, M. Marquis, K. B. Averyt, M. Tignor, and H. L. Miller, 434–97. Cambridge: Cambridge University Press.

Jansen, Mona Dverdal, Ha Thanh Dong, and Chadag Vishnumurthy Mohan. 2018. "Tilapia Lake Virus: A Threat to the Global Tilapia Industry?." *Reviews in Aquaculture* 46 (May): 120. https://doi.org/10.1111/raq .12254.

Jansson, Roland, and Mats Dynesius. 2002. "The Fate of Clades in a World of Recurrent Climatic Change: Milankovitch Oscillations and Evolution." *Annual Review of Ecology and Systematics* 3, no. 1: 741–77.

Janz, Niklas. 2011. "Ehrlich and Raven Revisited: Mechanisms Underlying Co-diversification of Plants and Enemies." *Annual Review of Ecology, Evolution, and Systematics* 42: 71–89.

Janz, Niklas, and Sören Nylin. 2008. "The Oscillation Hypothesis of Host-Plant Range and Speciation." In *Specialization, Speciation, and Radiation: The Evolutionary Biology of Herbivorous Insects*, edited by Kelley J. Tilmon, 203–15. Berkeley and Los Angeles: University of California Press.

Janz, Niklas, Sören Nylin, and Niklas Wahlberg. 2006. "Diversity Begets Diversity: Host Expansions and the Diversification of Plant-Feeding Insects." *BMC Evolutionary Biology* 6: 4.

Janzen, Daniel H. 1980. "When Is It Coevolution?" *Evolution* 34, no. 3: 611–12.

Janzen, Daniel H. 1985. "On Ecological Fitting." *Oikos* 45: 308–10.

Janzen, D. H., and W. Hallwachs. 1994. *All Taxa Biodiversity Inventory (ATBI) of Terrestrial Systems: A Generic Protocol for Preparing Wildland Biodiversity for Non-Damaging Use*. Washington, DC: Report of a National Science Foundation Workshop.

Janzen, Daniel H., and Winnie Hallwachs. 2011. "Joining Inventory by Parataxonomists with DNA Barcoding of a Large Complex Tropical Conserved Wildland in Northwestern Costa Rica." *PLoS ONE* 6, no. 7: e18123. https://doi.org/10.1371/journal.pone.0018123.

Janzen, Daniel H., Winnie Hallwachs, Jorge Jimenez, and Rodrigo Gámez. 1993. "The Role of the Parataxonomists, Inventory Managers, and Taxonomists in Costa Rica's National Biodiversity Inventory." In *Biodiversity Prospecting*, edited by W. V. Reid et al., 223–54. Washington, DC: World Resources Institute.

Janzen, Daniel H., and Paul S. Martin. 1982. "Neotropical Anachronisms: The Fruits the Gomphotheres Ate." *Science* 215: 19–27.

Jenkins, Emily J., Greg D. Appleyard, Eric P. Hoberg, Benjamin M. Rosenthal, Susan J. Kutz, Alasdair M. Veitch, Helen M. Schwantje, Brett T. Elkin, and Lydden Polley. 2005. "Geographic Distribution of the Muscle-Dwelling Nematode *Parelaphostrongylus odocoilei* in North America, Using Molecular Identification of First-Stage Larvae." *Journal of Parasitology* 91, no. 3: 574–84.

Jenkins, Emily J., Louisa J. Castrodale, S. J. de Rosemond, Brent R. Dixon, Stacey A. Elmore, Karen M. Gesy, Eric P. Hoberg, et al. 2013. "Tradition and Transition: Parasitic Zoonoses of People and Animals in Alaska, Northern Canada, and Greenland." *Advances in Parasitology* 82: 33–204. https://doi.org/10.1016/B978-0-12-407706-5.00002-2.

Jermy, Tibor. 1976. "Insect–Host-Plant Relationship—Co-Evolution or Sequential Evolution?" In *The Host-Plant in Relation to Insect Behaviour and Reproduction*, edited by Tibor Jermy, 109–13. Boston: Springer.

Jermy, Tibor. 1984. "Evolution of Insect/Host Plant Relationships." *American Naturalist* 124: 609–30.

Johnson, Christopher N., Andrew Balmford, Barry W. Brook, Jessie C. Buettel, Mauro Galetti, Lei Guangchun, and Janet M. Wilmshurst. 2017. "Biodiversity Losses and Conservation Responses in the Anthropocene." *Science* 356, no. 6335: 270–75. https://doi.org/10.1126/science.aam9317.

Johnson, Kevin P., R. J. Adams, and Dale H. Clayton. 2002. "The Phylogeny of the Louse Genus *Brueelia* Does Not Reflect Host Phylogeny." *Biological Journal of the Linnean Society* 77, no. 2: 233–47.

Johnson, Warren E., Eduardo Eizirik, Jill Pecon-Slattery, William J. Murphy, Agostinho Antunes, Emma Teeling, and Stephen J. O'Brien. 2006. "The Late Miocene Radiation of Modern Felidae: A Genetic Assessment." *Science* 311, no. 5757: 73–77.

Johnston, S. J. 1912. "On Some Trematode Parasites of Australian Frogs." *Proceedings of the Linnean Society of New South Wales* 37: 285–362.

Johnston, S. J. 1913. "Trematode Parasites and the Relationships and Distribution of Their Hosts." *Report of the Australian Association for the Advancement of Science Melbourne* 14: 272–78.

Johnston, S. J. 1914a. "Australian Trematodes and Cestodes: A Study in Zoogeography." *Proceedings of the British Association for the Advancement of Science Australia* 1: 3–7.

Johnston, S. J. 1914b. "Australian Trematodes and Cestodes." *Medical Journal of Australia* 1: 243–44.

Johnston, S. J. 1916. "On the Trematodes of Australian Birds." *Journal and Proceedings of the Royal Society, New South Wales* 50: 187–261.

Joly, D., C. Kreuder Johnson, T. Goldstein, S. J. Anthony, W. Karesh, P. Daszak, N. Wolfe, et al. 2016. "The First Phase of PREDICT: Surveillance for Emerging Infectious Zoonotic Diseases of Wildlife Origin (2009–2014)." *Inter-*

national Journal of Infectious Diseases 53: 31–32. https://doi.org/10.1016/j .ijid.2016.11.086.

Jones, Kate E., Nikkita G. Patel, Marc A. Levy, Adam Storeygard, Deborah Balk, John L. Gittleman, and Peter Daszak. 2008. "Global Trends in Emerging Infectious Diseases." *Nature* 451, no. 7181: 990–93. https://doi.org/10.1038/ nature06536.

Jones, Philip H., and Hugh B. Britten. 2010. "The Absence of Concordant Population Genetic Structure in the Black-Tailed Prairie Dog and the Flea, *Oropsylla hirsuta*, with Implications for the Spread of *Yersinia pestis*." *Molecular Ecology* 19, no. 10: 2038–49. https://doi.org/10.1111/j.1365-294X.2010.04634.x.

Jordan, David S. 1920. "Orthogenesis among Fishes." *Science* 52: 13–14.

Jordan, David S., and Vernon L. Kellogg. 1900. *Animal Life: A First Book of Zoology*. New York: Appleton.

Jordan, David S., and Vernon L. Kellogg. 1908. "The Law of Geminate Species." *American Naturalist* 42: 73–80.

Jordan, David S., Vernon L. Kellogg, and H. Heath. 1909. *Animal Studies: A Textbook of Elementary Zoology for Use in High Schools and Colleges*. New York: D. Appleton.

Jorge, Fátima, Ana Perera, Robert Poulin, Vicente Roca, and Miguel A. Carretero. 2018. "Getting There and Around: Host Range Oscillations during Colonization of the Canary Islands by the Parasitic Nematode *Spauligodon*." *Molecular Ecology* 2: 533–49. https://doi.org/10.1111/mec.14458.

Juarrero, Alicia. 1999. *Dynamics in Action*. Boston: MIT Press.

Juma, Gerald, Gilles Clément, Peter Ahuya, Ahmed Hassanali, Sylvie Derridj, Cyrile Gaertner, Romain Linard, et al. 2016. "Influence of Host-Plant Surface Chemicals on the Oviposition of the Cereal Stemborer *Busseola fusca*." *Journal of Chemical Ecology* 42, no. 5: 394–403.

Justiniano, Rebecca, John J. Schenk, Danilo S. Balete, Eric A. Rickart, Jacob A. Esselstyn, Lawrence R. Heaney, and Scott J. Steppan. 2014. "Testing Diversification Models of Endemic Philippine Forest Mice (*Apomys*) with Nuclear Phylogenies across Elevational Gradients Reveals Repeated Colonization of Isolated Mountain Ranges." *Journal of Biogeography* 42, no. 1: 51–64. https://doi.org/10.1111/jbi.12401.

Kang, Hae Ji, Shannon N. Bennett, Andrew G. Hope, Joseph A. Cook, and Richard Yanagihara. 2011. "Shared Ancestry between a Newfound Mole-Borne Hantavirus and Hantaviruses Harbored by Cricetid Rodents." *Journal of Virology* 85, no. 15: 7496–7503. https://doi.org/10.1128/JVI.02450-10.

Kappagoda, Shanthi, Upinder Singh, and Brian G. Blackburn. 2011. "Antiparasitic Therapy." *Mayo Clinic Proceedings* 86: 561–83.

Kardon, Gabrielle. 1998. "Evidence from the Fossil Record of an Antipredatory Exaptation: Conchiolin Layers in Corbulid Bivalves." *Evolution* 52, no. 1: 68–79.

Kasl, Emily L., Chris T. McAllister, Henry W. Robison, Matthew B. Connior, William F. Font, and Charles D. Criscione. 2015. "Evolutionary Consequence of a Change in Life Cycle Complexity: A Link Between Precocious Development and Evolution toward Female-Biased Sex Allocation in a Hermaphroditic Parasite." *Evolution* 69, no. 12: 3156–70. https://doi.org/10 .1111/evo.12805.

Kauffman, Stuart A. 1993. *The Origins of Order: Self-Organization and Selection in Evolution.* New York: Oxford University Press.

Kaufman, Darrell S., David P. Schneider, Nicholas P. McKay, Caspar M. Ammann, Raymond S. Bradley, Keith R. Briffa, Gifford H. Miller, et al. 2009. "Recent Warming Reverses Long-Term Arctic Cooling." *Science* 325, no. 5945: 1236–39.

Kearn, Graham C. 1986. "The Eggs of Monogeneans." *Advances in Parasitology* 25: 175–273.

Kedmi, M., M. Van Straten, E. Ezra, N. Galon, and E. Klement. 2010. "Assessment of the Productivity Effects Associated with Epizootic Hemorrhagic Disease in Dairy Herds." *Journal of Dairy Science* 93, no. 6: 2486–95. https:// doi.org/10.3168/jds.2009-2850.

Kellogg, Vernon L. 1896. "New Mallophaga, I, With Special Reference to a Collection from Maritime Birds of the Bay of Monterey, California." *Proceedings of the California Academy of Science,* 2nd ser., 6: 31–168.

Kellogg, Vernon L. 1907. *Darwinism Today.* New York: Holt.

Kellogg, Vernon L. 1913a. "Distribution and Species-Forming of Ectoparasites." *American Naturalist* 47: 129–58.

Kellogg, Vernon L. 1913b. "Ectoparasites of the Monkeys, Apes and Man." *Science* 38: 601–2.

Kellogg, Vernon L. 1914. "Ectoparasites of Mammals." *American Naturalist* 48: 257–79.

Kellogg, Vernon L., and Shinkai I. Kuwana. 1902. "Mallophaga from Birds." *Proceedings of the Washington Academy of Science* 4: 457–99.

Kethley, John B., and Donald E. Johnston. 1975. "Resource Tracking Patterns in Bird and Mammal Ectoparasites." *Miscellaneous Publications of the Entomoiogcal Society of America* 9: 231–36.

Khormi, Hassan M., and Lalit Kumar. 2014. "Climate Change and the Potential Global Distribution of *Aedes aegypti*: Spatial Modelling Using Geographical Information System and CLIMEX." *Geospatial Health* 8, no. 2: 405–15.

Kilpatrick, A. Marm, Andrew D. M. Dobson, Taal Levi, Daniel J. Salkeld, Andrea Swei, Howard S. Ginsberg, Anne Kjemtrup, et al. 2017. "Lyme Disease Ecology in a Changing World: Consensus, Uncertainty and Critical Gaps for Improving Control." *Philosophical Transactions of the Royal Society B: Biological Sciences* 372, no. 1722: 20160117–16. https://doi.org/10.1098/rstb .2016.0117.

Kindhauser, Mary Kay, Tomas Allen, Veronika Frank, Ravi Shankar Santhana, and Christopher Dye. 2017. "Zika: The Origin and Spread of a Mosquito-Borne Virus." *Bulletin of the World Health Organization* 94, no. 9: 675–86. http://dx.doi.org/10.2471/BLT.16.171082.

Kintisch, Eli. 2016. "The Lost Norse." *Science* 354: 696–701. https://doi.org/10.1126/science.354.6313.696.

Kintisch, Eli. 2017. "Meltdown." *Science* 355: 788–91. https://doi.org/10.1126/science.355.6327.788.

Kirby, Harold. 1937. "Host-Parasite Relations in the Distribution of Protozoa in Termites." *University of California Publications in Zoology* 41: 189–212.

Kirigia, Joses M., Luis G. Sambo, Allarangar Yokouide, Edoh Soumbey-Alley, Lenity K. Muthuri, and Doris G. Kirigia. 2009. "Economic Burden of Cholera in the WHO African Region." *BMC International Health and Human Rights* 9: 8. https://doi.org/10.1186/1472-698X-9-8.

Klein, Richard J. T., Saleemul Huq, Fatima Denton, Thomas E. Downing, Richard G. Richels, John B. Robinson, and Ferenc L. Toth. 2007. "Inter-Relationships between Adaptation and Mitigation." In *Climate Change 2007: Impacts, Adaptation and Vulnerability: Working Group II Contribution to the Fourth Assessment Report of the Intergovernmental Panel on Climate Change*, edited by M. L. Parry, O. F. Canziani, J. P. Palutikof, P. J. van der Linden, and C. E. Hanson, 745–77. Cambridge: Cambridge University Press.

Kloos, Helmut, and Rosalie David. 2002. "The Paleoepidemiology of Schistosomiasis in Ancient Egypt." *Human Ecology Review* 9: 14–25. https://doi.org/10.4324/9781315431734.

Knight-Jones, T. J. D., and J. Rushton. 2013. "The Economic Impacts of Foot and Mouth Disease—What Are They, How Big Are They and Where Do They Occur?" *Preventive Veterinary Medicine* 112, no. 3: 161–73.

Knowles, L. Lacey, Douglas J. Futuyma, Walter F. Eanes, and Bruce Rannala. 1999. "Insight into Speciation from Historical Demography in the Phytophagous Beetle Genus *Ophraella*." *Evolution* 53, no. 6: 1846–56.

Koch, R. 1876. "Untersuchungen über Bakterien: V. Die Ätiologie der Milzbrand-Krankheit, begründet auf die Entwicklungsgeschichte des *Bacillus anthracis*." *Beiträge zur Biologie der Pflanzen* 2: 277–310.

Koehler, Anson V. A., Eric P. Hoberg, Nikolai E. Dokuchaev, Nina A. Tranbenkova, Jackson S. Whitman, David W. Nagorsen, and Joseph A. Cook. 2009. "Phylogeography of a Holarctic Nematode, *Soboliphyme baturini*, among Mustelids: Climate Change, Episodic Colonization, and Diversification in a Complex Host-Parasite System." *Biological Journal of the Linnean Society* 96, no. 3: 651–63.

Koepfli, Klaus-Peter, Kerry A. Deere, Graham J. Slater, Colleen Begg, Keith Begg, Lon Grassman, Mauro Lucherini, et al. 2008. "Multigene Phylogeny of the Mustelidae: Resolving Relationships, Tempo and Biogeographic History of a Mammalian Adaptive Radiation." *BMC Biology* 6: 10. https://doi.org/10.1186/1741-7007-6-10.

Kohli, Brooks A., Kelly A. Speer, C. William Kilpatrick, Nyamsuren Batsaikhan, Darmaa Damdinbaza, and Joseph A. Cook. 2014. "Multilocus Systematics and Non-Punctuated Evolution of Holarctic Myodini (Rodentia: Arvicolinae)." *Molecular Phylogenetics and Evolution* 76: 18–29. https://doi.org/10 .1016/j.ympev.2014.02.019.

Kohli, Brooks A., Vadim B. Fedorov, Eric Waltari, and Joseph A. Cook. 2015. "Phylogeography of a Holarctic Rodent (*Myodes rutilus*): Testing High-Latitude Biogeographical Hypotheses and the Dynamics of Range Shifts." *Journal of Biogeography* 42, no. 2: 377–89. https://doi.org/10.1111/jbi.12433.

Köpf, Alfred, Nathan E. Rank, Heikki Roininen, Riitta Julkunen-Tiitto, Jacques M. Pasteels, and Jorma Tahvanainen. 1998. "The Evolution of Host-Plant Use and Sequestration in the Leaf Beetle *Phratora* (Coleoptera: Chrysomelidae)." *Evolution* 52, no. 2: 517–28.

Korhonen, Pasi K., Edoardo Pozio, Giuseppe La Rosa, Bill C. H. Chang, Anson V. Koehler, Eric P. Hoberg, Peter R. Boag, et al. 2016. "Phylogenomic and Biogeographic Reconstruction of the *Trichinella* Complex." *Nature Communications* 7: 10513.

Krech, Rüdiger, Ilona Kickbusch, Christian Franz, and Nadya Wells. 2018. "Banking for Health: The Role of Financial Sector Actors in Investing in Global Health." *BMJ Global Health* 3, suppl. 1: e000597. https://doi.org/10 .1136/bmjgh-2017-000597.

Kroll, J. S., and E. R. Moxon. 1990. "Capsulation in Distantly Related Strains of *Haemophilus* Influenzae Type b: Genetic Drift and Gene Transfer at the Capsulation Locus." *Journal of Bacteriology* 172, no. 3: 1374–79.

Kübler-Ross, Elisabeth. 1969. *On Death and Dying*. New York: Macmillan.

Kuiken, Thijs, Pascal Buijs, Peter van Run, Geert Van Amerongen, Marion Koopmans, and Bernadette van den Hoogen. 2017. "Pigeon Paramyxovirus Type 1 from a Fatal Human Case Induces Pneumonia in Experimentally Infected Cynomolgus Macaques (Macaca Fascicularis)." *Veterinary Research* 48, no. 1: 80. https://doi.org/10.1186/s13567-017-0486-6.

Kupferschmidt, Kai. 2017. "Bat Patrol." *Science* 356, no. 6341: 901–3.

Kuris, Armand M. 2012. "The Global Burden of Human Parasites: Who and Where Are They? How Are They Transmitted?" *Journal of Parasitology* 98, no. 6: 1056–64.

Kurtén, B., and E. Anderson. 1980. *Pleistocene Mammals of North America*. New York: Columbia University Press.

Kutz, Susan J., Ingrid Asmundsson, Eric P. Hoberg, Greg D. Appleyard, Emily J. Jenkins, Kimberlee Beckmen, Marsha Branigan, et al. 2007. "Serendipitous Discovery of a Novel Protostrongylid (Nematoda: Metastrongyloidea) Associated with Caribou, Muskoxen, and Moose from High Latitudes of North America Based on DNA Sequence Comparisons." *Canadian Journal of Zoology* 85, no. 11: 1143–56.

Kutz, Susan J., Eric P. Hoberg, Lydden Polley, and E. J. Jenkins. 2005. "Global Warming Is Changing the Dynamics of Arctic Host-Parasite Systems." *Proceedings of the Royal Society B: Biological Sciences* 272, no. 1581: 2571–76.

Kutz, Susan J., Eric P. Hoberg, Péter K. Molnár, Andy Dobson, and Guilherme G. Verocai. 2014. "A Walk on the Tundra: Host-Parasite Interactions in an Extreme Environment." *International Journal for Parasitology: Parasites and Wildlife* 3, no. 2: 198–208.

Kutz, Susan J., Julie Ducrocq, Guilherme G. Verocai, Bryanne M. Hoar, Doug D. Colwell, Kimberlee B. Beckmen, Lydden Polley, et al. 2012. "Parasites in Ungulates of Arctic North America and Greenland: A View of Contemporary Diversity, Ecology, and Impact in a World under Change." *Advances in Parasitology* 79: 99–252. https://doi.org/10.1016/B978-0-12-398457-9.00002-0.

Kutz, Susan J., Sylvia Checkley, Guilherme G. Verocai, Mathieu Dumond, Eric P. Hoberg, Rod Peacock, Jessica P. Wu, et al. 2013. "Invasion, Establishment, and Range Expansion of Two Parasitic Nematodes in the Canadian Arctic." *Global Change Biology* 19, no. 11: 3254–62.

Laaksonen, Sauli, Antti Oksanen, and Eric Hoberg. 2015. "A Lymphatic Dwelling Filarioid Nematode, *Rumenfilaria andersoni* (Filarioidea; Splendidofilariinae), Is an Emerging Parasite in Finnish Cervids." *Parasites & Vectors* 8, no. 1: 228.

Laaksonen, Sauli, Antti Oksanen, Susan Kutz, Pikka Jokelainen, Anniina Holma-Suutari, and Eric Hoberg. 2017. "Filarioid Nematodes, Threat to Arctic Food Safety and Security." In *Game Meat Hygiene: Food Safety and Security 2016*, edited by P. Paulsen, A. Bauer, and F. J. M. Smulders, 213–23. Netherlands: Wageningen Academic.

Labro, Marie-Thérèse. 2012. "Immunomodulatory Effects of Antimicrobial Agents, Part II: Antiparasitic and Antifungal Agents." *Expert Review of Anti-Infective Therapy* 10, no. 3: 341–57.

Laenen, Lies, Valentijn Vergote, Liana Eleni Kafetzopoulou, Tony Bokalanga Wawina, Despoina Vassou, Joseph A. Cook, Jean-Pierre Hugot, et al. 2017. "A Novel Hantavirus of the European Mole, Bruges Virus, Is Involved in Frequent Nova Virus Coinfections." *Genome Biology and Evolution* 10, no. 1: 45–55. https://doi.org/10.1093/gbe/evx268.

Lafferty, Kevin D. 2009. "The Ecology of Climate Change and Infectious Diseases." *Ecology* 90, no. 4: 888–900.

Lai, Yi-Te, Jukka Kekäläinen, and Raine Kortet. 2016. "Infestation with the Parasitic Nematode *Philometra ovata* Does Not Impair Behavioral Sexual Competitiveness or Odor Attractiveness of the Male European Minnow (*Phoxinus phoxinus*)." *Acta Ethologica* 19, no. 2: 103–11. https://doi.org/10.1007/s10211-015-0229-5.

Lajaunie, Claire, and Calvin Wai-Loon Ho. 2017. "Pathogens Collections, Biobanks and Related-Data in a One Health Legal and Ethical Perspective." *Parasitology* 7: 1–8. https://doi.org/10.1017/S0031182017001986.

Lam, Adriane R., Alycia L. Stigall, and Nicholas J. Matzke. 2018. "Dispersal in the Ordovician: Speciation Patterns and Paleobiogeographic Analyses of Brachiopods and Trilobites." *Palaeogeography, Palaeoclimatology, Palaeoecology* 489: 147–65. https://doi.org/10.1016/j.palaeo.2017.10.006.

Lam, Vicky W. Y., William W. L. Cheung, Gabriel Reygondeau, and U. Rashid Sumaila. 2016. "Projected Change in Global Fisheries Revenues under Climate Change." *Scientific Reports* 6: 32607.

Lamsdell, James C., Curtis R. Congreve, Melanie J. Hopkins, Andrew Z. Krug, and Mark E. Patzkowsky. 2017. "Phylogenetic Paleoecology: Tree-Thinking and Ecology in Deep Time." *Trends in Ecology & Evolution* 32: 452–63.

Larick, Roy R., and Russell L. Ciochon. 1996. "The African Emergence and Early Asian Dispersals of the Genus *Homo*." *American Scientist* 84: 538–51.

Larson, Greger, Keith Dobney, Umberto Albarella, Meiying Fang, Elizabeth Matisoo-Smith, Judith Robins, Stewart Lowden, et al. 2005. "Worldwide Phylogeography of Wild Boar Reveals Multiple Centers of Pig Domestication." *Science* 307, no. 5715: 1618–21.

Larson, Greger, Ranran Liu, Xingbo Zhao, Jing Yuan, Dorian Fuller, Loukas Barton, Keith Dobney, et al. 2010. "Patterns of East Asian Pig Domestication, Migration, and Turnover Revealed by Modern and Ancient DNA." *Proceedings of the National Academy of Sciences of the United States of America* 107, no. 17: 7686–91.

Larson, Greger, Umberto Albarella, Keith Dobney, Peter Rowley-Conwy, Joerg Schibler, Anne Tresset, Jean-Denis Vigne, et al. 2007. "Ancient DNA, Pig Domestication, and the Spread of the Neolithic into Europe." *Proceedings of the National Academy of Sciences of the United States of America* 104, no. 39: 15276–81.

Laundré, John W., Lucina Hernández, Perla López Medina, Andrea Campanella, Jorge López-Portillo, Alberto González-Romero, Karina M. Grajales-Tam, et al. 2014. "The Landscape of Fear: The Missing Link to Understand Top-Down and Bottom-Up Controls of Prey Abundance?" *Ecology* 95, no. 5: 1141–52. https://doi.org/10.1890/13-1083.1.

Lauron, Elvin J., Claire Loiseau, Raurie C. K. Bowie, Greg S. Spicer, Thomas B. Smith, Martim Melo, and Ravinder N. M. Sehgal. 2014. "Coevolutionary Patterns and Diversification of Avian Malaria Parasites in African Sunbirds (Family Nectariniidae)." *Parasitology* 142, no. 5: 635–47. https://doi.org/10.1017/S0031182014001681.

Lawler, Joshua J., Sarah L. Shafer, Denis White, Peter Kareiva, Edwin P. Maurer, Andrew R. Blaustein, and Patrick J. Bartlein. 2009. "Projected Climate-Induced Faunal Change in the Western Hemisphere." *Ecology* 90, no. 3: 588–97.

Lawton, Scott P., Hirohisa Hirai, Joe E. Ironside, David A. Johnston, and David Rollinson. 2011. "Genomes and Geography: Genomic Insights into the Evolution and Phylogeography of the Genus *Schistosoma*." *Parasites & Vectors* 4: 131. https://doi.org/10.1186/1756-3305-4-131.

Lebarbenchon, Camille, Audrey Jaeger, Chris Feare, Matthieu Bastien, Muriel Dietrich, Christine Larose, Erwan Lagadec, et al. 2015. "Influenza A Virus on Oceanic Islands: Host and Viral Diversity in Seabirds in the Western Indian Ocean." *PLoS Pathogens* 11, no. 5: e1004925. https://doi.org/10.1371/journal. ppat.1004925.

Lee, Bruce Y., Jorge A. Alfaro-Murillo, Alyssa S. Parpia, Lindsey Asti, Patrick T. Wedlock, Peter J. Hotez, and Alison P. Galvani. 2017. "The Potential

Economic Burden of Zika in the Continental United States." *PLoS Neglected Tropical Diseases* 11, no. 4: e0005531. https://doi.org/10.1371/journal.pntd .0005531.

Lee, Bruce Y., Kristina M. Bacon, Maria Elena Bottazzi, and Peter J. Hotez. 2013. "Global Economic Burden of Chagas Disease: A Computational Simulation Model." *Lancet: Infectious Diseases* 13, no. 4: P342–48. https://doi.org/10 .1016/S1473-3099(13)70002-1.

Lee, Richard B., and Irven DeVore, eds. 1968. *Man the Hunter*. Chicago: Aldine.

Lefoulon, Emilie, Alessio Giannelli, Benjamin L. Makepeace, Yasen Mutafchiev, Simon Townson, Shigehiko Uni, Guilherme G. Verocai, et al. 2017. "Whence River Blindness? The Domestication of Mammals and Host-Parasite Co-Evolution in the Nematode Genus *Onchocerca*." *International Journal for Parasitology* 47: 457–70.

León-Règagnon, V., S. Guillén-Hernández, and M. A. Arizmendi-Espinosa. 2005. "Intraspecific Variation of *Haematoloechus floedae* Harwood, 1932 (Digenea: Plagiorchiidae), from *Rana* spp. in North and Central America." *Journal of Parasitology* 91, no. 4: 915–21.

Leppik, E. E. 1953. "Some Viewpoints on the Phylogeny of Rust Fungi. I. Coniferous Rusts." *Mycologia* 45, no. 1: 46–74.

Leppik, E. E. 1955. "Evolution of Angiosperms as Mirrored in the Phylogeny of Rust Fungi." *Archivum Societatis Zoologicae Botanicae Fennicae "Vanamo"* 9: 149–60.

Leppik, E. E. 1959. "Some Viewpoints on the Phylogeny of Rust Fungi. III. Origin of Grass Rusts." *Mycologia* 51, no. 4: 512–28.

Leppik, E. E. 1965. "Some Viewpoints on the Phylogeny of Rust Fungi. V. Evolution of Biological Specialization." *Mycologia* 57, no. 1: 6–22.

Leppik, E. E. 1967. "Some Viewpoints on the Phylogeny of Rust Fungi. VI. Biogenic Radiation." *Mycologia* 59, no. 4: 568–79.

Leppik, E. E. 1972. "Evolution of Specialization of Rust Fungi (Uredinales) on the Leguminosae." *Annales Botanici Fennici* 9: 135–48.

Levina, Ellina, and Dennis Tirpak. 2006. "Key Adaptation Concepts and Terms." OECD/IEA Project for the Annex I Expert Group on the UNFCCC, 1–21. https://www.oecd.org/env/cc/36278739.pdf.

Levit, Georgy S., and Lennart Olsson. 2006. "Evolution on Rails: Mechanisms and Levels of Orthogenesis." *Annals of the History and Philosophy of Biology* 11: 99–138.

Lewontin, Richard C. 1978. "Adaptation." *Scientific American* 239: 212–30.

Lewontin, Richard C. 1983. "Gene, Organism, and Environment." In *Evolution from Molecules to Men*, edited by Derek S. Bendall, 273–28. Cambridge: Cambridge University Press.

Li, Yun-fang, De-biao Li, Hong-sheng Shao, Hong-jun Li, and Yue-dong Han. 2016. "Plague in China 2014—All Sporadic Case Report of Pneumonic Plague." *BMC Infectious Diseases* 16, no. 1: 49. https://doi.org/10.1186/ s12879-016-1403-8.

Lieberman, Bruce S. 2000. *Paleobiogeography: Using Fossils to Study Global Changes, Plate Tectonics, and Evolution.* Topics in Geobiology 16. New York: Kluwer Academic/Plenum Press.

Lieberman, Bruce S. 2003a. "Unifying Theory and Methodology in Biogeography." *Evolutionary Biology* 33: 1–25.

Lieberman, Bruce S. 2003b. "Paleobiogeography: The Relevance of Fossils to Biogeography." *Annual Review of Ecology and Systematics* 34: 51–69. https://doi.org/10.1146/annurev.ecolsys.34.121101.153549.

Liebherr, James K. 1988. "General Patterns in West Indian Insects, and Graphical Biogeographic Analysis of Some Circum-Caribbean *Platynus* Beetles (Carabidae)." *Systematic Biology* 37, no. 4: 385–409.

Liebherr, James K., and Ann E. Hajek. 1990. "A Cladistic Test of the Taxon Cycle and Taxon Pulse Hypotheses." *Cladistics* 6, no. 1: 39–59.

Lim, Burton K. 2008. "Historical Biogeography of New World Emballonurid Bats (Tribe Diclidurini): Taxon Pulse Diversification." *Journal of Biogeography* 35, no. 7: 1385–1401.

Lima, L. B., S. Bellay, H. C. Giacomini, A. Isaac, and D. P. Lima-Junior. 2015. "Influence of Host Diet and Phylogeny on Parasite Sharing by Fish in a Diverse Tropical Floodplain." *Parasitology* 143, no. 3: 343–49. https://doi.org/10.1017/S003118201500164X.

Lipman, Chas. B. 1922. "Orthogenesis in Bacteria." *American Naturalist* 56, no. 643: 105–15.

Lister, Adrian M. 2004. "The Impact of Quaternary Ice Ages on Mammalian Evolution." *Philosophical Transactions of the Royal Society B: Biological Sciences* 359, no. 1442: 221–41.

Lister, Adrian M., and A. V. Sher. 2015. "Evolution and Dispersal of Mammoths across the Northern Hemisphere." *Science* 350, no. 6262: 805–9.

Liu, M., N. Rodrigue, and J. Kolmer. 2014. "Population Divergence in the Wheat Leaf Rust Fungus *Puccinia triticina* Is Correlated with Wheat Evolution." *Heredity* 112, no. 4: 443–53. https://doi.org/10.1038/hdy.2013.123.

Líznarová, Eva, and Stano Pekár. 2016. "Metabolic Specialisation on Preferred Prey and Constraints in the Utilisation of Alternative Prey in an Ant-Eating Spider." *Zoology* 119, no. 5: 464–70. https://doi.org/10.1016/j.zool.2016.04.004.

Llambí, Luis D., Mario Fariñas, Julia K. Smith, Sandra M. Castañeda, and Benito Briceño. 2013. "Diversidad de la vegetación en dos paramos de Venezuela: Un enfoque multiescalar con fines de conservación." In *Avances en investigación para la conservación de los páramos andinos,* edited by Francisco Cuesta, Jan Sevink, Luis Daniel Llambí, Bert De Bièvre, and Joshua Posner, 41–68. Quito: CONDESAN.

Loftus, Ronan T., David E. MacHugh, Daniel G. Bradley, Paul M. Sharp, and Patrick Cunningham. 1994. "Evidence for Two Independent Domestications of Cattle." *Proceedings of the National Academy of Sciences* 91, no. 7: 2757–61.

Lombard, M., P. P. Pastoret, and A. M. Moulin. 2007. "A Brief History of Vaccines and Vaccination." *Revue scientifique et technique (International Office of Epizootics)* 26, no. 1: 29–48.

Lootvoet, Amélie, Simon Blanchet, Muriel Gevrey, Laetitia Buisson, Löic Tudesque, and Géraldine Loot. 2013. "Patterns and Processes of Alternative Host Use in a Generalist Parasite: Insights from a Natural Host-Parasite Interaction." *Functional Ecology* 27, no. 6: 1403–14. https://doi.org/10.1111/1365-2435.12140.

Lopman, Ben A., Mark H. Reacher, Ian B. Vipond, Dawn Hill, Christine Perry, Tracey Halladay, David W. Brown, et al. 2004. "Epidemiology and Cost of Nosocomial Gastroenteritis, Avon, England, 2002–2003." *Emerging Infectious Diseases* 10, no. 10: 1827–34.

Lundsgaard, T. 1996. "Filovirus-Like Particles Detected in Extracts from the Leafhopper *Psammotettix alienus*" [abstract]. Paper presented at the Tenth International Congress of Virology, Jerusalem.

Lynch, M. 2016. "Mutation and Human Exceptionalism: Our Future Genetic Load." *Genetics* 202, no. 3: 869–75. https://doi.org/10.1534/genetics.115.180471.

MacArthur, Robert H., and Edward O. Wilson. 1967. *The Theory of Island Biogeography*. Princeton, NJ: Princeton University Press.

Macedo, Renan, Lilian Patrícia Sales, Fernanda Yoshida, Lidianne Lemes Silva-Abud, and Murillo Lobo. 2017. "Potential Worldwide Distribution of Fusarium Dry Root Rot in Common Beans Based on the Optimal Environment for Disease Occurrence." Edited by Sabrina Sarrocco. *PLoS ONE* 12, no. 11: e0187770. https://doi.org/10.1371/journal.pone.0187770.

Machado, Antonio. 1912. "Proverbios y cantares XXIX." In *Campos de Castilla*. Madrid: Renacimiento.

MacIntyre, C. Raina, Thomas Edward Engells, Matthew Scotch, David James Heslop, Abba B. Gumel, George Poste, Xin Chen, et al. 2017. "Converging and Emerging Threats to Health Security." *Environment Systems and Decisions* 38, no. 2: 198–207. https://doi.org/10.1007/s10669-017-9667-0.

Makarikov, Arseny A., Kurt E. Galbreath, and Eric P. Hoberg. 2013. "Parasite Diversity at the Holarctic Nexus: Species of *Arostrilepis* (Eucestoda: Hymenolepididae) in Voles and Lemmings (Cricetidae: Arvicolinae) from Greater Beringia." *Zootaxa* 3608, no. 6: 401–39.

Makkonen, Jenny, Japo Jussila, Jörn Panteleit, Nina Sophie Keller, Anne Schrimpf, Kathrin Theissinger, Raine Kortet, et al. 2018. "MtDNA Allows the Sensitive Detection and Haplotyping of the Crayfish Plague Disease Agent *Aphanomyces astaci* Showing Clues about Its Origin and Migration." *Parasitology* 145, no. 9: 1210–18. https://doi.org/10.1017/S0031182018000227.

Makundi, Rhodes H., Apia W. Massawe, Benny Borremans, Anne Laudisoit, and Abdul Katakweba. 2015. "We Are Connected: Flea-Host Association Networks in the Plague Outbreak Focus in the Rift Valley, Northern

Tanzania." *Wildlife Research* 42, no. 2: 196–206. https://doi.org/10.1071/WR14254.

Malcicka, Miriama, Salvatore J. Agosta, and Jeffrey A. Harvey. 2015. "Multi Level Ecological Fitting: Indirect Life Cycles Are Not a Barrier to Host Switching and Invasion." *Global Change Biology* 21, no. 9: 3210–18.

Maldonado, Carla, Christopher J. Barnes, Claus Cornett, Else Holmfred, Steen H. Hansen, Claes Persson, Alexandre Antonelli, et al. 2017. "Phylogeny Predicts the Quantity of Antimalarial Alkaloids within the Iconic Yellow *Cinchona* Bark (Rubiaceae: *Cinchona calisaya*)." *Frontiers in Plant Science* 8: 391. https://doi.org/10.3389/fpls.2017.00391.

Malthus, Thomas Robert. 1798. *An Essay on the Principle of Population, as it affects the future Improvement of Society, with Remarks on the Speculations of Mr. Godwin, M. Condorcet, and Other Writers.* London: J. Johnson.

Mann, Michael E., Raymond S. Bradley, and Malcolm K. Hughes. 1999. "Northern Hemisphere Temperatures during the Past Millennium: Inferences, Uncertainties, and Limitations." *Geophysical Research Letters* 26, no. 6: 759–62.

Manter, Harold W. 1966. "Parasites of Fishes as Biological Indicators of Recent and Ancient Conditions." In *Host-Parasite Relationships*, edited by James E. McCauley, 59–71. Corvallis: Oregon State University Press.

Marinosci, Cassandra, Sara Magalhães, Emilie Macke, Maria Navajas, David Carbonell, Céline Devaux, and Isabelle Olivieri. 2015. "Effects of Host Plant on Life-History Traits in the Polyphagous Spider Mite *Tetranychus urticae*." *Ecology and Evolution* 5, no. 15: 3151–58. https://doi.org/10.1002/ece3.1554.

Mark, Justin T., Brian B. Marion, and Donald D. Hoffman. 2010. "Natural Selection and Veridical Perceptions." *Journal of Theoretical Biology* 266, no. 4: 504–15.

Marshall, Larry G., S. David Webb, J. John Sepkoski, and David M. Raup. 1982. "Mammalian Evolution and the Great American Interchange." *Science*, n.s., 215, no. 4538: 1351–57.

Martinez-Hernandez, Fernando, Diego Emiliano Jimenez-Gonzalez, Paola Chenillo, Cristina Alonso-Fernandez, Pablo Maravilla, and Ana Flisser. 2009. "Geographical Widespread of Two Lineages of *Taenia solium* due to Human Migrations: Can Population Genetic Analysis Strengthen This Hypothesis?" *Infection, Genetics and Evolution* 9, no. 6: 1108–14.

Mas-Coma, Santiago, Maria A. Valero, and Maria D. Bargues. 2008. "Effects of Climate Change on Animal and Zoonotic Helminthiases." *Review of Science and Technology* 27, no. 2: 443–57.

Mas-Coma, Santiago, Maria A. Valero, and Maria D. Bargues. 2009. "Climate Change Effects on Trematodiases, with Emphasis on Zoonotic Fascioliasis and Schistosomiasis." *Veterinary Parasitology* 163, no. 4: 264–80.

Mas-Coma, Santiago, María A. Valero, and María D. Bargues. 2009. "*Fasciola*, Lymnaeids and Human Fascioliasis, with a Global Overview on Disease

Transmission, Epidemiology, Evolutionary Genetics, Molecular Epidemiology and Control." *Advances in Parasitology* 69: 41–146.

Masson-Delmotte, V., M. Schulz, A. Abe-Ouchi, J. Beer, A. Ganopolski, J. F. González Rouco, E. Jansen, K. Lambeck, et al. 2013. "Information from Paleoclimate Archives." In *Climate Change 2013: The Physical Science Basis; Contribution of Working Group I to the Fifth Assessment Report of the Intergovernmental Panel on Climate Change*, 383–464. Cambridge: Cambridge University Press.

Masters, Judith C., and Richard J. Rayner. 1993. "Competition and Macroevolution: The Ghost of Competition Yet to Come?" *Biological Journal of the Linnaean Society* 49, no. 1: 87–98.

Matisz, Chelsea E., Cameron P. Goater, and Douglas Bray. 2010. "Migration and Site Selection of *Ornithodiplostomum ptychocheilus* (Trematoda: Digenea) Metacercariae in the Brain of Fathead Minnows (*Pimephales promelas*)." *Parasitology* 137, no. 4: 719–31. https://doi.org/10.1017/S0031182009991545.

Matamoros, Wilfredo, A., Caleb D. McMahan, Prosanta Chakrabarty, James S. Albert, and Jacob F. Schaefer. 2015. "Derivation of the Freshwater Fish Fauna of Central America Revisited: Myers's Hypothesis in the Twenty-First Century." *Cladistics* 31, no. 2: 177–88. https://doi.org/10.1111/cla.12081.

Matthews, Alix E., Pavel B. Klimov, Heather C. Proctor, Ashley P. G. Dowling, Lizzie Diener, Stephen B. Hager, Jeffery L. Larkin, et al. 2018. "Cophylogenetic Assessment of New World Warblers (Parulidae) and Their Symbiotic Feather Mites (Proctophyllodidae)." *Journal of Avian Biology* 49, no. 3: jav-01580. https://doi.org/10.1111/jav.01580.

Maubecin, C. C., A. Cosacov, A. N. Sérsic, J. Fornoni, and S. Benitez-Vieyra. 2016. "Drift Effects on the Multivariate Floral Phenotype of Calceolaria polyrhiza during a Post-Glacial Expansion in Patagonia." *Journal of Evolutionary Biology* 29, no. 7: 1523–34. https://doi.org/10.1111/jeb.12889.

Mavalankar, Dileep, Priya Shastri, Tathagata Bandyopadhyay, Jeram Parmar, and Karaikurichi V. Ramani. 2008. "Increased Mortality Rate Associated with Chikungunya Epidemic, Ahmedabad, India." *Emerging Infectious Diseases* 14, no. 3: 412–15.

May, Robert M., and Roy M. Anderson. 1985. "Epidemiology and Genetics in the Coevolution of Parasites and Hosts." *Proceedings of the Royal Society B: Biological Sciences* 219, no. 1216: 281–83.

Maynard Smith, John. 1976. "What Determines the Rate of Evolution?" *American Naturalist* 110, no. 973: 331–38.

Maynard Smith, John. 1982. *Evolution and the Theory of Games*. Cambridge: Cambridge University Press.

Maynard Smith, John, and Eörs Szathmáry. 1995. *The Major Transitions in Evolution*. Oxford: Oxford University Press.

Maynard Smith, John, and Eörs Szathmáry. 1999. *The Origins of Life: From the Birth of Life to the Origin of Language*. Oxford: Oxford University Press.

Mayr, Ernst. 1954. "Change of Genetic Environment and Evolution." In *Evolution as a Process*, edited by Julian Huxley, Alister C. Hardy, and Edmund B. Ford, 157–80. London: Allen & Unwin.

Mayr, Ernst. 1957. "Evolutionary Aspects of Host Specificity among Parasites of Vertebrates." In *Premier symposium sur la spécificité parasitaire des parasites des vertébrés/First Symposium of Host Specificity among Parasites of Vertebrates*, edited by J.-G. Baer, 7–14. Neuchatel: Paul Attinger S.A.

Mayr, Ernest. 1963. *Animal Species and Evolution*. Cambridge, MA: Harvard University Press.

Mazet, Jonna A. K., Qin Wei, Guoping Zhao, Derek A. T. Cummings, James S. Desmond, Joshua Rosenthal, Charles H. King, et al. 2015. "Joint China-US Call for Employing a Transdisciplinary Approach to Emerging Infectious Diseases." *EcoHealth* 12, no. 4: 555–59.

McCoy, Karen D., Muriel Dietrich, Audrey Jaeger, David A. Wilkinson, Matthieu Bastien, Erwan Lagadec, Thierry Boulinier, et al. 2016. "The Role of Seabirds of the Îles Eparses as Reservoirs and Disseminators of Parasites and Pathogens." *Acta Oecologica* 72: 98–109. https://doi.org/10.1016/j.actao.2015.12.013.

McHugh, Anne, Carol Yablonsky, Greta Binford, and Ingi Agnarsson. 2014. "Molecular Phylogenetics of Caribbean *Micrathena* (Araneae: Araneidae) Suggests Multiple Colonisation Events and Single Island Endemism." *Invertebrate Systematics* 28, no. 4: 337–49. https://doi.org/10.1071/IS13051

McIntosh, Allen. 1935. "A Progenetic Metacercaria of a *Clinostomum* in a West Indian Land Snail." *Proceedings of the Helminthological Society of Washington* 2: 79–80.

McLennan, Deborah A. 2008. "The Concept of Co-Option: Why Evolution Often Looks Miraculous." *Evolution: Education and Outreach* 1, no. 3: 247–58.

McMichael, Anthony J. 2004. "Environmental and Social Influences on Emerging Infectious Diseases: Past, Present and Future." *Philosophical Transactions of the Royal Society B: Biological Sciences* 359, no. 1447: 1049–58.

McMichael, Anthony J. 2013. "Globalization, Climate Change, and Human Health." *New England Journal of Medicine* 368: 1335–43.

McMichael, Anthony J., Alistair Woodward, and Cameron Muir. 2017. *Climate Change and the Health of Nations: Famines, Fevers, and the Fate of Populations*. Oxford: Oxford University Press.

McMichael, Anthony J., Rosalie E. Woodruff, and Simon Hales. 2006. "Climate Change and Human Health: Present and Future Risks." *Lancet* 367, no. 9513: 859–69.

McMichael, Celia. 2015. "Climate Change–Related Migration and Infectious Disease." *Virulence* 6, no. 6: 548–53.

McMullen, Donald B. 1938. "Observations on Precocious Metacercarial Development in the Trematode Superfamily Plagiorchioidea." *Journal of Parasitology* 24, no. 3: 273–80.

McNeill, W. H. 1976. *Plagues and Peoples*. New York: Anchor.

McVicar, A. H. 1979. "The Distribution of Cestodes within the Spiral Intestine of *Raja naevus* Muller & Henle." *International Journal for Parasitology* 9, no. 3: 165–76.

Meadows, Donella H., Dennis L. Meadows, Jorgen Randers, and William W. Behrens. 1972. *The Limits to Growth*. New York: Earthscan.

Mele, S., D. Macías, M. J. Gómez-Vives, G. Garippa, F. Alemany, and P. Merella. 2012. "Metazoan Parasites on the Gills of the Skipjack Tuna *Katsuwonus pelamis* (Osteichthyes: Scombridae) from the Alboran Sea (Western Mediterranean Sea)." *Diseases of Aquatic Organisms* 97, no. 3: 219–25. https://doi .org/10.3354/dao02421.

Meltofte, H., T. Barry, D. Berteaux, H. Bueltmann, J. S. Christiansen, J. A. Cook, F. J. A. Daniëls, et al. 2013. "Status and Trends in Arctic Biodiversity— Synthesis: Implications for Conservation." In *Arctic Biodiversity Assessment: Status and Trends in Arctic Biodiversity*, edited by H. Meltofte, 21–66. Akureyri, Iceland: Conservation of Arctic Flora and Fauna, Arctic Council.

Meltzer, M. I., N. J. Cox, and K. Fukuda. 1999. "The Economic Impact of Pandemic Influenza in the United States: Priorities for Intervention." *Emerging Infectious Diseases* 5, no. 5: 659–71. https://doi.org/10.3201/eid0505 .990507.

Menezes, Lucas, Clarissa Canedo, Henrique Batalha-Filho, Adrian Antonio Garda, Marcelo Gehara, and Marcelo Felgueiras Napoli. 2016. "Multilocus Phylogeography of the Treefrog *Scinax eurydice* (Anura, Hylidae) Reveals a Plio-Pleistocene Diversification in the Atlantic Forest." *PLoS ONE* 11, no. 6: e0154626. https://doi.org/10.1371/journal.pone.0154626.

Metcalf, Maynard M. 1920. "Upon an Important Method of Studying Problems of Relationship and Geographical Distribution." *Proceedings of the National Academy of Science of the United States of America* 6: 432–33.

Metcalf, Maynard M. 1922. "The Host Parasite Method of Investigation and Some Problems to Which It Gives Approach." *Anatomical Record* 23: 117.

Metcalf, Maynard M. 1923a. "The Opalinid Ciliate Infusorians." *Bulletin of the U.S. National Museum* 120: 1–484.

Metcalf, Maynard M. 1923b. "The Origin and Distribution of the Anura." *American Naturalist* 57: 385–411.

Metcalf, Maynard M. 1926. "Larval Stages in a Protozoon." *Proceedings of the National Academy of Science of the United States of America* 12: 734–737.

Metcalf, Maynard M. 1928a. "The Bell-Toads and Their Opalinid Parasites." *American Naturalist* 62: 5–21.

Metcalf, Maynard M. 1928b. "Trends in Evolution: A Discussion of Data Bearing upon 'Orthogenesis.'" *Journal of Morphology and Physiology* 45: 1–45.

Metcalf, Maynard M. 1929. "Parasites and the Aid They Give in Problems of Taxonomy, Geographic Distribution, and Paleogeography." *Smithsonian Institution Miscellaneous Collections* 81: 1–36.

Metcalf, Maynard M. 1934. "Frogs and Opalinids." *Science* 79: 213–14.

Metcalf, Maynard M. 1940. "Further Studies on the Opalinid Ciliate Infusorians." *Proceedings of the U.S. National Museum* 87: 465–634.

Michelet, Lorraine, Jean-François Carod, Mahenintsoa Rakontondrazaka, Laurence Ma, Frédérick Gay, and Catherine Dauga. 2010. "The Pig Tapeworm *Taenia solium*, the Cause of Cysticercosis: Biogeographic (Temporal and Spacial) Origins in Madagascar." *Molecular Phylogenetics and Evolution* 55, no. 2: 744–50.

Millán, Javier, Tatiana Proboste, Isabel G. Fernández de Mera, Andrea D. Chirife, José de la Fuente, and Laura Altet. 2016. "Molecular Detection of Vector-Borne Pathogens in Wild and Domestic Carnivores and Their Ticks at the Human-Wildlife Interface." *Ticks and Tick-Borne Diseases* 7, no. 2: 284–90. https://doi.org/10.1016/j.ttbdis.2015.11.003.

Miller, Alden H. 1949. "Some Ecologic and Morphologic Considerations in the Evolution of Higher Taxonomic Categories." In *Ornithologie als Biologische Wissenschaft: 28 Beiträge als Festschrift zum 60. Geburtstag von Erwin Stresemann*, edited by Erst Mayr and Ernst Schüz, 84–88. Heidelberg: C. Winter.

Miller, R. L., G. J. Armelagos, S. Ikram, N. de Jonge, F. W. Krijger, and A. M. Deelder. 1992. "Palaeoepidemiology of Schistosoma Infection in Mummies." *BMJ: British Medical Journal* 304, no. 6826: 555–56. https://doi.org/10.1136/bmj.304.6826.555.

Miller, R. L., S. Ikram, G. J. Armelagos, R. Walkers, W. B. Harer, C. J. Shiff, D. Bagget, M. Carrigan, and M. Maret. 1994. "Diagnosis of Plasmodium Falciparum Infections in Mummies Using the Rapid Manual ParaSight™-F Test." *Transactions of the Royal Society of Tropical Medicine and Hygiene* 88, no. 1: 31–32.

Millins, Caroline, Lucy Gilbert, Jolyon Medlock, Kayleigh Hansford, Des B. A. Thompson, and Roman Biek. 2017. "Effects of Conservation Management of Landscapes and Vertebrate Communities on Lyme Borreliosis Risk in the United Kingdom." *Philosophical Transactions of the Royal Society B: Biological Sciences* 372, no. 1722: 20160123–12. https://doi.org/10.1098/rstb.2016.0123.

Minter, Nic, Luis Buatois, Gabriela Mángano, Neil Davies, Martin Gibling, Robert MacNaughton, and Conrad Labandeira. 2017. "Early Bursts of Diversification Defined the Faunal Colonization of Land." *Nature Ecology & Evolution* 1: 0175. https://doi.org/10.17863/CAM.12604.

Mitchell, P. Chalmers. 1901. "VII. On the Intestinal Tract of Birds; with Remarks on the Valuation and Nomenclature of Zoological Characters." *Transactions of the Linnean Society of London* 8, no. 7: 173–275.

Mitchell, P. Chalmers. 1910. "Biogenesis." In *Encyclopædia Britannica*, 11th ed., 3:952. London: Encyclopædia Britannica.

Mitchell, P. Chalmers. 1905. "On the Intestinal Tract of Mammals." *Journal of Zoology* 17, no. 5: 437–536.

Mitter, Charles, and Daniel R. Brooks. 1983. "Phylogenetic Aspects of Coevolution." In *Coevolution*, edited by Douglas J. Futuyma and Montgomery Slatki, 65–98. Sunderland, MA: Sinauer Associates.

Mitter, Charles, Brian Farrell, and Brian Wiegmann. 1988. "The Phylogenetic Study of Adaptive Zones: Has Phytophagy Promoted Insect Diversification?" *American Naturalist* 132 , no. 1: 107–28.

Mize, Erica L., and Hugh B. Britten. 2016. "Detections of *Yersinia pestis* East of the Known Distribution of Active Plague in the United States." *Vector Borne and Zoonotic Diseases* 16: 88–95. https://doi.org/10.1089/vbz.2015 .1825.

Mize, Erica L., Shaun M. Grassel, and Hugh B. Britten. 2017. "Fleas of Black-Footed Ferrets (*Mustela nigripes*) and Their Potential Role in the Movement of Plague." *Journal of Wildlife Diseases* 53, no. 3: 521–31. https://doi.org/10 .7589/2016-09-202.

Mizukoshi, Atsushi, Kevin P. Johnson, and Kazunori Yoshizawa. 2012. "Co-Phylogeography and Morphological Evolution of Sika Deer Lice (*Damalinia sika*) with Their Hosts (*Cervus nippon*)." *Parasitology* 139, no. 12: 1614–29.

Mode, Charles J. 1958. "A Mathematical Model for the Co-Evolution of Obligate Parasites and Their Hosts." *Evolution* 12, no. 2: 158–65.

Mode, Charles J. 1961. "Generalized Model of a Host-Pathogen." *Biometrics* 17: 386–404.

Mode, Charles J. 1962. "Some Multi-Dimensional Birth and Death Processes and Their Applications in population genetics." *Biometrics* 18: 543–67.

Mode, Charles J. 1964. "A Stochastic Model of the Dynamics of Host-Pathogen Systems with Mutation." *Bulletin of Mathematical Biology* 26, no. 3: 205–33.

Moens, Michael A. J., and Javier Pérez-Tris. 2016. "Discovering Potential Sources of Emerging Pathogens: South America Is a Reservoir of Generalist Avian Blood Parasites." *International Journal for Parasitology* 46, no. 1: 41–49. http://dx.doi.org/10.1016/j.ijpara.2015.08.001.

Moore, Janice. 1981. "Asexual Reproduction and Environmental Predictability in Cestodes (Cyclophyllidea: Taeniidae)." *Evolution* 35, no. 4: 723–41.

Moore, Ruth E. 1966. *Niels Bohr: The Man, His Science, and the World They Changed*. Boston: MIT Press. http://cds.cern.ch/record/107731.

Morand, Serge. 2015. "Diversity and Origins of Human Infectious Diseases." In *Basics in Human Evolution*, edited by Michael P. Muehlenbein, 405–14. London: Academic Press. https://doi.org/10.1016/B978-0-12-802652-6 .00029-3.

Morand, Serge, and Claire Lajaunie. 2017. *Biodiversity and Health*. Amsterdam: Elsevier.

Moreira, Xoaquín, Colleen S. Nell, Angelos Katsanis, Sergio Rasmann and Kailen A. Mooney. 2016. "Herbivore Specificity and the Chemical Basis of Plant-Plant Communication in *Baccharis salicifolia* (Asteraceae)." *New Phytologist* https://doi.org/10.1111/nph.14164.

Morelli, Giovanna, Yajun Song, Camila J Mazzoni, Mark Eppinger, Philippe Roumagnac, David M Wagner, Mirjam Feldkamp, et al. 2010. "*Yersinia pestis* Genome Sequencing Identifies Patterns of Global Phylogenetic Diversity." *Nature Genetics* 42, no. 12: 1140–43. https://doi.org/10.1038/ng.705.

Morse, Stephen S. 1995. "Factors in the Emergence of Infectious Diseases." *Emerging Infectious Diseases* 1: 7–15. https://dx.doi.org/10.3201/eid0101.950102.

Mueller, Kai, Victoria Morin-Adeline, Katrina Gilchrist, Graeme Brown and Jan Šlapeta. 2015a. "High Prevalence of *Tritrichomonas foetus* 'Bovine Genotype' in Faecal Samples from Domestic Pigs at a Farm Where Bovine Trichomonosis Has Not Been Reported for over 30 Years." *Veterinary Parasitology* 212, no. 3–4: 105–10. https://doi.org/10.1016/j.vetpar.2015.08.010.

Müller, Anke, Peder K Bøcher, Christina Fischer, and Jens-Christian Svenning. 2018. "'Wild' in the City Context: Do Relative Wild Areas Offer Opportunities for Urban Biodiversity?" *Landscape and Urban Planning* 170: 256–65. https://doi.org/10.1016/j.landurbplan.2017.09.027.

Murphy, L., A. K. Pathak, and I. M. Cattadori. 2013. "A Co-Infection with Two Gastrointestinal Nematodes Alters Host Immune Responses and Only Partially Parasite Dynamics." *Parasite Immunology* 35, no. 12: 421–32. https://doi.org/10.1111/pim.12045.

Nadler, Steven A., Eugene T. Lyons, Christopher Pagan, Derek Hyman, Edwin E. Lewis, Kimberlee Beckmen, Cameron M. Bell, et al. 2013. "Molecular Systematics of Pinniped Hookworms (Nematoda: *Uncinaria*): Species Delimitation, Host Associations and Host-Induced Morphometric Variation." *International Journal for Parasitology* 43, no. 14: 1119–32.

Nakao, M., M. Okamoto, Y. Sako, H. Yamasaki, K. Nakaya, and A. Ito. 2002. "A Phylogenetic Hypothesis for the Distribution of Two Genotypes of the Pig Tapeworm *Taenia solium* Worldwide." *Parasitology* 124, no. 6: 657–62.

Nakao, Minoru, Antti Lavikainen, Takashi Iwaki, Voitto Haukisalmi, Sergey Konyaev, Yuzaburo Oku, Munehiro Okamoto, et al. 2013. "Molecular Phylogeny of the Genus *Taenia* (Cestoda: Taeniidae): Proposals for the Resurrection of *Hydatigera* Lamarck, 1816 and the Creation of a New Genus *Versteria*." *International Journal for Parasitology* 43, no. 6: 427–37.

Nakazawa, Yoshinori, Matthew R. Mauldin, Ginny L. Emerson, Mary G. Reynolds, R. Ryan Lash, Jinxin Gao, Hui Zhao, et al. 2015. "Phylogeographic Investigation of African Monkeypox." *Viruses* 7, no. 4: 2168–84. https://doi.org/10.3390/v7042168.

Navarro, Laetitia M., Néstor Fernández, Carlos Guerra, Rob Guralnick, W. Daniel Kissling, Maria Cecilia Londoño, Frank Muller-Karger, et al. 2017. "Monitoring Biodiversity Change through Effective Global Coordination." *Current Opinion in Environmental Sustainability* 29: 158–69. https://doi.org/10.1016/j.cosust.2018.02.005.

Neiman, Susan. 2002. *Evil in Modern Thought: An Alternative History of Philosophy*. Princeton, NJ: Princeton University Press.

Newman, C. M., J. E. Cohen, and C. Kipnis. 1985. "Neo-Darwinian Evolution Implies Punctuated Equilibria." *Nature* 315, no. 6018: 400–401.

Nieberding, Caroline, Serge Morand, Roland Libois, and Johan René Michaux. 2004. "A Parasite Reveals Cryptic Phylogeographic History of Its Host."

Proceedings of the Royal Society of London B: Biological Sciences 271, no. 1557: 2559–68.

Nielsen, Rasmus, Joshua M. Akey, Mattias Jakobsson, Jonathan K. Pritchard, Sarah Tishkoff, and Eske Willerslev. 2017. "Tracing the Peopling of the World through Genomics." *Nature* 541, no. 7637: 302–10.

Niles, Meredith T., Margaret Brown, and Robyn Dynes. 2016. "Farmer's Intended and Actual Adoption of Climate Change Mitigation and Adaptation Strategies." *Climatic Change* 135, no. 2: 277–95. https://doi.org/10.1007/s10584-015-1558-0.

Norman, Janette A., and Les Christidis. 2016. "Ecological Opportunity and the Evolution of Habitat Preferences in an Arid-Zone Bird: Implications for Speciation in a Climate-Modified Landscape." *Scientific Reports* 6: 19613. https://doi.org/10.1038/srep19613.

Nutting, W. B. 1968. "Host Specificity in Parasitic Acarines." *Acarologia* 10, no. 1: 165–80. https://doi.org/10.5962/bhl.title.59487.

Nylin, Sören. 1988. "Host Plant Specialization and Seasonality in a Polyphagous Butterfly, *Polygonia c-album* (Nymphalidae)." *Oikos* 53: 381–86.

Nylin, Sören, and Niklas Janz. 2009. "Butterfly Host Plant Range: An Example of Plasticity as a Promoter of Speciation?" *Evolutionary Ecology* 23, no. 1: 137–46.

Nylin, Sören, Georg H. Nygren, Jack J. Windig, Niklas Janz, and Anders Bergström. 2005. "Genetics of Host-Plant Preference in the Comma Butterfly *Polygonia c-Album* (Nymphalidae), and Evolutionary Implications." *Biological Journal of the Linnean Society* 84, no. 4: 755–65.

Nylin, Sören, Salvatore Agosta, Staffan Bensch, Walter Boeger, Mariana P. Braga, Daniel R. Brooks, Matthew L. Forister, et al. 2018. "Embracing Colonizations: A New Paradigm for Species Association Dynamics." *Trends in Ecology & Evolution* 33, no. 1: 4–14. https://doi.org/10.1016/j.tree.2017.10.005.

Ochoa-Ochoa, Leticia M., Jonathan A. Campbell, and Oscar A. Flores-Villela. 2014. "Patterns of Richness and Endemism of the Mexican Herpetofauna, a Matter of Spatial Scale?" *Biological Journal of the Linnean Society* 111, no. 2: 305–36.

Odling-Smee, F. John, Kevin N. Laland, and Marcus W. Feldman. 1996. "Niche Construction." *American Naturalist* 147, no. 4: 641–48.

Ogden, Nick H., and L. Robbin Lindsay. 2016. "Effects of Climate and Climate Change on Vectors and Vector-Borne Diseases: Ticks Are Different." *Trends in Parasitology* 32, no. 7: 646–56. https://doi.org/10.1016/j.pt.2016.04.015.

Ogden, N. H., A. Maarouf, I. K. Barker, M. Bigras-Poulin, L. R. Lindsay, M. G. Morshed, C. J. O'Callaghan, F. Ramay, D. Waltner-Toews, and D. F. Charron. 2006. "Climate Change and the Potential for Range Expansion of the Lyme Disease Vector *Ixodes scapularis* in Canada." *International Journal for Parasitology* 36, no. 1: 63–70. https://doi.org/10.1016/j.ijpara.2005.08.016.

Ogden, Nicholas H., Radojević Milka, Cyril Caminade, and Philippe Gachon. 2014. "Recent and Projected Future Climatic Suitability of North America for the Asian Tiger Mosquito *Aedes albopictus.*" *Parasites & Vectors* 7, no. 1: 532. https://doi.org/10.1186/s13071-014-0532-4.

Olival, Kevin J., Parviez R. Hosseini, Carlos Zambrana-Torrelio, Noam Ross, Tiffany L. Bogich, and Peter Daszak. 2017. "Host and Viral Traits Predict Zoonotic Spillover from Mammals." *Nature* 546: 646–50. https://doi.org/10.1038/nature22975.

Oliveira, Ubirajara, Adriano Pereira Paglia, Antonio D. Brescovit, Claudio J. B. de Carvalho, Daniel Paiva Silva, Daniella T. Rezende, Felipe Sá Fortes Leite, et al. 2016. "The Strong Influence of Collection Bias on Biodiversity Knowledge Shortfalls of Brazilian Terrestrial Biodiversity." *Diversity and Distributions* 22, no. 12: 1232–44. https://doi.org/10.1111/ddi.12489.

Olivero, Jesús, John E. Fa, Raimundo Real, Miguel Ángel Farfán, Ana Luz Márquez, J. Mario Vargas, J. Paul Gonzalez, Andrew A. Cunningham, and Robert Nasi. 2016. "Mammalian Biogeography and the Ebola Virus in Africa." *Mammal Review* 47, no. 1: 24–37. https://doi.org/10.1111/mam.12074.

Olivero, Jesús, John E. Fa, Raimundo Real, Ana L. Márquez, Miguel A. Farfán, J. Mario Vargas, David Gaveau, et al. 2017. "Recent Loss of Closed Forests Is Associated with Ebola Virus Disease Outbreaks." *Scientific Reports* 7: 14291. https://doi.org/10.1038/s41598-017-14727-9.

Ono, Yoshitaka, and Joe F. Hennen. 1983. "Taxonomy of the *Chaconiaceous* Genera (Uredinales)." *Transactions of the Mycological Society of Japan* 24, no. 4: 369–402.

Oppenheimer, Michael, and Richard B. Alley. 2016. "How High Will the Seas Rise?" *Science* 354, no. 6318: 1375–77.

Orlando, Ludovic. 2016. "Back to the Roots and Routes of Dromedary Domestication." *Proceedings of the National Academy of Sciences of the United States of America* 113, no. 24: 6588–90.

Orsucci, M., P. Audiot, A. Pommier, C. Raynaud, B. Ramora, A. Zanetto, D. Bourguet, et al. 2016. "Host Specialization Involving Attraction, Avoidance and Performance, in Two Phytophagous Moth Species." *Journal of Evolutionary Biology* 29, no. 1: 114–25.

Osadebe, L. U., K. Manthiram, A. M. McCollum, Y. Li, G. L. Emerson, N. F. Gallardo-Romero, J. B. Doty, et al. 2014. "Novel Poxvirus Infection in 2 Patients from the United States." *Clinical Infectious Diseases* 60, no. 2: 195–202. https://doi.org/10.1093/cid/ciu790.

Osborn, Henry F. 1910. *The Age of Mammals in Europe, Asia and North America.* New York: Macmillan.

Osborn, Henry F. 1918. *The Origin and Evolution of Life: On the Theory of Action, Reaction and Interaction of Energy.* London: G. Bell & Sons.

Osborn, Henry F. 1932. "The Nine Principles of Evolution Revealed by Paleontology." *American Naturalist* 66, no. 702: 52–60.

Osborn, Henry F. 1934. "Aristogenesis, the Creative Principle in the Origin of Species." *American Naturalist* 68, no. 716: 193–235.

Osche, Günther. 1958. "Beiträge zur Morphologie, Ökologie und Phylogenie der Ascaridoidea (Nematoda)." *Zeitschrift für Parasitenkunde* 18, no. 6: 479–572.

Osche, Günther. 1960. "Systematische, morphologische und parasitophy-letische studien an parasitischen Oxyuroidea (Nematoda) exotischer Diplopoden (ein Beitrag zur Morphologie des Sexualdimorphismus)." *Zoologischer Jahrbuch* 87: 395–440.

Osche, Günther. 1963. "Morphological, Biological, and Ecological Consider-ations in the Phylogeny of Parasitic Nematodes." In *The Lower Metazoa: Comparative Biology and Phylogeny*, edited by Ellsworth C. Dougherty, 283–302. Berkeley and Los Angeles: University of California Press.

Othman, Ahmad A., and Rasha H. Soliman. 2015. "Schistosomiasis in Egypt: A Never-Ending Story?" *Acta Tropica* 148: 179–90. https://doi.org/10.1016/j.actatropica.2015.04.016.

Ottersen, Trygve, Steven J. Hoffman, and Gaëlle Groux. 2016. "Ebola Again Shows the International Health Regulations Are Broken." *American Journal of Law & Medicine* 42, no. 2–3: 356–92. https://doi.org/10.1177/0098858816658273.

Page, Roderic D. M. 2003. *Tangled Trees: Phylogeny, Cospeciation, and Coevolution.* Chicago: University of Chicago Press.

Page, Scott E. 2011. *Diversity and Complexity.* Princeton, NJ: Princeton University Press.

Pálinkás, Melinda. 2018. "Ecological Responses to Climate Change at Biogeo-graphical Boundaries." In *Pure and Applied Biogeography*, edited by Levente Hufnagel, 165–211. https://doi.org/10.5772/intechopen.69514.

Palms, Danielle L., Elisabeth Mungai, Taniece Eure, Angela Anttila, Nicola D. Thompson, Margaret A. Dudeck, Jonathan R. Edwards, et al. 2017. "The National Healthcare Safety Network Long-Term Care Facility Component Early Reporting Experience: January 2013–December 2015." *American Journal of Infection Control* 46, no. 6: 637–42. https://doi.org/10.1016/j.ajic.2018.01.003.

Panzer, J., P. Saavedra, et al. 2016. "The Short-Term Economic Costs of Zika in Latin America and the Caribbean." February 18. World Bank Group.

Papadopoulou, Anna, and L. Lacey Knowles. 2015. "Species-Specific Responses to Island Connectivity Cycles: Refined Models for Testing Phylogeo-graphic Concordance across a Mediterranean Pleistocene Aggregate Island Complex." *Molecular Ecology* 24, no. 16: 4252–68. https://doi.org/10.1111/mec.13305.

Pardey, Philip, Yuan Chai, Jason Beddow, Terry Hurley, and Darren Kriticos. 2014. "The Global Occurrence and Economic Consequences of Stripe Rust in Wheat." Paper presented at the. BGRI 2014 Technical Workshp, Ciudad Obregón, Mexico, 24 March.

Pardi, Norbert, Michael J. Hogan, Frederick W. Porter, and Drew Weissman. 2018. "mRNA Vaccines—A New Era in Vaccinology." *Nature Reviews: Drug Discovery* 17, no. 4: 261–79. https://doi.org/10.1038/nrd.2017.243.

Parmesan, C. 2006. "Ecological and Evolutionary Responses to Recent Climate Change." *Annual Review of Ecology, Evolution and Systematics* 37: 637–69.

Parrish, Colin R., Edward C. Holmes, David M. Morens, Eun-Chung Park, Donald S. Burke, Charles H. Calisher, Catherine A. Laughlin, et al. 2008. "Cross-Species Virus Transmission and the Emergence of New Epidemic Diseases." *Microbiology and Molecular Biology Reviews* 72: 457–70.

Paseka, Rachel E., Carrie C. Heitman, and Karl J. Reinhard. 2018. "New Evidence of Ancient Parasitism among Late Archaic and Ancestral Puebloan Residents of Chaco Canyon." *Journal of Archaeological Science: Reports* 18: 51–58. https://doi.org/10.1016/j.jasrep.2018.01.001.

Pasquini, Lorena, Gina Ziervogel, Richard M. Cowling, and Clifford Shearing. 2014. "What Enables Local Governments to Mainstream Climate Change Adaptation? Lessons Learned from Two Municipal Case Studies in the Western Cape, South Africa." *Climate and Development* 7, no. 1: 60–70. https://doi.org/10.1080/17565529.2014.886994.

Patella, L., Daniel R. Brooks, and Walter Antonio Boeger. 2017. "Phylogeny and Ecology Illuminate the Evolution of Associations under the Stockholm Paradigm." *Vie et milieu* 67, no. 2: 91–102.

Patz, J. A., A. K. Githeko, J. P. McCarty, S. Hussein, U. Confalonieri, N. De Wet, et al. 2003. "Climate Change and Infectious Diseases." *Climate Change and Human Health: Risks and Responses* 6: 103–37.

Patz, J. A., D. Campbell-Lendrum, T. Holloway, and J. A. Foley. 2005. "Impact of Regional Climate Change on Human Health." *Nature* 438, no. 7066: 310–17. https://doi.org/10.1038/nature04188.

Patz, Jonathan A., and Sarah H. Olson. 2006. "Climate Change and Health: Global to Local Influences on Disease Risk." *Annals of Tropical Medicine & Parasitology* 100, no. 5–6: 535–49. https://doi.org/10.1179/136485906X97426.

Patz, Jonathan A., Holly K. Gibbs, Jonathan A. Foley, Jamesine V. Rogers, and Kirk R. Smith. 2007. "Climate Change and Global Health: Quantifying a Growing Ethical Crisis." *EcoHealth* 4, no. 4: 397–405. https://doi.org/10.1007/s10393-007-0141-1.

Patz, Jonathan A., Sarah H. Olson, Christopher K. Uejio, and Holly K. Gibbs. 2008. "Disease Emergence from Global Climate and Land Use Change." *Medical Clinics of North America* 92, no. 6: 1473–91. https://doi.org/10.1016/j.mena.2008.07.007.

Patz, Jonathan A., Thaddeus K. Graczyk, Nina Geller, and Amy Y. Vittor. 2000. "Effects of Environmental Change on Emerging Parasitic Diseases." *International Journal for Parasitology* 30, no. 12: 1395–1405.

Pauly, Daniel, Reg Watson, and Jackie Alder. 2005. "Global Trends in World Fisheries: Impacts on Marine Ecosystems and Food Security." *Philosophi-*

cal Transactions of the Royal Society of London B: Biological Sciences 360, no. 1453: 5–12. https://doi.org/10.1098/rstb.2004.1574.

Pearce-Duvet, Jessica M. C. 2006. "The Origin of Human Pathogens: Evaluating the Role of Agriculture and Domestic Animals in the Evolution of Human Disease." *Biological Reviews* 81, no. 3: 369–382.

Pecl, Gretta T., Miguel B. Araújo, Johann D. Bell, Julia Blanchard, Timothy C. Bonebrake, I-Ching Chen, Timothy D. Clark, et al. 2017. "Biodiversity Redistribution under Climate Change: Impacts on Ecosystems and Human Well-Being." *Science* 355, no. 6332: https://doi.org/10.1126/science.aai9214.

Pellegrino, Irene, Alessandro Negri, Marco Cucco, Nadia Mucci, Marco Pavia, Martin Šálek, Giovani Boano, et al. 2014. "Phylogeography and Pleistocene Refugia of the Little Owl *Athene noctua* Inferred from mtDNA Sequence Data." *Ibis* 156, no. 3: 639–57.

Peñuelas, Josep, Jordi Sardans, Constantí Stefanescu, Teodor Parella, and Iolanda Filella. 2006. "*Lonicera implexa* Leaves Bearing Naturally Laid Eggs of the Specialist Herbivore *Euphydryas aurinia* Have Dramatically Greater Concentrations of Iridoid Glycosides than Other Leaves." *Journal of Chemical Ecology* 32, no. 9: 1925–33.

Pérez-Rodríguez, R., O. Domínguez-Domínguez, I. Doadrio, E. Cuevas-García, and G. Pérez-Ponce de León. 2015. "Comparative Historical Biogeography of Three Groups of Nearctic Freshwater Fishes across Central Mexico." *Journal of Fish Biology* 86, no. 3: 993–1015. https://doi.org/10.1111/jfb.12611.

Perkins, S. L. 2001. "Phylogeography of Caribbean Lizard Malaria: Tracing the History of Vector-Borne Parasites." *Journal of Evolutionary Biology* 14, no. 1: 34–45.

Perry, B. D., and T. F. Randolph. 1999. "Improving the Assessment of the Economic Impact of Parasitic Diseases and of Their Control in Production Animals." *Veterinary Parasitology* 84, no. 3–4: 145–68. https://doi.org/10.1016/S0304-4017(99)00040-0.

Peterson, A. Townsend. 2006. "Ecologic Niche Modeling and Spatial Patterns of Disease Transmission." *Emerging Infectious Diseases* 12, no. 12: 1822–26. https://doi.org/10.3201/eid1212.060373

Peterson, A. Townsend, and David A. Vieglais. 2001. "Predicting Species Invasions Using Ecological Niche Modeling." *BioScience* 51, no. 5: 363–71.

Peterson, A. Townsend, Hanqin Tian, Enrique Martinez-Meyer, Jorge Soberon, Victor Sanchez-Cordero, and Brian Huntley. 2005. "Modeling Distributional Shifts of Individual Species and Biomes." In *Climate Change and Biodiversity*, edited by Thomas E. Lovejoy and Lee Hannah, 211–28. New Haven, CT: Yale University Press.

Peterson, A. Townsend, Miguel A. Ortega-Huerta, Jeremy Bartley, Victor Sánchez-Cordero, Jorge Soberón, Robert H. Buddemeier, and David R. B. Stockwell. 2002. "Future Projections for Mexican Faunas under Global Climate Change Scenarios." *Nature* 416, no. 6881: 626–29.

Pfennig, David W., Matthew A. Wund, Emilie C. Snell-Rood, Tami Cruick-shank, Carl D. Schlichting, and Armin P. Moczek. 2010. "Phenotypic Plasticity's Impacts on Diversification and Speciation." *Trends in Ecology & Evolution* 25, no. 7: 459–67. https://doi.org/10.1016/j.tree.2010.05.006.

Pichler, Verena, Romeo Bellini, Rodolfo Veronesi, Daniele Arnoldi, Annapaola Rizzoli, Riccardo Paolo Lia, Domenico Otranto, et al. 2018. "First Evidence of Resistance to Pyrethroid Insecticides in Italian *Aedes albopictus* Populations 26 Years after Invasion." *Pest Management Science* 14 (February): 1319–27. https://doi.org/10.1002/ps.4840.

Pickles, Rob S. A., Daniel Thornton, Richard Feldman, Adam Marques, and Dennis L. Murray. 2013. "Predicting Shifts in Parasite Distribution with Climate Change: A Multitrophic Level Approach." *Global Change Biology* 19, no. 9: 2645–54. https://doi.org/10.1111/gcb.12255.

Pimentel, David, Rodolfo Zuniga, and Doug Morrison. 2005. "Update on the Environmental and Economic Costs Associated with Alien-Invasive Species in the United States." *Ecological Economics* 52, no. 3: 273–88. https://doi.org/10.1016/j.ecolecon.2004.10.002.

Pinsent, Amy, Kim M. Pepin, Huachen Zhu, Yi Guan, Michael T. White, and Steven Riley. 2017. "The Persistence of Multiple Strains of Avian Influenza in Live Bird Markets." *Proceedings of the Royal Society of London B: Biological Sciences* 28, no. 1868: 20170715. https://doi.org/10.1098/rspb.2017.0715.

Pinsky, Malin L., Boris Worm, Michael J. Fogarty, Jorge L. Sarmiento, and Simon A. Levin. 2013. "Marine Taxa Track Local Climate Velocities." *Science* 341, no. 6151: 1239–42.

Poe, Edgar Allan. 1842. "The Mask of the Red Death: A Fantasy." In *Graham's Lady's and Gentleman's Magazine*, 257–59. May.

Pohlmann, Anne, Elke Starick, Christian Grund, Dirk Höper, Günter Strebelow, Anja Globig, Christoph Staubach, et al. 2018. "Swarm Incursions of Reassortants of Highly Pathogenic Avian Influenza Virus Strains H5N8 and H5N5, Clade 2.3.4.4b, Germany, Winter 2016/17." *Scientific Reports* 8, no. 1. https://doi.org/10.1038/s41598-017-16936-8.

Post, Eric, Uma S. Bhatt, Cecilia M. Bitz, Jedediah F. Brodie, Tara L. Fulton, Mark Hebblewhite, Jeffrey Kerby, et al. 2013. "Ecological Consequences of Sea-Ice Decline." *Science* 341, no. 6145: 519–24.

Poulin, Robert. 1992. "Determinants of Host-Specificity in Parasites of Freshwater Fishes." *International Journal for Parasitology* 22, no. 6: 753–58. https://doi.org/10.1016/0020-7519(92)90124-4.

Pozio, Edoardo, Eric Hoberg, Giuseppe La Rosa, and Dante S. Zarlenga. 2009. "Molecular Taxonomy, Phylogeny and Biogeography of Nematodes Belonging to the *Trichinella* Genus." *Infection, Genetics and Evolution* 9, no. 4: 606–16.

President's Council of Advisors on Science and Technology [PCAST]. 1998. *Teaming with Life: Investing in Science to Understand and Use America's Living Capital*. Washington, DC: PCAST, Biodiversity and Ecosystems Panel.

President's Council of Advisors on Science and Technology [PCAST]. 2005. "Global Risks of Infectious Animal Diseases." Issue Paper 28 (January): 1–16. http://www.castscience.org/download.cfm?PublicationID=2900& File=f030f5b5845ecc35 e2b0631a124043596147.

Prairie Climate Centre. 2018. "Will Climate History Repeat Itself? Lessons From Another Global Warming Event." February 5. http:// prairieclimatecentre.ca/2018/02/will-climate- history-repeat-itself-lessons -from-another-global-warming-event/.

Prates, Ivan, Danielle Rivera, Miguel T. Rodrigues, and Ana C. Carnaval. 2016. "A Mid-Pleistocene Rainforest Corridor Enabled Synchronous Invasions of the Atlantic Forest by Amazonian Anole Lizards." *Molecular Ecology* 25, no. 20: 5174–86. https://doi.org/10.1111/mec.13821.

Preece, Noel D., Sandra E. Abell, Lauran Grogan, Adrian Wayne, Lee F. Skerratt, Penny van Oosterzee, Amy L. Shima, et al. 2017. "A Guide for Ecologists: Detecting the Role of Disease in Faunal Declines and Managing Population Recovery." *Biological Conservation* 214: 136–46. https://doi.org/10 .1016/j.biocon.2017.08.014.

Prevéy, Janet, Mark Vellend, Nadja Rüger, Robert D. Hollister, Anne D. Bjorkman, Isla H. Myers-Smith, Sarah C. Elmendorf, et al. 2017. "Greater Temperature Sensitivity of Plant Phenology at Colder Sites: Implications for Convergence across Northern Latitudes." *Global Change Biology* 23, no. 7: 2660–71. https://doi.org/10.1111/gcb.13619.

Price, Peter W. 1980. *Evolutionary Biology of Parasites*. Vol. 15. Princeton, NJ: Princeton University Press.

Prosser, Sean W. J., Maria G. Velarde-Aguilar, Virginia Léon-Règagnon, and Paul D. N. Hebert. 2013. "Advancing Nematode Barcoding: A Primer Cocktail for the Cytochrome Coxidase Subunit I Gene from Vertebrate Parasitic Nematodes." *Molecular Ecology Resources* 13, no. 6: 1108–15. https://doi.org/ 10.1111/1755-0998.12082.

Pullin, Andrew S., William Sutherland, Toby Gardner, Valerie Kapos, and John E. Fa. 2013. *Conservation Priorities: Identifying Need, Taking Action and Evaluating Success*. In *Key Topics in Conservation Biology 2*, edited by David W. Macdonald and Katherine J. Willis, 1–22. New York: Wiley. https:// doi.org/10.1002/9781118520178.

Råberg, Lars, Elisabet Alacid, Esther Garces, and Rosa Figueroa. 2014. "The Potential for Arms Race and Red Queen Coevolution in a Protist Host-Parasite System." *Ecology and Evolution* 4, no. 24: 4775–85. https://doi.org/ 10.1002/ece3.1314.

Racloz, Vanessa, Gert Johannes Venter, C. Griot, and Katharina D. C. Stärk. 2008. "Estimating the Temporal and Spatial Risk of Bluetongue Related to the Incursion of Infected Vectors into Switzerland." *BioMed Central Veterinary Research* 4: 42. https://doi.org/10.1186/1746-6148-4-42.

Raeymaekers, Joost A. M., Anurag Chaturvedi, Pascal I. Hablützel, Io Verdonck, Bart Hellemans, Gregory E. Maes, Luc De Meester, and Filip A. M.

Volckaert. 2017. "Adaptive and Non-Adaptive Divergence in a Common Landscape." *Nature Communications* 8: 267. https://doi.org/10.1038/s41467 -017-00256-6.

Raff, Jennifer. 2016. "Ancient DNA Reveals Genetic Legacy of Pandemics in the Americas." *Guardian*, November 12. https://doi.org/10.1038/ncomms13175.

Ragazzo, Leo J., Sarah Zohdy, Mamitiana Velonabison, James Herrera, Patricia C. Wright, and Thomas R. Gillespie. 2018. "*Entamoeba histolytica* Infection in Wild Lemurs Associated with Proximity to Humans." *Veterinary Parasitology* 249: 98–101. https://doi.org/10.1016/j.vetpar.2017.12.002.

Raghavan, M., Pontus Skoglund, Kelly E. Graf, Mait Metspalu, Anders Albrechtsen, Ida Moltke, Simon Rasmussen, et al. 2013. "Upper Palaeolithic Siberian Genome Reveals Dual Ancestry of Native Americans." *Nature* 505, no. 7481: 87–91. https://doi.org/10.1038/nature12736.

Rahfeld, Peter, Wiebke Haeger, Roy Kirsch, Gerhard Pauls, Tobias Becker, Eva Schulze, Natalie Wielsch, et al. 2015. "Glandular β-Glucosidases in Juvenile *Chrysomelina* Leaf Beetles Support the Evolution of a Host-Plant-Dependent Chemical Defense." *Insect Biochemistry and Molecular Biology* 58: 28–38. https://doi.org/10.1016/j.ibmb.2015.01.003.

Ramalli, Lauriane, Stephen Mulero, Harold Noël, Jean-Dominique Chiappini, Josselin Vincent, Hélène Barré-Cardi, Philippe Malfait, et al. 2018. "Persistence of Schistosomal Transmission Linked to the Cavu River in Southern Corsica since 2013." *Eurosurveillance* 23, no. 4: 30100. https://doi.org/10 .2807/1560-7917.ES.2018.23.4.18-00017.

Randhawa, Haseeb S., and Michael D. B. Burt. 2008. "Determinants of Host Specificity and Comments on Attachment Site Specificity of Tetraphyllidean Cestodes Infecting Rajid Skates from the Northwest Atlantic." *Journal of Parasitology* 94, no. 2: 436–61. https://doi.org/10.1645/ge-1180.1.

Rasmussen, S. O., Matthias Bigler, Simon P. Blockley, Thomas Blunier, Susanne L. Buchardt, Henrik B. Clausen, Ivana Cvijanovic, et al. 2014. "A Stratigraphic Framework for Abrupt Climatic Changes during the Last Glacial Period based on Three Synchronized Greenland Ice-Core Records: Refining and Extending the Intimate Event Stratigraphy." *Quaternary Science Reviews* 106: 14–28.

Rausch, Robert L. 1994. "Transberingian Dispersal of Cestodes in Mammals." *International Journal for Parasitology* 24, no. 7: 1203–12.

Rausher, Mark D. 2001. "Co-Evolution and Plant Resistance to Natural Enemies." *Nature* 411: 857–64.

Razgour, Orly, Irene Salicini, Carlos Ibáñez, Ettore Randi, and Javier Juste. 2015. "Unravelling the Evolutionary History and Future Prospects of Endemic Species Restricted to Former Glacial Refugia." *Molecular Ecology* 24, no. 20: 5267–83. https://doi.org/10.1111/mec.13379.

Reiczigel, Jenö, and Lajos Rózsa. 1998. "Host-Mediated Site Segregation of Ectoparasites: An Individual-Based Simulation Study." *Journal of Parasitology* 84, no. 3: 491–98.

Reinhard, Karl J., and Elisa Pucu de Araújo. 2014. "Comparative Parasitological Perspectives on Epidemiologic Transitions: The Americas and Europe." In *Modern Environments and Human Health: Revisiting the Second Epidemiologic Transition*, edited by Molly K. Zuckerman, 311–26. New York: Wiley.

Renema, Willem, D. R. Bellwood, J. C. Braga, K. Bromfield, R. Hall, K. G. Johnson, P. Lunt, et al. 2008. "Hopping Hotspots: Global Shifts in Marine Biodiversity." *Science* 321, no. 5889: 654–57.

Repenning, Charles A. 1980. "Faunal Exchanges between Siberia and North America." *Canadian Journal of Anthropology* 1, no. 1: 37–44.

Repenning, Charles A. 2001. "Beringian Climate during Intercontinental Dispersal: A Mouse Eye View." *Quaternary Science Reviews* 20, no. 1: 25–40.

Respicio-Kingry, Laurel B., Brook M. Yockey, Sarah Acayo, John Kaggwa, Titus Apangu, Kiersten J. Kugeler, Rebecca J. Eisen, et al. 2016. "Two Distinct *Yersinia pestis* Populations Causing Plague among Humans in the West Nile Region of Uganda." *PLoS Neglected Tropical Diseases* 10, no. 2: e0004360. https://doi.org/10.1371/journal.pntd.0004360.

Reynolds, Mary G., Sarah Anne J. Guagliardo, Yoshinori J. Nakazawa, Jeffrey B. Doty, and Matthew R. Mauldin. 2017. "Understanding Orthopoxvirus Host Range and Evolution: From the Enigmatic to the Usual Suspects." *Current Opinion in Virology* 28: 108–15. https://doi.org/10.1016/j.coviro.2017.11.012.

Richards, Lora A., Lee A. Dyer, Matthew L. Forister, Angela M. Smilanich, Craig D. Dodson, Michael D. Leonard, and Christopher S. Jeffrey. 2015. "Phytochemical Diversity Drives Plant-Insect Community Diversity." *Proceedings of the National Academy of Sciences of the United States of America* 112, no. 35: 10973–78.

Richgels, Katherine L. D., Robin E. Russell, Gebbiena M. Bron, and Tonie E. Rocke. 2016. "Evaluation of *Yersinia pestis* Transmission Pathways for Sylvatic Plague in Prairie Dog Populations in the Western U.S." *EcoHealth* 13, no. 2: 415–27. https://doi.org/10.1007/s10393-016-1133-9.

Richter, Veronika, Karin Lebl, Walter Baumgartner, Walter Obritzhauser, Annemarie Käsbohrer, and Beate Pinior. 2017. "A Systematic Worldwide Review of the Direct Monetary Losses in Cattle due to Bovine Viral Diarrhoea Virus Infection." *Veterinary Journal* 220 (February): 80–87. https://doi.org/10.1016/j.tvjl.2017.01.005.

Rios, Liliam, Liani Coronado, Dany Naranjo-Feliciano, Orlando Martínez-Pérez, Carmen L. Perera, Lilian Hernandez-Alvarez, Heidy Díaz de Arce, et al. 2017. "Deciphering the Emergence, Genetic Diversity and Evolution of Classical Swine Fever Virus." *Scientific Reports* 7: 17887. https://doi.org/10.1038/s41598-017-18196-y.

Rizzoli, Annapola, Cornelia Silaghi, Anna Obiegala, Ivo Rudolf, Zdeněk Hubálek, Gábor Földvári, Olivier Plantard, Muriel Vayssier-Taussat, et al. 2014. "*Ixodes ricinus* and Its Transmitted Pathogens in Urban and Peri-Urban Areas in Europe: New Hazards and Relevance for Public Health." *Frontiers in Public Health* 2: 251. https://doi.org/10.3389/fpubh.2014.00251.

Robertson, Gemma J., Anson V. Koehler, Robin B. Gasser, Matthew Watts, Robert Norton, and Richard S. Bradbury. 2017. "Application of PCR-Based Tools to Explore *Strongyloides* Infection in People in Parts of Northern Australia." *Tropical Medicine and Infectious Disease* 2, no. 4: 62. https://doi .org/10.3390/tropicalmed2040062.

Robertson, Lucy J., Hein Sprong, Ynes R. Ortega, Joke W. B. van der Giessen, and Ron Fayer. 2014. "Impacts of Globalisation on Foodborne Parasites." *Trends in Parasitology* 3, no. 1: 37–52.

Robertson, Lucy J., Joke W. B. van der Giessen, Michael B. Batz, Mina Kojima, and Sarah Cahill. 2013. "Have Foodborne Parasites Finally Become a Global Concern?" *Trends in Parasitology* 29, no. 3: 101–3.

Robles, Maria del Rosario. 2010. "La importanciade los nematodes Syphaciini (Syphaciinae-Oxyuridae) como marcedores especifícos de sus hospeda- dores." *Mastozoología neotropical* 17, no. 2: 305–15.

Roche, Benjamin, Serge Morand, Eric Elguero, Thomas Balenghien, Jean- François Guégan, and Nicolas Gaidet. 2015. "Does Host Receptivity or Host Exposure Drives Dynamics of Infectious Diseases? The Case of West Nile Virus in Wild Birds." *Infection, Genetics and Evolution* 33: 11–19.

Roeber, Florian, Aaron R. Jex, and Robin B. Gasser. 2013. "Impact of Gastrointesti- nal Parasitic Nematodes of Sheep, and the Role of Advanced Molecular Tools for Exploring Epidemiology and Drug Resistance—An Australian Perspec- tive." *Parasites & Vectors* 6: 153. https://doi.org/10.1186/1756-3305-6-153.

Roelfs, A. P. 1989. "Epidemiology of the Cereal Rusts in North America." *Cana- dian Journal of Plant Pathology* 11, no. 1: 86–89.

Rogers, William P. 1962. *The Nature of Parasitism: The Relationship of Some Meta- zoan Parasites to Their Hosts*. New York: Academic Press.

Rohde, Klaus. 1979. "A Critical Evaluation of Intrinsic and Extrinsic Factors Responsible for Niche Restriction in Parasites." *American Naturalist* 114, no. 5: 648–71.

Rohde, Klaus. 1981. "Niche Width of Parasites in Species-Rich and Species-Poor Communities." *Cellular and Molecular Life Sciences* 37, no. 4: 359–61.

Rohde, Klaus. 1984. "Ecology of Marine Parasites." *Helgoländer Meeresuntersu- chungen* 37, no. 1: 5.

Rohde, Klaus. 1994. "Niche Restriction in Parasites: Proximate and Ultimate Causes." In "Parasites and Behaviour," edited by M. V. K. Sukhdeo. Supplement, *Parasitology* 109, no. S1: S69–84. https://doi.org/10.1017/ S0031182000085097.

Rohde, Klaus, and R. P. Hobbs. 1986. "Species Segregation: Competition or Re- inforcement of Reproductive Barriers?" In *Parasite Lives: Papers on Parasites, Their Hosts and Their Associations to Honour J. F. A. Sprent*, edited by Mary Cremin, Colin Dobson, and Douglas E. Moorehouse, 189–99. St. Lucia, Queensland: University of Queensland Press.

Rolland, Jonathan, Marc W. Cadotte, Jonathan Davies, Vincent Devic- tor, Sebastien Lavergne, Nicolas Mouquet, Sandrine Pavoine, et al.

2012. *Using Phylogenies in Conservation: New Perspectives*. London: Royal Society.

Rooney, N., K. S. McCann, and D. L. G. Noakes, eds. 2006. *From Energetics to Ecosystems: The Dynamics and Structure of Ecological Systems*. Peter Yodzis Fundamental Ecology Series 1. New York: Springer Science and Business Media.

Roossinck, Marilyn J. 2013. "Plant Virus Ecology." *PLoS Pathogens* 9, no. 5: e1003304. https://doi.org/10.1371/journal.ppat.1003304.

Roossinck, Marilyn J. 2015a. "Plants, Viruses and the Environment: Ecology and Mutualism." *Virology* 479–80: 271–77. https://doi.org/10.1016/j.virol.2015.03.041.

Roossinck, Marilyn J. 2015b. "Move Over, Bacteria! Viruses Make Their Mark as Mutualistic Microbial Symbionts." *Journal of Virology* 89, no. 13: 6532–35. https://doi.org/10.1128/JVI.02974-14.

Roossinck, Marilyn J., and Fernando García-Arenal. 2015. "Ecosystem Simplification, Biodiversity Loss and Plant Virus Emergence." *Current Opinion in Virology* 10: 56–62. https://doi.org/10.1016/j.coviro.2015.01.005.

Roossinck, Marilyn J., Darren P. Martin, and Philippe Roumagnac. 2015. "Plant Virus Metagenomics: Advances in Virus Discovery." *Phytopathology* 105, no. 6: 716–27. https://doi.org/10.1094/PHYTO-12-14-0356-RVW.

Rosa, Daniele. 1918. *Ologenesi*. Florence: R. Bemparad.

Rosa, Daniele. 1931. *L'ologénèse: Nouvelle théorie de l'évolution et de la distribution géographique des êtres vivants*. Paris: F. Alcan.

Rosà, Roberto, Veronica Andreo, Valentina Tagliapietra, Ivana Baráková, Daniele Arnoldi, Heidi Hauffe, Mattia Manica, et al. 2018. "Effect of Climate and Land Use on the Spatio-Temporal Variability of Tick-Borne Bacteria in Europe." *International Journal of Environmental Research and Public Health* 15, no. 4: 732. https://doi.org/10.3390/ijerph15040732.

Rosen, Julia. 2017. "California Rains Put Spotlight on Atmospheric Rivers." *Science Magazine Online*, February 22.

Rosenthal, Benjamin M. 2008. "How Has Agriculture Influenced the Geography and Genetics of Animal Parasites?" *Trends in Parasitology* 25, no. 2: 67–70.

Rosenthal, Benjamin M., Giuseppe LaRosa, Dante Zarlenga, Detiger Dunams, Yao Chunyu, Liu Mingyuan and Edoardo Pozio. 2008. "Human Dispersal of *Trichinella spiralis* in Domesticated Pigs." *Infection, Genetics and Evolution* 8, no. 6: 799–805.

Rosenthal, Joshua. 2009. "Climate Change and the Geographic Distribution of Infectious Diseases." *EcoHealth* 6, no. 4: 489–95.

Rosenzweig, Cynthia, Ana Iglesias, X. B. Yang, Paul R. Epstein, and Eric Chivian. 2001. "Climate Change and Extreme Weather Events; Implications for Food Production, Plant Diseases, and Pests." *Global Change & Human Health* 2, no. 2: 90–104.

Ross, Herbert H. 1972a. "The Origin of Species Diversity in Ecological Communities." *Taxon* 21: 253–59.

Ross, Herbert H. 1972b. "An Uncertainty Principle in Ecological Evolution." *University of Arkansas Museum Occasional Papers* 4: 133–57.

Rothman, Daniel H. 2017. "Thresholds of Catastrophe in the Earth System." *Science Advances* 3, no. 9: 31770906. https://doi.org/10.1126/sciadv.1700906.

Rowe, Kevin C., Michael L. Reno, Daniel M. Richmond, Ronald M. Adkins, and Scott J. Steppan. 2008. "Pliocene Colonization and Adaptive Radiations in Australia and New Guinea (Sahul): Multilocus Systematics of the Old Endemic Rodents (Muroidea: Murinae)." *Molecular Phylogenetics and Evolution* 47, no. 1: 84–101.

Ruddiman, William F. 2013. "The Anthropocene." *Annual Review of Earth and Planetary Sciences* 41: 45–68.

Ruedi, Manuel, Nicole Friedli-Weyeneth, Emma C. Teeling, Sébastien J. Puechmaille, and Steven M. Goodman. 2012. "Biogeography of Old World Emballonurine Bats (Chiroptera: Emballonuridae) Inferred with Mitochondrial and Nuclear DNA." *Molecular Phylogenetics and Evolution* 64, no. 1: 204–11. https://doi.org/10.1016/j.ympev.2012.03.019.

Russell, Robert J. 1968. "Evolution and Classification of the Pocket Gophers of the Subfamily Geomyinae." *University of Kansas Publications, Museum of Natural History* 16: 473–579.

Rutschmann, Sereina, Harald Detering, Sabrina Simon, David H. Funk, Jean-Luc Gattolliat, Samantha J. Hughes, Pedro M. Raposeiro, et al. 2017. "Colonization and Diversification of Aquatic Insects on Three Macaronesian Archipelagos Using 59 Nuclear Loci Derived from a Draft Genome." *Molecular Phylogenetics and Evolution* 107 (February): 27–38.

Ryss, A. Y., K. S. Polyanina, B. G. Popovichev, S. A. Krivets, and I. A. Kerchev. 2018. "Plant Host Range Specificity of *Bursaphelenchus mucronatus* Mamiya et Enda, 1979 Tested in the Laboratory Experiments." *Parazitologiya* 52, no. 1: 32–40.

Sachs, Jeffrey, and Pia Malaney. 2002. "The Economic and Social Burden of Malaria." *Nature* 415, no. 6872: 680–85.

Sady, Hany, Hesham M. Al-Mekhlafi, Bonnie L. Webster, Romano Ngui, Wahib M. Atroosh, Ahmed K. Al-Delaimy, Nabil A. Nasr, Kek Heng Chua, et al. 2015. "New Insights into the Genetic Diversity of *Schistosoma mansoni* and *S. haematobium* in Yemen." *Parasites & Vectors* 8, no. 1: 544.

Sala, Osvaldo E., F. Stuart Chapin, Juan J. Armesto, Eric Berlow, Janine Bloomfield, Rodolfo Dirzo, Elisabeth Huber-Sanwald, et al. 2000. "Global Biodiversity Scenarios for the Year 2100." *Science* 287, no. 5459: 1770–74. https://doi.org/10.1126/science.287.5459.1770.

Salkeld, Dan J., and Paul Stapp. 2008. "Prevalence and Abundance of Fleas in Black-Tailed Prairie Dog Burrows: Implications for the Transmission of Plague (*Yersinia pestis*)." *Journal of Parasitology* 94, no. 3: 616–21. https://doi.org/10.1645/GE-1368.1.

Salkeld, Daniel J., Marcel Salathé, Paul Stapp, and James Holland Jones. 2010. "Plague Outbreaks in Prairie Dog Populations Explained by Percolation

Thresholds of Alternate Host Abundance." *Proceedings of the National Academy of Sciences of the United States* 107, no. 32: 14247–50. https://doi.org/10.1073/pnas.1002826107.

Samy, Abdallah M., and A. Townsend Peterson. 2016. "Climate Change Influences on the Global Potential Distribution of Bluetongue Virus." *PLoS ONE* 11, no. 3: e0150489. https://doi.org/10.1371/journal.pone.0150489.

Samy, Abdallah M., A. Townsend Peterson, and Matthew Hall. 2017. "Phylogeography of Rift Valley Fever Virus in Africa and the Arabian Peninsula." *PLoS Neglected Tropical Diseases* 11, no. 1: e0005226. https://doi.org/10.1371/journal.pntd.0005226.

Santoro, M., V. V. Tkach, S. Mattiucci, J. M. Kinsella, and G. Nascetti. 2011. "*Renifer aniarum* (Digenea: Reniferidae), an Introduced North American Parasite in Grass Snakes *Natrix natrix* in Calabria, Southern Italy." *Diseases of Aquatic Organisms* 95, no. 3: 233–40. https://doi.org/10.3354/dao02365.

Sasal, P., S. Trouvé, C. Müller-Graf, and S. Morand. 1999. "Specificity and Host Predictability: A Comparative Analysis among Monogenean Parasites of Fish." *Journal of Animal Ecology* 68, no. 3: 437–44. https://doi.org/10.2307/778809.

Sato, Masaaki, Kentaro Honda, Wilfredo H. Uy, Darwin I. Baslot, Tom G. Genovia, Yohei Nakamura, Lawrence Patrick C. Bernardo, et al. 2017. "Marine Protected Area Restricts Demographic Connectivity: Dissimilarity in a Marine Environment Can Function as a Biological Barrier." *Ecology and Evolution* 7, no. 19: 7859–71. https://doi.org/10.1002/ece3.3318.

Sayers, Dorothy L. 1935. *Gaudy Night.* London: Victor Gollancz.

Schad, G. A. 1963. "Niche Diversification in a Parasitic Species Flock." *Nature* 198, no. 4878: 404–6.

Schade, Franziska M., Michael J. Raupach, and K. Mathias Wegner. 2016. "Seasonal Variation in Parasite Infection Patterns of Marine Fish Species from the Northern Wadden Sea in Relation to Interannual Temperature Fluctuations." *Journal of Sea Research* 113: 73–84. https://doi.org/10.1016/j.seares.2015.09.002.

Schäpers, Alexander, Mikael A. Carlsson, Gabriella Gamberale-Stille, and Niklas Janz. 2015. "The Role of Olfactory Cues for the Search Behavior of a Specialist and Generalist Butterfly." *Journal of Insect Behavior* 28, no. 1: 77–87. https://doi.org/10.1007/s10905-014-9482-0.

Scheffers, Brett R., Luc De Meester, Tom C. L. Bridge, Ary A. Hoffmann, John M. Pandolfi, Richard T. Corlett, Stuart H. M. Butchart, et al. 2016. "The Broad Footprint of Climate Change from Genes to Biomes to People." *Science* 354, no. 6313: aaf7671.

Schindewolf, Otto H. 1969. *Über den "Typus" in morphologischer und phylogenetischer Biologie.* N.p.: Verlag der Akademie der Wissenschaften und der Literatur.

Schmid, Boris V., Ulf Büntgen, W. Ryan Easterday, Christian Ginzler, Lars Walløe, Barbara Bramanti, and Nils Chr. Stenseth. 2015. "Climate-Driven

Introduction of the Black Death and Successive Plague Reintroductions into Europe." *Proceedings of the National Academy of Sciences of the United States of America* 112, no. 10: 3020–25.

Schmiedel, Ute, Yoseph Araya, Maria Ieda Bortolotto, Linda Boeckenhoff, Winnie Hallwachs, Daniel Janzen, Shekhar S. Kolipaka, et al. 2016. "Contributions of Paraecologists and Parataxonomists to Research, Conservation, and Social Development." *Conservation Biology* 30, no. 3: 506–19. https://doi.org/10.1111/cobi.12661.

Schoonhoven, L. M. 1996. "After the Verschaffelt-Dethier Era: The Insect-Plant Field Comes of Age." *Entomologia Experimentalis et Applicata* 80, no. 1: 1–5.

Schumann, Gay L., and Kurt J. Leonard. 2000. "Stem Rust of Wheat." http://www.apsnet.org/edcenter/intropp/lessons/fungi/Basidiomycetes/Pages/StemRust.as px.

Schwab, Samantha R., Chris M. Stone, Dina M. Fonseca, and Nina H. Fefferman. 2017. "The Importance of Being Urgent: The Impact of Surveillance Target and Scale on Mosquito-Borne Disease Control." *Epidemics* 23 (June): 55–63. https://doi.org/10.1016/j.epidem.2017.12.004.

Scientific Collections International. 2015. "Scientific Collections and Emerging Infectious Diseases: Report of an Interdisciplinary Workshop." *SciColl Report* 1 (March). https://doi.org/10.5479/si.SciColl.

Semenza, Jan C., Susanne Herbst, Andrea Rechenburg, Jonathan E. Suk, Christoph Höser, Christiane Schreiber, and Thomas Kistemann. 2012. "Climate Change Impact Assessment of Food- and Waterborne Diseases." *Critical Reviews in Environmental Science and Technology* 42, no. 7: 857–90.

Shafer, Aaron B., Catherine I. Cullingham, Steeve D. Côté, and David W. Coltman. 2010. "Of Glaciers and Refugia: A Decade of Study Sheds New Light on the Phylogeography of Northwestern North America." *Molecular Ecology* 19, no. 21: 4589–4621. https://doi.org/10.1111/j.1365-294X.2010.04828.x.

Shaikh, Alanna. 2018. "The World Health Organization Wants You to Worry About 'Disease X.'" UN Dispatch. February 28.

Shalaby, Hatem A. 2013. "Anthelmintics Resistance: How to Overcome It?" *Iranian Journal of Parasitology* 8, no. 1: 18–32.

Shannon, L. M., Ryan H. Boyko, Marta Castelhano, Elizabeth Corey, Jessica J. Hayward, Corin McLean, Michelle E. White, et al. 2015. "Genetic Structure in Village Dogs reveals a Central Asia Domestication Origin." *Proceedings of the National Academy of Sciences of the United States of America* 112, no. 44: 13639–44.

Shepard, Donald S. 2010. "Cost and Burden of Dengue and Chikungunya from the Americas to Asia." *Dengue Bulletin* 34: 1–5.

Shepard, Donald S., Eduardo A. Undurraga, Yara A. Halasa, and Jeffrey D. Stanaway. 2016. "The Global Economic Burden of Dengue: A Systematic Analysis." *Lancet: Infectious Diseases* 16, no. 7: 935–41.

Shepard, Donald S., Laurent Coudeville, Yara A. Halasa, Betzana Zambrano, and Gustavo H. Dayan. 2011. "Economic Impact of Dengue Illness in the Americas." *American Journal of Tropical Medicine and Hygiene* 84, no. 2: 200–207.

Sher, Andrei. 1999. "Traffic Lights at the Beringian Crossroads." *Nature* 397, no. 6715: 103–4.

Shi, Suhua, Yelin Huang, Kai Zeng, Fengxiao Tan, Hanghang He, Jianzi Huang, and Yunxin Fu. 2005. "Molecular Phylogenetic Analysis of Mangroves: Independent Evolutionary Origins of Vivipary and Salt Secretion." *Molecular Phylogenetics and Evolution* 34, no. 1: 159–66. https://doi.org/10.1016/j .ympev.2004.09.002.

Shinn, A. P., J. Pratoomyot, J. E. Bron, G. Paladini, E. E. Brooker, and A. J. Brooker. 2014. "Economic Costs of Protistan and Metazoan Parasites to Global Mariculture." *Parasitology* 142, no. 1: 196–270. https://doi.org/10 .1017/S0031182014001437.

Shoop, Wesley L. 1988. "Trematode Transmission Patterns." *Journal of Parasitology* 74: 46–59.

Shrestha, S., and M. Maharjan. 2017. "Cross Infection with Gastro-Intestinal Tract Parasites between Red Panda (*Ailurus fulgens* Cuvier, 1825) and Livestocks in Community Forest of Ilam, Nepal." *International Journal of Research Studies in Zoology* 3, no. 4: 15–24. https://doi.org/10.20431/2454 -941X.0304003.

Shurkin, Joel. 2014. "News Feature: Animals That Self-Medicate." *Proceedings of the National Academy of Sciences of the United States of America* 111, no. 49: 17339–41.

Siepielski, A. M., Michael B. Morrissey, Mathieu Buoro, Stephanie M. Carlson, Christina M. Caruso, Sonya M. Clegg, Tim Coulson, et al. 2017. "Precipitation Drives Global Variation in Natural Selection." *Science* 355: 959–62.

Silaghi, Cornelia, Relja Beck, José A. Oteo, Martin Pfeffer, and Hein Sprong. 2015. "Neoehrlichiosis: An Emerging Tick-Borne Zoonosis Caused by *Candidatus neoehrlichia mikurensis*." *Experimental and Applied Acarology* (June): 1–20. https://doi.org/10.1007/s10493-015-9935-y.

Simberloff, Daniel, Kenneth L. Heck, Earl D. McCoy, and Edward F. Connor. 1980. "There Have Been Statistical Tests of Cladistic Biogeographic Hypotheses." In *Vicariance Biogeography: A Critique*, edited by Gareth Nelson and Donn E. Rosen, 40–63. New York: Columbia University Press.

Simberloff, Daniel. 1987. "Calculating Probabilities That Cladograms Match: A Method of Biogeographical Inference." *Systematic Zoology* 36, no. 2: 175–95.

Simões, M., L. Breitkreuz, M. Alvarado, S. Baca, J. C. Cooper, L. Heins, K. Herzog, and B. S. Lieberman. 2016. "The Evolving Theory of Evolutionary Radiations." *Trends in Ecology & Evolution* 31, no. 1: 27–34.

Simpson, George G. 1944. *Tempo and Mode in Evolution*. New York: Columbia University Press.

Simpson, George G. 1954. *The Major Features of Evolution*. New York: Columbia University Press.

Singer, Michael C. 2017. "Shifts in Time and Space Interact as Climate Warms." *Proceedings of the National Academy of Sciences of the United States of America* 114, no. 49: 12848–50. https://doi.org/10.1073/pnas.1718334114.

Singh, Ravi P., David P. Hodson, Julio Huerta-Espino, Yue Jin, Sridhar Bhavani, Peter Njau, Sybil Herrera-Foessel, Pawan K. Singh, Sukhwinder Singh, and Velu Govindan. 2011. "The Emergence of Ug99 Races of the Stem Rust Fungus Is a Threat to World Wheat Production." *Annual Review of Phytopathology* 49, no. 1: 465–81. https://doi.org/10.1146/annurev-phyto-072910 -095423.

Singh, S. K., M. Hodda, and G. J. Ash. 2013. "Plant-Parasitic Nematodes of Potential Phytosanitary Importance, Their Main Hosts and Reported Yield Losses." *EPPO Bulletin* 43, no. 2: 334–74. https://doi.org/10.1111/epp.12050.

Skoglund, P., Helena Malmström, Ayça Omrak, Maanasa Raghavan, Cristina Valdiosera, Torsten Günther, Per Hall, et al. 2014. "Genomic Diversity and Admixture Differs for Stone-Age Scandinavian Foragers and Farmers." *Science* 344: 747–50.

Skórka, Piotr, Magdalena Lenda, Dawid Moroń, Rafał Martyka, Piotr Tryjanowski, and William J. Sutherland. 2015. "Biodiversity Collision Blackspots in Poland: Separation Causality from Stochasticity in Roadkills of Butterflies." *Biological Conservation* 187: 154–63. https://doi.org/10.1016/ j.biocon.2015.04.017.

Smilanich, Angela M., R. Malia Fincher, and Lee A. Dyer. 2016. "Does Plant Apparency Matter? Thirty Years of Data Provide Limited Support but Reveal Clear Patterns of the Effects of Plant Chemistry on Herbivores." *New Phytologist* 210, no. 3: 1044–57. https://doi.org/10.1111/nph.13875.

Smith, Adam B. 2018. "2017 U.S. Billion Dollar Weather and Climate Disasters: A Historic Year in Context." https://www.climate.gov/news-features/blogs/ beyond-data/2017-us-billion-dollar-weather-and-climate-disasters-historic -year.

Smout, Felicity A., Lee F. Skerratt, James R. A. Butler, Christopher N. Johnson, Bradley C. Congdon, and R. C. Andrew Thompson. 2017. "The Hookworm *Ancylostoma ceylanicum*: An Emerging Public Health Risk in Australian Tropical Rainforests and Indigenous Communities." *One Health* 3: 66–69. https://doi.org/10.1016/j.onehlt.2017.04.002.

Smythe, Ashleigh B., and William F. Font. 2001. "Phylogenetic Analysis of *Alloglossidium* (Digenea: Macroderoididae) and Related Genera: Life-Cycle Evolution and Taxonomic Revision." *Journal of Parasitology* 87, no. 2: 386–91.

Snow, Robert W., Benn Sartorius, David Kyalo, Joseph Maina, Punam Amratia, Clara W. Mundia, Philip Bejon, and Abdisalan M. Noor. 2017. "The Prevalence of *Plasmodium falciparum* in Sub-Saharan Africa since 1900." *Nature* 78: 342–45. https://doi.org/10.1038/nature24059.

Snyder, Scott D., and Eric S. Loker. 2000. "Evolutionary Relationships among the Schistosomatidae (Platyhelminthes: Digenea) and an Asian Origin for *Schistosoma*." *Journal of Parasitology* 86, no. 2: 283–88.

Soberón, J., and B. Arroyo-Peña. 2017. "Are Fundamental Niches Larger than the Realized? Testing a 50-Year-Old Prediction by Hutchinson." *PLoS ONE* 12, no. 4: e0175138.

Sokolov, A., S. Paltsev, H. Chen, M. Haigh, R. Prinn and E. Monier. 2017. "Climate Stabilization at 2°C and Net Zero Carbon Emissions." *MIT Joint Program on the Science and Policy of Global Change*, Report 309. http://globalchange@mit.edu/publication/16629.

Solà, Eduard, Marta Álvarez-Presas, Cristina Frías-López, D. Timothy J. Littlewood, Julio Rozas, and Marta Riutort. 2015. "Evolutionary Analysis of Mitogenomes From Parasitic and Free-Living Flatworms." *PLoS ONE* 10, no. 3: e0120081. https://doi.org/10.1371/journal.pone.0120081.

Soranno, Patricia A., Kendra S. Cheruvelil, Edward G. Bissell, Mary T. Bremigan, John A. Downing, Carol E. Fergus, Christopher T. Filstrup, et al. 2014. "Cross-Scale Interactions: Quantifying Multi-Scaled Cause–Effect Relationships in Macrosystems." *Frontiers in Ecology and the Environment* 12, no. 1: 65–73.

Spector, Benjamin. 1959. "Darwin—Down House—Dawn." *New England Journal of Medicine* 260 (May 28): 1119–24. https://doi.org/10.1056/NEJM195905282602205.

Speed, Michael P., Andy Fenton, Meriel G. Jones, Graeme D. Ruxton, and Michael A. Brockhurst. 2015. "Coevolution Can Explain Defensive Secondary Metabolite Diversity in Plants." *New Phytologist* 208, no. 4: 1251–63. https://doi.org/10.1111/nph.13560.

Spencer, H. 1898–99. *The Principles of Biology.* 2 vols. New York: D. Appleton.

Sperling, Felix A. H., and P. Feeny. 1995. "*Umbellifer* and Composite Feeding in *Papilio*: Phylogenetic Frameworks and Constraints on Caterpillars." In *Swallowtail Butterflies: Their Ecology and Evolutionary Biology*, edited by J. Mark Scriber, Yoshitaka Tsubaki, and Robert C. Lederhouse, 299–306. Gainesville, FL: Scientific Publishers.

Spironello, Mike, and Daniel R. Brooks. 2003. "Dispersal and Diversification: Macroevolutionary Implications of the MacArthur-Wilson Model, Illustrated by *Simulium* (*Inseliellum*) Rubstov (Diptera: Simuliidae)." *Journal of Biogeography* 30, no. 10: 1563–73.

Sponheimer, Matt, and Julia A. Lee-Thorp. 1999. "Isotopic Evidence for the Diet of an Early Hominid, *Australopithecus africanus*." *Science* 283, no. 5400: 368–70.

Spradling, T. A., S. V. Brant, M. S. Hafner, and C. J. Dickerson. 2004. "DNA Data Support a Rapid Radiation of Pocket Gopher Genera (Rodentia: Geomyidae)." *Journal of Mammalian Evolution* 11, no. 2: 105–25.

Sprent, J. F. A. 1982. "Host-Parasite Relationships of Ascaridoid Nematodes and Their Vertebrate Hosts in Time and Space." *Memoires du Muséum national d'histoire naturelle*, n.s., sér. A, *Zoologie* 123: 255–63.

Springer, Mark S., Anthony V. Signore, Johanna L. A. Paijmans, Jorge Vélez-Juarbe, Daryl P. Domning, Cameron E. Bauer, Kai He, et al. 2015. "Interordinal Gene Capture, the Phylogenetic Position of Steller's Sea Cow Based on Molecular and Morphological Data, and the Macroevolutionary History of Sirenia." *Molecular Phylogenetics and Evolution* 91: 178–93. https://doi.org/10.1016/j.ympev.2015.05.022.

Springer, Yuri P., Christopher H. Hsu, Zachary R. Werle, Link E. Olson, Michael P. Cooper, Louisa J. Castrodale, Nisha Fowler, et al. 2017. "Novel Orthopoxvirus Infection in an Alaska Resident." *Clinical Infectious Diseases* 64, no. 12: 1737–41. https://doi.org/10.1093/cid/cix219.

Stacy, Brian A., Phoebe A. Chapman, Allen M. Foley, Ellis C. Greiner, Lawrence H. Herbst, Alan B. Bolten, Paul A. Klein, et al. 2017. "Evidence of Diversity, Site, and Host Specificity of Sea Turtle Blood Flukes (Digenea: Schistosomatoidea: 'Spirorchiidae'): A Molecular Prospecting Study." *Journal of Parasitology* 103, no. 6: 756–67. https://doi.org/10.1645/16-31.

Stakman, E. C. 1915. "Relation between *Puccinia graminis* and Plants Highly Resistant to Its Attack." *Journal of Agricultural Research* 4: 193–200.

Stakman, E. C., and F. J. Piemeisel. 1917. "Biologic Forms of *Puccinia grammis* on Cereals and Grasses." *Journal of Agricultural Research* 10: 429–95.

Stammer, H. J. 1955. "Ökologische Wechselbeziehungen zwischen Insekten und anderen Tiergruppen." *Wanderversammlung deutsches Entomologie* 7: 12–61.

Stammer, H. J. 1957. "Gedanken zu den parasitophyletischen Regeln und zur Evolution der Parasiten." *Zoologischer Anzeiger* 159, no. 11–12: 255–67.

Stanley, Steven M. 1979. *Macroevolution: Pattern and Process*. Baltimore, MD: Johns Hopkins University Press.

Staples, J. Erin, Manjunath B. Shankar, James J. Sejvar, Martin I. Meltzer, and Marc Fischer. 2014. "Initial and Long-Term Costs of Patients Hospitalized with West Nile Virus Disease." *American Journal of Tropical Medicine and Hygiene* 90, no. 3: 402–9.

Stark, S., M. Väisänen, H. Ylänne, R. Julkunen-Tiitto, and F. Martz. 2015. "Decreased Phenolic Defence in Dwarf Birch (*Betula nana*) after Warming in Subarctic Tundra." *Polar Biology* 38, no. 12: 1993–2005.

Steffen, Will, Jacques Grinevald, Paul Crutzen, and John McNeill. 2011. "The Anthropocene: Conceptual and Historical Perspectives." *Philosophical Transactions of the Royal Society of London A: Mathematical, Physical and Engineering Sciences* 369, no. 1938: 842–67.

Steffen, Will, Johan Rockström, Katherine Richardson, Timothy M. Lenton, Carl Folke, Diana Liverman, Colin P. Summerhayes, et al. 2018. "Trajectories of the Earth System in the Anthropocene." *Proceedings of the National Academy of Sciences on the United States of America*. https://doi.org/10.1073/pnas.1810141115.

Stenseth, Nils Chr., and J. Maynard Smith. 1984. "Coevolution in Ecosystems: Red Queen Evolution or Stasis?" *Evolution* 38, no. 4: 870–80.

Stenseth, Nils C., Noelle I. Samia, Hildegunn Viljugrein, Kyrre Linné Kausrud, Mike Begon, Stephen Davis, Herwig Leirs, et al. 2006. "Plague Dynamics

Are Driven by Climate Variation." *Proceedings of the National Academy of Sciences of the United States of America* 103, no. 35: 13110–15.

Stephens, Patrick R., Sonia Altizer, Katherine F. Smith, A. Alonso Aguirre, James H. Brown, Sarah A. Budischak, James E. Byers, et al. 2016. "The Macroecology of Infectious Diseases: A New Perspective on Global-Scale Drivers of Pathogen Distributions and Impacts." *Ecology Letters* 19, no. 9: 1159–71. https://doi.org/10.1111/ele.12644.

Stevens, G., B. McCluskey, A. King, E. O'Hearn, and G. Mayr. 2015. "Epizootic Hemorrhagic Disease Outbreak in Domestic Ruminants in the United States." *PLoS ONE* 10, no. 7: e0133359. https://doi.org/10.1371/journal.pone.0133359.

Stigall, Alycia L. 2010. "Invasive Species and Biodiversity Crises: Testing the Link in the Late Devonian." *PLoS ONE* 5, no. 12: e15584. https://doi.org/10.1371/journal.pone.0015584.

Stigall, Alycia L. 2012a. "Speciation Collapse and Invasive Species Dynamics during the Late Devonian 'Mass Extinction.'" *Geological Socety of America Today* 22: 4–9.

Stigall, Alycia L. 2012b. "Using Ecological Niche Modelling to Evaluate Niche Stability in Deep Time." *Journal of Biogeography* 39: 772–81.

Stigall, Alycia L. 2014. "When and How Do Species Achieve Niche Stability over Long Time Scales?" *Ecography* 37, no. 11: 1123–32.

Stigall, Alycia L. 2017. "How Is Biodiversity Produced? Examining Speciation Processes during the GOBE." *Lethaia* 51, no. 2: 165–72. https://doi.org/10.1111/let.12232.

Stigall, Alycia L., Jennifer E. Bauer, Adriane R. Lam, and David F. Wright. 2017. "Biotic Immigration Events, Speciation, and the Accumulation of Biodiversity in the Fossil Record." *Global and Planetary Change* 148: 242–57.

Stocker, Thomas F., Dahe Qin, G.-K. Plattner, Lisa V. Alexander, Simon K. Allen, Nathaniel L. Bindoff, F.-M. Bréon, et al. 2013. "Technical Summary." In *Climate Change 2013: The Physical Science Basis; Contribution of Working Group I to the Fifth Assessment Report of the Intergovernmental Panel on Climate Change*, 33–115. Cambridge: Cambridge University Press.

Stone, Brandee L, Yvonne Tourand, and Catherine A Brissette. 2017. "Brave New Worlds: the Expanding Universe of Lyme Disease." *Vector-Borne and Zoonotic Diseases* 17, no. 9: 619–29. https://doi.org/10.1089/vbz.2017.2127.

Stone, Richard. 2016. "Dam-Building Threatens Mekong Fisheries." *Science* 354: 1084–85.

Stott, Peter. 2016. "How Climate Change Affects Extreme Weather Events." *Science* 352, no. 6293: 1517–18. https://doi.org/10.1126/science.aaf7271.

Stout, Rex. 1947. *Too Many Women*. New York: Penguin Random House.

St. Romain, Krista, Daniel W. Tripp, Daniel J. Salkeld, and Michael F. Antolin. 2013. "Duration of Plague (*Yersinia pestis*) Outbreaks in Black-Tailed Prairie Dog (*Cynomys ludovicianus*) Colonies of Northern Colorado." *EcoHealth* 10, no. 3: 241–45. https://doi.org/10.1007/s10393-013-0860-4.

Strauss, Sharon Y., Jennifer A. Rudgers, Jennifer A. Lau, and Rebecca E. Irwin. 2002. "Direct and Ecological Costs of Resistance to Herbivory." *Trends in Ecology & Evolution* 17, no. 6: 278–85.

Streicker, Daniel G., Jamie C. Winternitz, Dara A. Satterfield, Rene Edgar Condori-Condori, Alice Broos, Carlos Tello, Sergio Recuenco, et al. 2016. "Host–Pathogen Evolutionary Signatures Reveal Dynamics and Future Invasions of Vampire Bat Rabies." *Proceedings of the National Academy of Sciences of the United States* 113, no. 39: 10926–31. https://doi.org/10.1073/pnas.1606587113.

Strona, Giovanni, and Claudio Castellano. 2018. "Rapid Decay in the Relative Efficiency of Quarantine to Halt Epidemics in Networks." *Physical Review E* 97, no. 2: 022308. https://doi.org/10.1103/PhysRevE.97.022308.

Strona, Giovanni, Paolo Galli, and Simone Fattorini. 2013. "Fish Parasites Resolve the Paradox of Missing Coextinctions." *Nature Communications* 4: 1718. https://doi.org/10.1038/ncomms2723.

Stunkard, Horace W. 1959. "The Morphology and Life-History of the Digenetic Trematode, *Asymphylodora amnicolae* n. sp.; The Possible Significance of Progenesis for the Phylogeny of the Digenea." *Biological Bulletin* 117, no. 3: 562–81.

Suinyuy, Terence N., John S. Donaldson, and Steven D. Johnson. 2015. "Geographical Matching of Volatile Signals and Pollinator Olfactory Responses in a Cycad Brood-Site Mutualism." *Proceedings of the Royal Society of London B: Biological Sciences* 282, no. 1816: 20152053.

Sukhdeo, M. V. K. 1994. "Parasites and Behaviour." In "Parasites and Behaviour," edited by M. V. K. Sukhdeo. Supplement, *Parasitology* 109, no. S1: S1. Cambridge: Cambridge University Press.

Sullivan, J., J. R. Demboski, K. C. Bell, S. Hird, B. Sarver, N. Reid, and J. M. Good. 2014. "Divergence with Gene Flow within the Recent Chipmunk Radiation (*Tamias*)." *Heredity* 113, no. 3: 185–94.

Sun Tzu. 2017. *The Art of War*. New York: Knickerbocker Classics.

Sunny, Anupam, Swati Diwakar, and Gyan Prakash Sharma. 2015. "Native Insects and Invasive Plants Encounters." *Arthropod-Plant Interactions* 9, no. 4: 323–31.

Sutherland, William J., Stuart H. M. Butchart, Ben Connor, Caroline Culshaw, Lynn V. Dicks, Jason Dinsdale, Helen Doran, et al. 2018. "A 2018 Horizon Scan of Emerging Issues for Global Conservation and Biological Diversity." *Trends in Ecology & Evolution* 33, no. 1: 47–58. https://doi.org/10.1016/j.tree.2017.11.006.

Suydam, E. Lynn. 1971. "The Micro-Ecology of Three Species of Monogenetic Trematodes of Fishes from the Beaufort-Cape Hatteras Area." *Proceedings of the Helminthological Society of Washington* 38, no. 2: 240–46. https://doi.org/10.2307/3272789.

Svitálková, Zuzana, Danka Haruštiaková, Lenka Mahríková, Lenka Berthová, Mirko Slovák, Elena Kocianová, and Mária Kazimírová. 2015. "*Anaplasma *phagocytophilum* Prevalence in Ticks and Rodents in an Urban and Natural Habitat in South-Western Slovakia." *Parasites & Vectors* 8: 276. https://doi.org/10.1186/s13071-015-0880-8.

Swenson, Ulf, Stephan Nylinder, and Jérôme Munzinger. 2014. "Sapotaceae Biogeography Supports New Caledonia Being an Old Darwinian Island." *Journal of Biogeography* 41, no. 4: 797–809. https://doi.org/10.1111/jbi.12246.

Swenson, Ulf, Jérôme Munzinger, Porter P. Lowry II, Bodil Cronholm, and Stephan Nylinder. 2015. "Island Life—Classification, Speciation and Cryptic Species of *Pycnandra* (Sapotaceae) in New Caledonia." *Botanical Journal of the Linnean Society* 179, no. 1: 57–77. https://doi.org/10.1111/boj.12308.

Sydeman, William J., Elvira Poloczanska, Thomas E. Reed, and Sarah Ann Thompson. 2015. "Climate Change and Marine Vertebrates." *Science* 350, no. 6262: 772–77.

Syed, Zainulabeuddin. 2015. "Chemical Ecology and Olfaction in Arthropod Vectors of Diseases." *Current Opinion in Insect Science* 10: 83–89.

Szathmáry, Eörs. 2015. "Toward Major Evolutionary Transitions Theory 2.0." *Proceedings of the National Academy of Sciences of the United States of America* 112, no. 33: 10104–111.

Szekeres, Sándor, Viktória Majláthová, Igor Majláth, and Gábor Földvári. 2016. "Neglected Hosts: The Role of Lacertid Lizards and Medium-Sized Mammals in the Ecoepidemiology of Lyme Borreliosis." In *Ecology and Control of Vector-Borne Diseases*, vol. 4, *Ecology and Prevention of Lyme Borreliosis*, edited by Marieta A. H. Braks, Sipke E. van Wieren, WillemTakken, and Hein Sprong, 103–26. Netherlands: Wageningen Academic. https://doi.org/10.3920/978-90-8686-838-4_8.

Taber, Stephen W., and Craig M. Pease. 1990. "Paramyxovirus Phylogeny: Tissue Tropism Evolves Slower than Host Specificity." *Evolution* 44, no. 2: 435–38.

Tagliacollo, Victor A., Fábio Fernandes Roxo, Scott M. Duke-Sylvester, Claudio Oliveira, and James S. Albert. 2015a. "Biogeographical Signature of River Capture Events in Amazonian Lowlands." *Journal of Biogeography* 42, no. 12: 2349–62. https://doi.org/10.1111/jbi.12594.

Tagliacollo, Victor A., Scott M. Duke-Sylvester, Wilfredo A. Matamoros, Prosanta Chakrabarty, and James S. Albert. 2015b. "Coordinated Dispersal and Pre-Isthmian Assembly of the Central American Ichthyofauna." *Systematic Biology* 66, no. 2: 183–96. https://doi.org/10.1093/sysbio/syv064.

Taylor, D. L., and T. D. Bruns. 1999. "Community Structure of Ectomycorrhizal Fungi in a *Pinus muricata* Forest: Minimal Overlap between the Mature Forest and Resistant Propagule Communities." *Molecular Ecology* 8, no. 11: 1837–50.

Taylor, Louise H., Sophia M. Latham, and E. J. Mark. 2001. "Risk Factors for Human Disease Emergence." *Philosophical Transactions of the Royal Society of London B: Biological Sciences* 356, no. 1411: 983–89.

Tenzin, Tenzin, Chador Wangdi, and Purna Bdr Rai. 2017. "Biosecurity Survey in Relation to the Risk of HPAI Outbreaks in Backyard Poultry Holdings in Thimphu City Area, Bhutan." *BMC Veterinary Research* 13, no. 1: 491. https://doi.org/10.1186/s12917-017-1033-4.

Terefe, Yitagele, Zerihun Hailemariam, Sissay Menkir, Minoru Nakao, Antti Lavikainen, Voitto Haukisalmi, Takashi Iwaki, et al. 2014. "Phylogenetic Characterisation of *Taenia* Tapeworms in Spotted Hyenas and Reconsideration of the 'Out of Africa' Hypothesis of *Taenia* in Humans." *International Journal for Parasitology* 44, no. 7: 533–41.

ter Hofstede, Hannah M., M. Brock Fenton, and John O. Whitaker, Jr. 2004. "Host and Host-Site Specificity of Bat Flies (Diptera: Streblidae and Nycteribiidae) on Neotropical Bats (Chiroptera)." *Canadian Journal of Zoology* 82, no. 4: 616–26. https://doi.org/10.1139/z04-030.

Thomas, D. C., M. Hughes, T. Phutthai, W. H. Ardi, S. Rajbhandary, R. Rubite, A. D. Twyford, et al. 2011. "West to East Dispersal and Subsequent Rapid Diversification of the Mega-Diverse Genus *Begonia* (Begoniaceae) in the Malesian Archipelago." *Journal of Biogeography* 39, no. 1: 98–113. https://doi.org/10.1111/j.1365-2699.2011.02596.x.

Thomaz, Andréa T., Luiz R. Malabarba, Sandro L Bonatto, and L. Lacey Knowles. 2015. "Testing the Effect of Palaeodrainages versus Habitat Stability on Genetic Divergence in Riverine Systems: Study of a Neotropical Fish of the Brazilian Coastal Atlantic Forest." *Journal of Biogeography* 4, no. 12: 2389–2401. https://doi.org/10.1111/jbi.12597.

Thomaz-Soccol, Vanete, André Luiz Gonçalves, Claudio Adriano Piechnik, Rafael Antunes Baggio, Walter Antonio Boeger, Themis Leão Buchman, Mario Sergio Michaliszyn, et al. 2018. "Hidden Danger: Unexpected Scenario in the Vector-Parasite Dynamics of Leishmaniases in the Brazil Side of Triple Border (Argentina, Brazil and Paraguay)." *PLoS Neglected Tropical Diseases* 12, no. 4: e0006336. https://doi.org/10.1371/journal.pntd.0006336.

Thompson, John N. 1988. "Coevolution and Alternative Hypotheses on Insect/Plant Interactions." *Ecology* 69, no. 4: 893–95.

Thompson, John N. 1994. *The Coevolutionary Process*. Chicago: University of Chicago Press.

Thompson, John N. 1997. "Evaluating the Dynamics of Coevolution among Geographically Structured Populations." *Ecology* 78: 1619–23.

Thompson, John N. 1999a. "Specific Hypotheses on the Geographic Mosaic of Coevolution." *American Naturalist* 153, suppl.: 1–14.

Thompson, John N. 1999b. "Coevolution and Escalation: Are Ingoing Coevolutionary Meanderings Important?" *American Naturalist* 153, suppl.: 92–93.

Thompson, John N. 1999c. "The Evolution of Species Interactions." *Science* 284: 2116–18.

Thompson, John N. 2005. *The Geographic Mosaic of Coevolution*. Chicago: University of Chicago Press.

Thoreau, Henry David. 2009. *The Journal of Henry David Thoreau, 1837–1861*. Edited by Damion Searls. Preface by John R. Stilgoe. New York: Penguin Random House.

Thorington R. W., and Hoffmann, R. S. 2005. "Family Sciuridae." In *Mammal Species of the World: A Taxonomic and Geographic Reference*, edited by Don E. Wilson and DeeAnn M. Reeder, 754-818 Baltimore, MD: Johns Hopkins University Press.

Timmermann, Axel, and Tobias Friedrich. 2016. "Late Pleistocene Climate Drivers of Early Human Migration." *Nature* 538, no. 7623: 92–95.

Tinsley, R. C. 1983. "Ovoviviparity in Platyhelminth Life-Cycles." *Parasitology* 86, no. 4: 161–96.

Tolkien, J. R. R. 1954. *The Lord of the Rings*. London: Allen & Unwin.

Tong, Shilu, and Wenbiao Hu. 2001. "Climate Variation and Incidence of Ross River Virus in Cairns, Australia: A Time-Series Analysis." *Environmental Health Perspectives* 109, no. 12: 1271–73.

Topley, W. W. C., and G. S. Wilson. 1923. "The Spread of Bacterial Infection: The Problem of Herd-Immunity." *Epidemiology & Infection* 21, no. 3: 243–49.

Travers, Marie-Agnès, Katherine Boettcher Miller, Ana Roque, and Carolyn S. Friedman. 2015. "Bacterial Diseases in Marine Bivalves." *Journal of Invertebrate Pathology* 131: 11–31.

Trenberth, Kevin E., Philip D. Jones, Peter Ambenje, Roxana Bojariu, David Easterling, Albert Klein Tank, David Parker, et al. 2007. "Observations: Surface and Atmospheric Climate Change." In *Climate Change 2007: The Physical Science Basis*. Contribution of Working Group I to the Fourth Assessment Report of the Intergovernmental Panel on Climate Change, edited by S. Solomon, D. Qin, M. Manning, Z. Chen, M. Marquis, K. B. Averyt, M. Tignor, and H. L. Miller, 236–335. Cambridge and New York: Cambridge University Press.

"The Triumvirate of Heredity." 1958. *Lancet* 272, no. 7051 (October 18): 838. https://doi.org/10.1016/S0140-6376(58)90386-6.

Trouvé, Sandrine, Pierre Sasal, Joseph Jourdane, François Renaud, and Serge Morand. 1998. "The Evolution of Life-History Traits in Parasitic and Free-Living Platyhelminthes: A New Perspective." *Oecologia* 115, no. 3: 370–78.

Trtanj, J., L. Jantarasami, J. Brunkard, T. Collier, J. Jacobs, E. Lipp, S. McLellan, et al. 2016. "Chapter 6: Climate Impacts on Water-Related Illness." In *The Impacts of Climate Change on Human Health in the United States: A Scientific Assessment*, 157–88. Washington, DC: U.S. Global Change Research Program.

Turner, Graham M. 2014. "Is Global Collapse Imminent? An Updated Comparison of the Limits to Growth with Historical Data." In *MSSI Research Paper 4*. Melbourne: Melbourne Sustainable Society Institute, Melbourne University.

Tzedakis, P. C., E. W. Wolff, L. C. Skinner, Victor Brovkin, D. A. Hodell, Jerry F. McManus, and D. Raynaud. 2012. "Can We Predict the Duration of an Interglacial?" *Climate of the Past* 8, no. 5: 1473–85.

Ulanowicz, Robert E. 1997. *Ecology: The Ascendent Perspective.* New York: Columbia University Press.

Ulett, M. A. 2014. "Making the Case for Orthogenesis: The Popularization of Definitely Directed Evolution (1890–1926)." *Studies in History and Philosophy of Biological and Biomedical Sciences* 45: 124–32.

United Nations Development Programme. 2017. "Social and Economic Costs of Zika Can Reach up to US$18 Billion in Latin America and the Caribbean." http://www.undp.org/content/undp/en/home/presscenter/pressreleases/2017/04/06/social-and-economic-costs-of-zika-can-reach-up-to-us-18-billion-in-latin-america-and-the-caribbean.html.

United States Agency for International Development [USAID]. 2014. "Reducing Pandemic Risk, Promoting Global Health." PREDICT 1 (2009–2014) Final Report.

United States Agency for International Development [USAID]. 2016. *Reducing Pandemic Risk, Promoting Global Health, Supporting the Global Health Security Agenda.* Washington, DC: PREDICT Annual Report.

United States Department of Agriculture [USDA]. 2016. "Plant Protection and Quarantine: Helping U.S. Agriculture Thrive—Across the Country and around the World." *Annual Report, Animal and Plant Health Inspection Service.* APHIS 81-05-020. https://www.aphis.usda.gov.

US Biologic. 2018. "Lyme Disease." http://usbiologic.com/lyme-disease/.

United StatesGlobal Change Research Program [USGCRP]. 2016. *The Impacts of Climate Change on Human Health in the United States: A Scientific Assessment.* Edited by A. Crimmins, J. Balbus, J. L. Gamble, C. B. Beard, J. E. Bell, D. Dodgen, R. J. Eisen, N. Fann, M. D. Hawkins, S. C. Herring, L. Jantarasami, D. M. Mills, S. Saha, M. C. Sarofim., J. Trtanj, and L. Ziska. Washington, DC: The Program. http://dx.doi.org/10.7930/J0R49NQX. https://doi.org/10.1111/pan.12888.

U.S. Global Change Research Program [USGCRP]. 2017. *Climate Science Special Report: Fourth National Climate Assessment (NCA4).* Vol. 1. [Edited by Donald J. Wuebbles, David W. Fahey, Kathy A. Hibbard, David J. Dokken, Brooke C. Stewart, and Thomas K. Maycock.] Washington, DC: The Program. https://doi.org/10.7930/J0J964J6.

Utaaker, Kjersti S., and Lucy J. Robertson. 2015. "Climate Change and Foodborne Transmission of Parasites: A Consideration of Possible Interactions and Impacts for Selected Parasites." *Food Research International* 68: 16–23.

Vandermark, Cailey, Elliott Zieman, Esmerie Boyles, Clayton K. Nielsen, Cheryl Davis, and Francisco Augustín Jiménez. 2018. "*Trypanosoma cruzi* Strain TcIV Infects Raccoons From Illinois." *Memórias do Instituto Oswaldo Cruz* 2018: 1–8. https://doi.org/10.1590/0074-02760170230.

van Soest, Rob W. M., and Eduardo Hajdu. 1997. "Marine Area Relationships from Twenty Sponge Phylogenies. A Comparison of Methods and Coding Strategies." *Cladistics* 13, no. 1–2: 1–20.

Van Valen, Leigh. 1973. "A New Evolutionary Law." *Evolutionary Theory* 1: 1–30.

Vasbinder, Jan Wouter. 2019. *Disrupted Balance: Society at Risk.* Exploring Complexity 6. Singapore: World Scientific.

Vellend, Mark, Lander Baeten, Antoine Becker-Scarpitta, Véronique Boucher-Lalonde, Jenny L. McCune, Julie Messier, Isla H. Myers-Smith, et al. 2017. "Plant Biodiversity Change across Scales during the Anthropocene." *Annual Review of Plant Biology* 68: 563–86. https://doi.org/10.1146/annurev-arplant-042916-040949.

Vercruysse, Jozef, Marco Albonico, Jerzy M. Behnke, Andrew C. Kotze, Roger K. Prichard, James S. McCarthy, Antonio Montresor, et al.. 2011. "Is Anthelmintic Resistance a Concern for the Control of Human Soil-Transmitted Helminths?" *International Journal for Parasitology: Drugs and Drug Resistance* 1, no. 1: 14–27.

Vermeij, Geerat J. 1991a. "When Biotas Meet: Understanding Biotic Unterchange." *Science* 253: 1099–1104.

Vermeij, Geerat J. 1991b. "Anatomy of an Invasion: The Trans-Arctic Interchange." *Paleobiology* 17: 281–307.

Verneau, Olivier, Sophie Bentz, Neeta D. Sinnappah, Louis du Preez, Ian Whittington, and Claude Combes. 2002. "A View of Early Vertebrate Evolution Inferred from the Phylogeny of Polystome Parasites (Monogenea: Polystomatidae)." *Proceedings of the Royal Society of London B: Biological Sciences* 269, no. 1490: 535–43.

Verocai, Guilherme G. 2015. "Contributions to the Biodiversity and Biogeography of the Genus *Varestrongylus* Bhalerao, 1932 (Nematoda: Protostrongylidae), Lungworms of Ungulates, with Emphasis on a New Nearctic Species." Ph.D. diss., University of Calgary.

Verocai, Guilherme G., Susan J. Kutz, Manon Simard, and Eric P. Hoberg. 2014. "*Varestrongylus eleguneniensis* sp. n. (Nematoda: Protostrongylidae): A Widespread, Multi-Host Lungworm of Wild North American Ungulates, with an Emended Diagnosis for the Genus and Explorations of Biogeography." *Parasites & Vectors* 7, no. 1: 556.

Verschaffelt, E. 1910. "The Cause Determining the Selection of Food in Some Herbivorous Insects." *Proceedings of the Koninklijke Nederlandse Akademie van Wetenschappen* 13: 536–42.

Vezzulli, Luigi, Ingrid Brettar, Elisabetta Pezzati, Philip C. Reid, Rita R. Colwell, Manfred G. Höfle, and Carla Pruzzo. 2012. "Long-Term Effects of Ocean Warming on the Prokaryotic Community: Evidence from the Vibrios." *International Society for Microbial Ecology Journal* 6, no. 1: 21–30.

Vignieri, Sacha, and Julia Fahrenkamp-Uppenbrink. 2017. "Ecosystem Earth." *Science* 356, no. 6335: 258–59. https://doi.org/10.1126/science.356.6335.258.

Vihakas, Matti, Isrrael Gómez, Maarit Karonen, Petri Tähtinen, Ilari Sääksjärvi, and Juha-Pekka Salminen. 2015. "Phenolic Compounds and Their Fates in Tropical Lepidopteran Larvae: Modifications in Alkaline Conditions." *Journal of Chemical Ecology* 41, no. 9: 822–36. https://doi.org/10.1007/s10886-015-0620-8.

Vittecoq, Marion, Sylvain Godreuil, Franck Prugnolle, Patrick Durand, Lionel Brazier, Nicolas Renaud, Audrey Arnal, et al. 2016. "Antimicrobial Resistance in Wildlife." *Journal of Applied Ecology* 53, no. 2: 519–29. https://doi.org/10.1111/1365-2664.12596.

Voegele, Ralf T., Matthias Hahn, and Kurt Mendgen. 2009. "The Uredinales: Cytology, Biochemistry, and Molecular Biology." In *The Mycota*, vol. 5, *Plant Relationships*, 2nd ed., edited by Holger B. Deising, 69–98. Berlin: Springer.

Volf, Martin, Jan Hrcek, Riitta Julkunen-Tiitto, and Vojtech Novotny. 2015a. "To Each Its Own: Differential Response of Specialist and Generalist Herbivores to Plant Defence in Willows." *Journal of Animal Ecology* 84, no. 4: 1123–32. https://doi.org/10.1111/1365-2656.12349.

Volf, Martin, Riitta Julkunen-Tiitto, Jan Hrcek, and Vojtech Novotny. 2015b. "Insect Herbivores Drive the Loss of Unique Chemical Defense in Willows." *Entomologia Experimentalis et Applicata* 156, no. 1: 88–98. https://doi.org/10.1111/eea.12312.

von Ihering, Hermann. 1891. "On the Ancient Relations between New Zealand and South America." *Transactions and Proceedings of the New Zealand Institute* 24: 431–45.

von Ihering, Hermann. 1902. "Die Helminthen als Hilfsmittel der zoogeographischen Forschung." *Zoologischer Anzeiger* 26: 42–51.

von Kéler, S. 1938. "Baustoffe zu einer Monographie der Mallophagen. I. Teil: Überfamilie Trichodectoidea." *Nova Acta Leopoldina* 5: 393–467.

von Kéler, S. 1939. "Baustoffe zur einer Monographie der Mallophagen. II. Teil: Überfamilie Nirmoidea." *Nova Acta Leopoldina* 8: 1–254.

von Nägeli, Carl. 1884. *Mechanisch-physiologische Theorie der Abstammungslehre.* Moscow: Ripol Classics.

von Son-de Fernex, Elke, Miguel Ángel Alonso-Díaz, Pedro Mendoza-de-Gives, Braulio Valles-de la Mora, Enrique Liébano-Hernández, María Eugenia López-Arellano, and Liliana Aguilar-Marcelino. 2014. "Reappearance of *Mecistocirrus digitatus* in Cattle from the Mexican Tropics: Prevalence, Molecular, and Scanning Electron Microscopy Identification." *Journal of Parasitology* 100, no. 3: 296–301. https://doi.org/10.1645/13-377.1.

Vrba, Elisabeth S. 1985. "African Bovidae; Evolutionary Events since the Miocene." *South African Journal of Science* 81: 263–66.

Vrba, Elisabeth S. 1995a. "On the Connections between Paleoclimate and Evolution." In *Paleoclimate and Evolution with Emphasis on Human Origins*, edited by Elisabeth S. Vrba, George H. Denton, Timothy C. Partridge, and Lloyd H. Burckle, 24–48. New Haven, CT: Yale University Press.

Vrba, Elisabeth S. 1995b. "The Fossil Record of African Antelopes (Mammalia: Bovidae) in Relation to Human Evolution and Paleoclimate." In *Paleoclimate and Evolution with Emphasis on Human Origins*, edited by Elisabeth S. Vrba, George H. Denton, Timothy C. Partridge, and Lloyd H. Burckle, 385–424. New Haven, CT: Yale University Press.

Vrba, Elisabeth S., and George B. Schaller. 2000. "Phylogeny of Bovidae (Mammalia) Based on Behavior, Glands, Skull, and Postcrania." In *Antelopes, Deer, and Relatives: Fossil Record, Behavioral Ecology, Systematics, and Conservation*, edited by Elisabeth S. Vrba and George B. Schaller, 203–22. New Haven, CT: Yale University Press.

Vucinich, Alexander. 1988. *Darwin in Russian Thought*. Berkeley and Los Angeles: University of California Press.

Wagner, Moritz. 1868. *Die Darwin'sche Theorie und das Migrationsgesetz der Organismen*. Leipzig: Duncker & Humboldt.

Wahlberg, Niklas. 2001. "The Phylogenetics and Biochemistry of Host-Plant Specialization in Melitaeine Butterflies (Lepidoptera: Nymphalidae)." *Evolution* 55, no. 3: 522–37.

Wallace, Alfred R. 1878. *Tropical Nature and Other Essays*. London and New York: Macmillan.

Waltari, Eric, Eric P. Hoberg, Enrique P. Lessa, and Joseph A. Cook. 2007a. "Eastward Ho: Phylogeographical Perspectives on Colonization of Hosts and Parasites across the Beringian Nexus." *Journal of Biogeography* 34, no. 4: 561–74.

Waltari, Eric, Robert J. Hijmans, A. Townsend Peterson, Árpád S. Nyári, Susan L. Perkins, and Robert P. Guralnick. 2007b. "Locating Pleistocene Refugia: Comparing Phylogeographic and Ecological Niche Model Predictions." *PLoS ONE* 2, no. 7: e563.

Walter, Katharine S., Giovanna Carpi, Adalgisa Caccone, and Maria A. Diuk-Wasser. 2017. "Genomic Insights into the Ancient Spread of Lyme Disease across North America." *Nature Ecology & Evolution* 1, no. 10: 1569–76. https://doi.org/10.1038/s41559-017-0282-8.

Wang, Houshuai, Jeremy D. Holloway, Niklas Janz, Mariana P. Braga, Niklas Wahlberg, Min Wang, and Soren Nylin. 2017. "Polyphagy and Diversification in Tussock Moths: Support for the Oscillation Hypothesis from Extreme Generalists." *Ecology and Evolution* 7: 7975–86. https://doi.org/10.1002/ece3.3350.

Wang, Shuai, Sen Wang, Yingfeng Luo, Lihua Xiao, Xuenong Luo, Shenghan Gao, Yongxi Dou1, et al. 2016. "Comparative Genomics Reveals Adaptive Evolution of Asian Tapeworm in Switching to a New Intermediate Host." *Nature Communications* 7: 12845.

Wanntorp, Hans-Erik. 1983. "Historical Constraints in Adaptation Theory: Traits and Non-Traits." *Oikos* 41: 157–60.

Wanntorp, Hans-Erik, Daniel R. Brooks, Thomas Nilsson, Sören Nylin, Fredrik Ronquist, Stephen C. Stearns, and Nina Wedell. 1990. "Phylogenetic Approaches in Ecology." *Oikos* 57: 119–32.

Wappes, Jim. 2015. "Report Finds $1.2 Billion in Iowa Avian Flu Damage." *Center for Infectious Disease Research and Policy*. http://www.cidrap.umn .edu/news-perspective/2015/08/report-finds-12-billion-iowa-avian-flu -damage.

Warburton, Elizabeth M., Michael Kam, Enav Bar-Shira, Aharon Friedman, Irina S. Khokhlova, Lee Koren, Mustafa Asfur, et al. 2016. "Effects of Parasite Pressure on Parasite Mortality and Reproductive Output in a Rodent-Flea System: Inferring Host Defense Trade-Offs." *Parasitology Research* 115, no. 9: 3337–44. https://doi.org/10.1007/s00436-016-5093-3.

Waters, Colin N., Jan Zalasiewicz, Colin Summerhayes, Anthony D. Barnosky, Clément Poirier, Agnieszka Galuszka, Alejandro Cearreta, et al. 2016. "The Anthropocene Is Functionally and Stratigraphically Distinct from the Holocene." *Science* 351, no. 6269: aad2622.

Watertor, Jean L. 1967. "Intraspecific Variation of Adult *Telorchis bonnerensis* (Trematoda: Telorchiidae) in Amphibian and Reptilian Hosts." *Journal of Parasitology* 53, no. 5: 962–68.

Watts, Nick, Markus Amann, Sonja Ayeb-Karlsson, Kristine Belesova, Timothy Bouley, Maxwell Boykoff, Prof Peter Byass, et al. 2017a. "The *Lancet* Countdown on Health and Climate Change: From 25 Years of Inaction to a Global Transformation for Public Health." *Lancet* 391, no. 10120: 1–50. https://doi.org/10.1016/S0140-6736(17)32464-9.

Watts, Nick, W. Neil Adger, Sonja Ayeb-Karlsson, Yuqi Bai, Peter Byass, Diarmid Campbell-Lendrum, Tim Colbourn, et al. 2017b. "The *Lancet* Countdown: Tracking Progress on Health and Climate Change." *Lancet* 389, no. 10074: 1151–64. https://doi.org/10.1016/ S0140-6736(16)32124-9

Watts, Peter. 2006. *Blindsight*. New York: Tor Books.

Weaver, Haylee J., John M. Hawdon, and Eric P. Hoberg. 2010. "Soil-Transmitted Helminthiases: Implications of Climate Change and Human Behavior." *Trends in Parasitology* 26, no. 12: 574–81.

Webb, S. David. 1995. "Biological Implications of the Middle Miocene Amazon Seaway." *Science* 269, no. 5222: 361–62.

Wegener, A. 1912. "Die Entstehung der Kontinente." *Petermanns Mitteilungsblatt* 58: 185–95.

Weigelt, Patrick, W. Daniel Kissling, Yael Kisel, Susanne A. Fritz, Dirk Nikolaus Karger, Michael Kessler, Samuli Lehtonen, Jens-Christian Svenning, and Holger Kreft. 2016. "Global Patterns and Drivers of Phylogenetic Structure in Island Floras." *Scientific Reports* 5, no. 1: 5925. https://doi.org/10.1038/ srep12213.

Wells, Konstans, David I. Gibson, Nicholas J. Clark, Alexis Ribas, Serge Morand, and Hamish I. McCallum. 2018. "Global Spread of Helminth Parasites at the Human–Domestic Animal–Wildlife Interface." *Global Change Biology* 11: 1123. https://doi.org/10.1111/gcb.14064.

Wernberg, Thomas, Scott Bennett, Russell C. Babcock, Thibaut de Bettignies, Katherine Cure, Martial Depczynski, et al. 2016. "Climate-Driven Regime

Shift of a Temperate Marine Ecosystem." *Science* 353, no. 6295: 169–72. https://doi.org/10.1126/science.aad8745.

West-Eberhard, Mary J. 2003. *Developmental Plasticity and Evolution.* New York: Oxford University Press.

Whiteman, Noah K., and Kailen A. Mooney. 2012. "Evolutionary Biology: Insects Converge on Resistance." *Nature* 489, no. 7416: 376–77. https://doi .org/10.1038/489376a.

Whiteman, Noah K., and Naomi E. Pierce. 2008. "Delicious Poison: Genetics of *Drosophila* Host Plant Preference." *Trends in Ecology & Evolution* 23, no. 9: 473–78.

Whiteman, Noah K., Rebecca T. Kimball, and Patricia G. Parker. 2007. "Co-Phylogeography and Comparative Population Genetics of the Threatened Galápagos Hawk and Three Ectoparasite Species: Ecology Shapes Population Histories within Parasite Communities." *Molecular Ecology* 16, no. 22: 4759–73.

Whitman, Charles Otis. 1899. *Animal Behavior.* Boston: Ginn.

Whitman, Charles Otis. 1919. *Orthogenetic Evolution in Pigeons.* Edited by O. Riddle. Washington, DC: Carnegie Institution.

Whitmee, Sarah, Andy Haines, Chris Beyrer, Frederick Boltz, Anthony G. Capon, Braulio Ferreira de Souza Dias, Alex Ezeh, et al. 2015. "Safeguarding Human Health in the Anthropocene Epoch: Report of the Rockefeller Foundation–*Lancet* Commission on Planetary Health." *Lancet* 386, no. 10007: 1973–2028.

Whittaker, Robert J. 2000. "Scale, Succession and Complexity in Island Biogeography: Are We Asking the Right Questions?" *Global Ecology and Biogeography* 9: 75–85.

Whittington, I. D., and I. Ernst. 2002. "Migration, Site-Specificity and Development of *Benedenia lutjani* (Monogenea: Capsalidae) on the Surface of Its Host, *Lutjanus carponotatus* (Pisces: Lutjanidae)." *Parasitology* 124, no. 4: 423–43. https://doi.org/10.1017/S0031182001001287.

Wickström, L. M., V. Haukisalmi, S. Varis, J. Hantula, V. B. Fedorov, and H. Henttonen. 2003. "Phylogeography of the Circumpolar *Paranoplocephala arctica* Species Complex (Cestoda: Anoplocephalidae) Parasitizing Collared Lemmings (*Dicrostonyx* spp.)." *Molecular Ecology* 12, no. 12: 3359–71.

Wickström, Lotta M., Voitto Haukisalmi, Saila Varis, Jarkko Hantula, and Heikki Henttonen. 2005. "Molecular Phylogeny and Systematics of Anoplocephaline Cestodes in Rodents and Lagomorphs." *Systematic Parasitology* 62, no. 2: 83–99.

Wiens, John J., David D. Ackerly, Andrew P. Allen, Brian L. Anacker, Lauren B. Buckley, Howard V. Cornell, Ellen I. Damschen, et al. 2010. "Niche Conservatism as an Emerging Principle in Ecology and Conservation Biology." *Ecology Letters* 13, no. 10: 1310–24. https://doi.org/10.1111/j.1461-0248 .2010.01515.x.

Wiens, John J., Richard T. Lapoint, and Noah K. Whiteman. 2015. "Herbivory Increases Diversification across Insect Clades." *Nature Communications* 6: 8370. https://doi.org/10.1038/ncomms9370.

Wijová, Martina, František Moravec, Aleš Horák, and Julius Lukeš. 2006. "Evolutionary Relationships of Spirurina (Nematoda: Chromadorea: Rhabditida) with Special Emphasis on Dracunculoid Nematodes Inferred from SSU rRNA Gene Sequences." *International Journal for Parasitology* 36, no. 9: 1067–75.

Wiklund, Christer. 1973. "Host Plant Suitability and the Mechanism of Host Selection in Larvae of *Papilio machaon*." *Entomologia Experimentalis et Applicata* 16, no. 2: 232–42.

Wiklund, Christer, and Carl Åhrberg. 1978. "Host Plants, Nectar Source Plants, and Habitat Selection of Males and Females of *Anthocharis cardamines* (Lepidoptera)." *Oikos* 31, no. 2: 169–83. https://doi.org/10.2307/3543560.

Wilder, Aryn P., Rebecca J. Eisen, Scott W. Bearden, John A. Montenieri, Kenneth L. Gage, and Michael F. Antolin. 2008. "*Oropsylla hirsuta* (Siphonaptera: Ceratophyllidae) Can Support Plague Epizootics in Black-Tailed Prairie Dogs (*Cynomys ludovicianus*) by Early-Phase Transmission of *Yersinia pestis*." *Vector-Borne and Zoonotic Diseases* 8, no. 3: 359–68. https://doi.org/10.1089/vbz.2007.0181.

Wiley, Edward O. 1981. *Phylogenetics: The Theory and Practice of Phylogenetic Systematics*. New York: Wiley-Interscience.

Wiley, Edward O. 1986. "Methods in Vicariance Biogeography." In *Systematics and Evolution: A Matter of Diversity*, edited by Peter Hovenkamp, E. Gittenberger, E. Hennipman, R. de Jong, Marco C Roos, Ronald Sluys, and Marinus Zandee, 283–306. Utrecht: Utrecht University.

Wiley, Edward O. 1988a. "Parsimony Analysis and Vicariance Biogeography." *Systematic Zoology* 37: 271–90.

Wiley, Edward O. 1988b. "Vicariance Biogeography." *Annual Review of Ecology and Systematics* 19: 513–42.

Williams, H. Harford, A. H. McVicar, R. Ralph, et al. 1970. "The Alimentary Canal of Fish as an Environment for Helminth Parasites." In *Aspects of Fish Parasitology. Symposium of the British Society for Parasitology*, edited by Angela E. R. Taylor and Ralph Louis Muller, 43–77. Oxford: Blackwell Scientific Publications.

Williams, H. Harford. 1960. "The Intestine in Members of the Genus *Raja* and Host-Specificity in the Tetraphyllidea." *Nature* 188: 514–16.

Williams, H. Harford. 1966. "The Ecology, Functional Morphology and Taxonomy of *Echeneibothrium* Beneden, 1849 (Cestoda: Tetraphyllidea), a Revision of the Genus and Comments on *Discobothrium* Beneden, 1870, *Pseudanthobothrium* Baer, 1956, and *Phormobothrium* Alexander, 1963." *Parasitology* 56: 227–85.

Williams, H. Harford. 1968. "*Phyllobothrium piriei* sp. nov. (Cestoda: Tetraphyllidea) from *Raja naevus* with a Comment on Its Habit and Mode of Attachment." *Parasitology* 58: 929–37.

Williams, John W., and Stephen T. Jackson. 2007. "Novel Climates, No-Analog Communities, and Ecological Surprises." *Frontiers in Ecology and Environment* 5, no. 9: 475–82. https://doi.org/ 10.1890/070037.

Wilson, Edward O. 1959. "Adaptive Shift and Dispersal in a Tropical and Fauna." *Evolution* 13: 122–44.

Wilson, Edward O. 1961. "The Nature of the Taxon Cycle in the Melanesian Ant Fauna." *American Midland Naturalist* 95: 169–93.

Wilson, Edward O. 1984. *Biophilia*. Cambridge, MA: Harvard University Press.

Wilson, R. A., G. Smith, and M. R. Thomas. 1982. "Fascioliasis." In *Population Dynamics of Infectious Diseases: Theory and Applications*, edited by Robert M. Anderson, 262–319. London: Chapman & Hall.

Winemiller, K. O., P. B. McIntyre, L. Castello, E. Fluet-Chouinard, T. Giarrizzo, S. Nam, I. G. Baird, et al. 2016. "Balancing Hydropower and Biodiversity in the Amazon, Congo, and Mekong." *Science* 351, no. 6269: 128–29.

Winkworth, Richard C., and Michael J. Donoghue. 2005. "*Vibrunum* Phylogeny Based on Combined Molecular Data: Implications for Taxonomy and Biogeography." *American Journal of Botany* 92, no. 4: 653–66.

Woese, Carl R. 2000. "Interpreting the Universal Phylogenetic Tree." *Proceedings of the National Academy of Sciences of the United States of America* 97, no. 15: 8392–96.

Wolfe, Nathan D., Claire Panosian Dunavan, and Jared Diamond. 2007. "Origins of Major Human Infectious Diseases." *Nature* 447, no. 7142: 279–83.

Woodburne, Michael O. 2010. "The Great American Biotic Interchange: dispersals, tectonics, climate, sea level and holding pens." *Journal of Mammalian Evolution* 17, no. 4: 245–64.

Woolhouse, Mark E. J. 2008. "Emerging Diseases Go Global." *Nature* 451: 898–99.

Woolhouse, Mark E. J., and Sonya Gowtage-Sequeria. 2005. "Host Range and Emerging and Reemerging Pathogens." *Emerging Infectious Diseases* 11, no. 12: 1842-47.

World Economic Forum. 2018. *The Global Risks Report 2018*. 13th ed. https://www.weforum.org/reports/the-global-risks-report-2018.

World Health Organization [WHO]. 2005. "Control of Neglected Zoonotic Diseases: A Route to Poverty Alleviation." http://www.who.int/zoonoses.

World Health Organization [WHO]. 2015. "Connecting Global Priorities: Biodiversity and Human Health: A State of Knowledge Review." https://www.cbd.int/health/SOK-biodiversity.

World Health Organization [WHO]. 2018. "Middle East Respiratory Syndrome Coronavirus (MERS-CoV)." http://www.who.int/mediacentre/factsheets/mers-cov/en/.

Wright, Sewall. 1984. *Evolution and the Genetics of Populations*. Vol. 1, *Genetics and Biometric Foundations*. Chicago: University of Chicago Press.

Wright, W. H. 1972. "A Consideration of Economic Impact of Schistosomiasis." *Bulletin of the World Health Organization* 47, no. 5: 559–65.

Wu, Xiaoxu, Yongmei Lu, Sen Zhou, Lifan Chen, and Bing Xu. 2016. "Impact of Climate Change on Human Infectious Diseases: Empirical Evidence and Human Adaptation." *Environment International* 86: 14–23. https://doi.org/10.1016/j.envint.2015.09.007.

Xu, L., Qiyong Liuc, Leif Chr. Stige, Tamara Ben Ari, Xiy Fang, Kung-Sik Chan, Shuchun Wange, et al. 2011. "Nonlinear Effect of Climate on Plague during the Third Pandemic in China." *Proceedings of the National Academy of Sciences of the United States of America* 108: 10214–19.

Yabsley, Michael J., and Barbara C. Shock. 2013. "Natural History of Zoonotic *Babesia*: Role of Wildlife Reservoirs." *International Journal for Parasitology: Parasites and Wildlife* 2: 18–31. https://doi.org/10.1016/j.ijppaw.2012.11.003.

Yanagida, Tetsuya, Jean-François Carod, Yasuhito Sako, Minoru Nakao, Eric P. Hoberg, and Akira Ito. 2014. "Genetics of the Pig Tapeworm in Madagascar Reveal a History of Human Dispersal and Colonization." *PLoS ONE* 9, no. 10: e109002.

Yanagihara, Richard, Se Hun Gu, and Jin-Won Song. 2015. "Expanded Host Diversity and Global Distribution of Hantaviruses: Implications for Identifying and Investigating Previously Unrecognized Hantaviral Diseases." In *Global Virology I: Identifying and Investigating Viral Diseases*, edited by Paul Shapshak, John T. Sinnott, Charurut Somboonwit, and Jens H. Kuhn, 161–98. New York: Springer.

Yap, Tiffany A., Natalie T. Nguyen, Megan Serr, Alexander Shepack, and Vance T. Vredenburg. 2017. "*Batrachochytrium salamandrivorans* and the Risk of a Second Amphibian Pandemic." *EcoHealth* 14, no. 4: 851–64. https://doi.org/10.1007/s10393-017-1278-1.

Yates, Terry L., James N. Mills, Cheryl A. Parmenter, Thomas G. Ksiazek, Robert R. Parmenter, John R. Vande Castle, Charles H. Calisher, et al. 2002. "The Ecology and Evolutionary History of an Emergent Disease: Hantavirus Pulmonary Syndrome." *AIBS Bulletin* 52, no. 11: 989–98.

Yellen, John E., Alison S. Brooks, E. Cornelissen, Michael J. Mehlman, and Kathlyn Stewart. 1995. "A Middle Stone Age Worked Bone Industry from Katanda, Upper Semliki Valley, Zaire." *Science* 268: 553–56.

Young, P. J., A. T. Archibald, K. W. Bowman, J. F. Lamarque, V. Naik, D. S. Stevenson, S. Tilmes, et al. 2013. "Pre-Industrial to End 21st Century Projections of Tropospheric Ozone from the Atmospheric Chemistry and Climate Model: Intercomparison Project (ACCMIP)." *Atmospheric Chemistry and Physics* 13: 2063–90. https://doi.org/10.5194/acp-13-2063-2013.

Young, Cristin C. W., and Kevin J. Olival. 2016. "Optimizing Viral Discovery in Bats." *PLoS ONE* 11, no. 2: e0149237. https://doi.org/10.1371/journal.pone.0149237.

Young, Hillary S., Chelsea L. Wood, A. Marm Kilpatrick, Kevin D. Lafferty, Charles L. Nunn, and Jeffrey R. Vincent, eds. 2017. "Conservation, Biodiversity and Infectious Disease: Scientific Evidence and Policy Implications." Special issue, *Philosophical Transactions of the Royal Society B: Biological Sciences* 372, no. 1722 (5 June). https://doi.org/10.1098/rstb.2016.0124.

Yue, Ricci P. H., Harry F. Lee, and Connor Y. H. Wu. 2016. "Navigable Rivers Facilitated the Spread and Recurrence of Plague in Pre-Industrial Europe." *Scientific Reports* 6: 34867 https://doi.org/10.1038/srep34867.

Zamparo, David, Daniel R. Brooks, Eric P. Hoberg, and Deborah A. McLennan. 2001. "Phylogenetic Analysis of the Rhabdocoela (Platyhelminthes) with Emphasis on the Neodermata and Relatives." *Zoologica Scripta* 30, no. 1: 59–77.

Zarlenga, D. S., B. M. Rosenthal, G. La Rosa, E. Pozio, and Eric P. Hoberg. 2006. "Post-Miocene Expansion, Colonization, and Host Switching Drove Speciation among Extant Nematodes of the Archaic Genus *Trichinella*." *Proceedings of the National Academy of Sciences of the United States of America* 103, no. 19: 7354–59.

Zarlenga, Dante S., Eric Hoberg, Benjamin Rosenthal, Simonetta Mattiucci, and Giuseppe Nascetti. 2014. "Anthropogenics: Human Influence on Global and Genetic Homogenization of Parasite Populations." *Journal of Parasitology* 100, no. 6: 756–72.

Zhang, Guanyang, Usmaan Basharat, Nicholas Matzke, and Nico M. Franz. 2017. "Model Selection in Statistical Historical Biogeography of Neotropical Insects—The *Exophthalmus* Genus Complex (Curculionidae: Entiminae)." *Molecular Phylogenetics and Evolution* 109: 226–39. https://doi.org/10.1016/j.ympev.2016.12.039.

Ziętara, Marek S., and Jaakko Lumme. 2002. "Speciation by Host-Switch and Adaptive Radiation in a Fish Parasite Genus *Gyrodactylus* (Monogenea, Gyrodactylidae)." *Evolution* 56, no. 12: 2445–58.

Zinsser, Hans. 1935. *Rats, Lice, and History.* Boston: Little, Brown.

Zinsstag, Jakob, Esther Schelling, David Waltner-Toews, and Marcel Tanner. 2011. "From 'One Medicine' to 'One Health' and Systemic Approaches to Health and Well-Being." *Preventive Veterinary Medicine* 101, no. 3: 148–56.

Zschokke, Fritz. 1904. "Die Darmcestoden der amerikanischen Beuteltiere." *Centralblatt für Bakteriologie und Parasitenkunde* 36: 51–61.

Zschokke, Fritz. 1933. "Die Parasiten als Zeugen für die geologische Vergangenheit ihrer Trager." *Forschung und Fortschrift deutsche Wissenschaft* 9: 466–67.

Züst, Tobias, and Anurag A. Agrawal. 2016. "Population Growth and Sequestration of Plant Toxins along a Gradient of Specialization in Four Aphid Species on the Common Milkweed *Asclepias syriaca*." *Functional Ecology* 30, no. 4: 547–56. https://doi.org/10.1111/1365-2435.12523.

Index

co-option: definitions, 78–81; exaptation and co-optation, 80–81; as key to understanding origins of host-pathogen systems, 81; origins of new associations, 78; in parasitic flatworms, 81; preadaptation paradox and, 79
Cope, Edward Drinker, 76
coping with climate change and emerging disease: contemporary approaches to, 37, 47, 65, 70, 217, 226, 249, 257, 263; denial spectrum syndrome and, 222, 223, 260; living with fear and, 260; proactive measures for, 263–66
cospeciation and cophylogeny: chipmunks and pinworms, 93–95; classic case reanalysis, 131–33; cophylogeny analysis, 93; cryptic cospeciation, 93; estimating host fundamental fitness space, 93; general comparisons, 91–95; lice on rodents, 132–33; relation to Stockholm Paradigm, 93; representative studies, 282n20
Cracraft, Joel, 173
Cuénot, Lucien, 79
Curtice, Cooper, 37; and tick biology, 37

DAMA protocol, "Act": "Act" recommendations, 243–47; "anticipate to mitigate" as action plan, 226–27; applying the precautionary principle, 244–45; proof of concept, 247; roles of health practitioners, 244; sense of urgency, 243–48; urban threats, 247–48
DAMA protocol, "Assess" (triage): *Haemonchus* example, 236–37; hantaviruses in mammals example, 238–40; importance of archival biological collections, 238–40; infraspecific phylogenies and variants, 236; infrastructure and baselines for change, 239–40; phylogenetic triage, 235–38; risk assessment and decisions, 235–40; zoonotic viruses in bats example, 238
DAMA protocol, "Document": Área de Conservación Guanacaste, Costa Rica example, 231–32; *Borrelia* and Lyme disease pathogen example, 231; definition of, 229–35; ecological interfaces, 233; "finding them before they find us" initiative, 227, 272;

genetic barcoding, 234; monkeypox virus in North America example, 231; parataxonomists, 231–32; reservoirs of infection, 233; taxonomic inventories, 234
DAMA protocol, "Monitor": as early warning systems, 242; geography and natural history, 240–43; linked to models of anthropogenic climate forcing, 241; parataxonomists and local citizens, 241–42; snapshots of pathogen distribution, 242–43; traditional ecological knowledge, 243
DAMA protocol for EID: "anticipate to mitigate," 226–27; definition of, 227–48; as policy imperative, 270–71; sequential components of a DAMA program, 227–48
Darlington, Philip, 138
Darwin, Charles, 27–33, 50, 71–83, 86–87, 97, 120, 122, 170, 176–78, 257
Darwin Declaration, 230
Darwinism: as complex systems theory, 75–76; conditions of existence as higher law, 71, 73; contrasting nature of organism and nature of conditions, 71–76; Darwinian centenary and health professions, 51–52; Darwin's necessary misfit, 72–73; evolution as selective accumulation of diversity, 75, 137–38; natural selection as an emergent property, 76; nature of the conditions, 72–73, 97, 109, 180, 196; nature of the organism, 72–73, 75, 77, 86–87, 90, 97–98, 105, 109, 135, 137, 178, 180, 182, 194, 196, 218; significance of phylogenetic trees, 74–75; tangled bank, 73–74; tree of life, 74–75
death by a thousand cuts scenario, 21, 217, 226, 262
Dethier, Vincent, 42, 61, 281nn39–40
Diamond, Jared, 4, 21, 268
disease alert portals, 302n5
Dobzhansky, Theodosius, 79
Dogiel, Valentin Alexandrovitch, 84, 287n33
domestication (animal), 5, 17, 19, 53, 68, 100, 129, 136, 150, 151, 153–60, 165, 167–68, 209–10, 213, 222, 228, 236, 253, 285n48

ecological fitting: and colonization,
113; and coping with changes, 78;
ectomycorrhizal communities on
trees example, 178; and evolutionary
anachronisms, 62, 83, 286n30; and
functional traits, 297n31; historical
flexibility or evolutionary conservatism, 45, 82, 178; and invasive species,
2, 122, 156, 165, 172, 178–79, 182, 199,
230, 281n48, 290n24; origin of concept, 61–62; and oscillation, 129; and
parasite paradox, 64, 78, 105, 137; and
pathogen communities, 100–104; and
sloppy fitness space, 192; and spillback,
124; and spillover, 115, 124
ecological fitting as functional capacity, 78, 113; co-option (situational
flexibility), 78–81, 177; microhabitat
preference, 85–86; phylogenetic conservatism (historical flexibility), 81–83,
177; plasticity (contextual flexibility),
83–84, 177; transmission dynamics, 85
ecological fitting in sloppy fitness space:
community structure, 104; definitions, 97–100; examples, 100–104;
exploitation-biased activities, 97–98;
exploration-biased activities, 98–100;
exploring and exploiting, 98; host
fitness space, 178; invasions, 201;
operational aspect of capacity, 98; organisms as oscillators, 98–100; research
programs, 192
ecological niche modeling, 91, 93, 179
economic cost of emerging disease, 8,
14–19, 53, 55, 97, 129, 167, 214, 217–18,
222, 230, 249–50, 254, 256, 269,
286n30
Ehrlich, Paul, 51
Eichler, Wolfdietrich, 40
Eimer, Theodor, 32
El Niño Southern Oscillation (ENSO), 162,
240–41, 295n103
Elton, Charles, 1–2, 55, 201, 212, 249,
305n9
emerging infectious disease (EID): "anticipate to mitigate," 22, 70, 226–27;
black elephants, 262; black swans, 20,
276n32; business as usual, 10, 13, 69,
220–21, 224, 226, 228, 264–65; and
climate change, 165; climate change
challenge to human species, 69;

contemporary examples, 2–3; costs of
eradicating biodiversity, 206–7; crisis
and definition, 3; crisis as evolutionary
and ecological issue, 23–24; databases
and archives, 303n30; death by a
thousand cuts scenario, 21, 217, 226,
262; evolutionary minefield, 226–27;
humanity's responses to, as existential threat, 8–9, 25, 249, 262, 271;
frequency of colonization, 113–15;
globalization, 6, 8, 22, 162, 167, 211,
215–18, 221–22, 250, 253, 264; global
monetary costs, 14; global pandemics,
161–62, 222, 261–62; high probability/
low impact, 20, 217, 229, 249–50; host
range expansion, 145–46; implications
of taxon pulses, 145–46; inextricable
link to climate, 69; and landscape
perturbation, 207–8; links to climate,
6; low probability/high impact, 20–21;
monkeypox and ecological fitting,
231; monkeypox history in Africa,
289n23; as national security threat,
8, 246, 269–71; origins and risk of, 5;
as outcome of density, urbanization,
and globalization, 211–22; pathogen
morbidity and preventive chemotherapy examples, 284n43; perspective of
pathogen, 105; portals for disease alert
information, 302n5; recognition and
warning, 4; responses as economically
unsustainable, 225; and science and
solutions, 10; scientists and climate,
12; and socioeconomic collapse, 69–70;
and threat multipliers and climate, 8,
161, 168, 209, 212, 264; threat space
and opportunity space, 264
environmental harshness and species
radiation, 173
equilibrium theory of island biogeography,
139–40
eradication programs, consequences of,
54–55
Ericksson, Jakob, 35
Erwin, Terry, 140
evolutionary and functional capacity
(ecological fitting), 78; co-option
(situational flexibility), 78–81, 177;
microhabitat preference, 61, 63, 84–86,
94, 100, 103, 178, 192, 204, 236, 249;
phylogenetic conservatism (histori-

cal flexibility), 81–83, 177; plasticity
(contextual flexibility), 83–84, 177;
transmission dynamics, 45, 55, 61–63,
84–86, 90–91, 94, 100, 102–4, 151, 178,
186, 192, 236, 249
evolutionary capacity, general concepts, 77
evolutionary conservatism (conservative
nature of inheritance): ecological homologues, 82–83; evolutionary anachronisms, 83; functional traits, 177;
functional traits and speciation events,
182; phylogenetic conservatism, 63,
297n31; phylogenetic distribution of
host resources, 91–92
evolutionary opportunity: as filter and
facilitator, 88, 96–98; and fitness
space, 87; general concepts of, 86–91;
geography as fundamental element of,
90–91; host range, 61, 64, 70, 86, 90,
100, 107–8, 113–14, 116, 119, 121–22,
124–26, 131, 133–35, 143, 145–47,
151, 156, 158, 160, 163, 166, 178, 183,
192–93, 200–201, 204–5, 207–8, 211,
227–28, 233, 236–37, 241, 248, 279n27,
289n22; operative environments, 87; for
pathogens, 88, 93–94, 96–99, 111–12,
117, 127–28, 136–38, 142, 144, 170, 174,
179–81, 189, 191, 202, 233, 238, 257,
263–65; phylogenetic comparisons, 91
evolutionary oscillations: in altered fitness
space, 138; cospeciation and coextinction, 134; demographic outcomes,
121–22; directionality, 124–25; and
ecological fitting, 114, 129–30, 179–81;
evolutionary time scales, 131; and
external perturbations, 135, 307; generalizing and specializing, 106–7, 114,
179–81; hosts as samplers and selectors,
126–27; insect virus and primate example, 126–28; organisms as oscillators,
135; reanalysis of classic case of cospeciation, 132–33; and stepping stones,
125–29; taxon pulses, 140–44, 146–48,
182–83, 298n46; and Zika virus, 124–25
evolutionary transitions: equivalence of
host and pathogen evolution, 181;
information/capacity limited, 180;
macroevolution and microevolution,
290; *Ophraella* example, 180–81; op-
portunity/selection limited, 180
evolution of hominids and humans, 148–50

Fahrenholz, Heinrich, 40
Filipchenko, Aleksandr Alexandrovich, 84,
287n33
fitness space: cophylogeny studies, 93;
ecological niche modeling, 91; fundamental fitness space (FFS), 89–90, 111;
generalizing in, 97, 107–8, 134–35,
138, 178–79, 181, 190, 192, 252, 258,
299n62; realized fitness space (RFS),
88–90, 111; sloppy fitness space, 65,
88–91, 93, 95–99, 103, 106–7, 111–12,
130, 134–35, 138, 177–81, 190–92, 194,
207–8, 250, 284n40; specializing in,
106–7, 179–81; technological fitness
space, 216, 218
Fleming, Alexander, 43
Földvári, Gábor, 247–48
Fontana, Felice, 30
Food and Agricultural Organization (FAO),
227
food security and EID: agriculture, aquaculture, fisheries, 208; climate and
pathogen opportunity, 209; threat
multipliers and climate, 209; water
distribution as threat, 209; founder
effect, 115–17
fungal pathogens: *Puccinia* rust fungi,
162–63; *Puccinia gramini* stem rust, 29,
163–65; *Puccinia striiformis* yellow leaf
rust, 162–63; *Puccinia triticina* wheat
leaf rust, 163
fungal pathogens, wheat rusts: coevolution
between wheat and rust fungus, 49;
colonization from *Aegilopus speltoides*,
163; complex life cycles of, 30, 163–64;
discovery of, 29, 49; epidemics and
accidental introductions of, 163–65;
epidemics and climate envelopes, 30,
165; and global spread on trade routes,
163–64; and host range in wild grasses,
163; origins of, 163

Gambler's Ruin, 217–18; humanity's survival in climate change and, 252–53;
on Planet B, 263; policy makers and,
254–55; politicians and bureaucrats
and, 254; responses, five stages of grief,
223; response to threats and, 257; unit-
ing humanity, 272
Gamgee, John, 36
Garrett, Laurie, 4

modelling simulations (*continued*)
of selection and virulence, 120; network analysis, 109; pathogenicity, 120; potential reservoir hosts, 120; stepping stone host range changes, 125–26, 129, 146, 158, 240
modelling simulations, propagule pressure: continuous colonization, 122–23; definition of, 122; discontinuous colonization, 123; and ecological fitting, 122–23; multiple colonization events, 123–24
molluscan vectors of disease: snails and slugs for ungulate lungworms, 165; snails for *Fasciola*, 188; snails for *Fascioloides*, 100; snails for *Haematoloechus*, 101–3; snails for schistosomes, 147, 207, 211

Nägeli, Carl von, 32
natural selection, 31–33, 38, 43, 50, 57, 71, 73, 75–76, 78, 80, 127, 170–71, 177, 180, 196, 278n21, 296n2
nature of the biosphere: balance of nature concepts, 201–2; myth of a perfectly adapted biosphere, 76; Stockholm Paradigm, 65–66, 134, 190–94, 209, 284n40; survival of the fittest vs. origin of the fittest, 76; tangled bank, 73–74
nature of the conditions, 72–73, 75, 85–86, 88, 96–98, 104–5, 109, 134, 137–38, 144, 176, 190, 201, 218
nature of the organism, 72–73, 75, 77, 86–87, 89–90, 97–98, 104–5, 109, 135, 137, 178, 180, 182, 194, 196, 218
neglected tropical diseases, 285, 291
Neiman, Susan, 145
neo-Darwinism, 33, 41, 50, 71, 76, 172, 176, 296
Neodermata, 81, 184–90, 299n49
Newton, Isaac, and Newtonian Revolution, 258–59
North Atlantic Oscillation (NAO), 295n103

ocean-atmosphere regime shifts, 295n103
Office International des Epizooties (OIE), 227
One Health initiatives, 250–51
organisms as explorers and exploiters, 77, 88, 163

Origin of Species (Darwin), 27–28, 50, 71, 73–76, 177
orthogenesis, 32–34, 39–41, 51, 170, 172, 278n13, 280n34
Osborn, Henry Fairfield, 170, 172, 175, 296nn1–2, 296nn4–5
oscillation hypothesis, 107–8, 289n1
Osler, William, 251

Pacific Decadal Oscillation (PDO), 295n103
Pandora's box, 267
parasite paradox, 64
parasitism as an ecological phenomenon, 84
paratenic host, 126
Parmenides, 258
Pasteur, Louis, 28, 34–36, 276n4
pathogen biodiversity, global estimates, 21
pathogenicity, 119
pathogen pollution, 20
penicillin, 43, 46, 52, 67
Persoon, Christiaan Hendrick, 30
Petri, Julius Richard, 35
phenotypic plasticity, 177
phenotypic variation, 113
philosophy and nature of life, 258–59
phylogenetic analysis, 55–56
planetary health initiatives, 250–51
planetary state thresholds and tipping points, 281n48
plant hosts: *Ambrosia*, 181; *Artemisiae*, 181; barberry, 30, 163–64; barley, 162; *Chrysopsis*, 181; corn (maize), 2; *Eupatorium*, 181; *Helianthus*, 181; *Iva*, 181; mistletoe (Santanales), 62; Poaceae (grasses), 162; *Solidago*, 181; tomato, 2; wheat, 3, 30, 162
Poe, Edgar Allan, 214
policy and lifestyle solutions: building cooperation, 250; climate change as urgent risk, 248–49; DAMA protocols to buy time, 249–50
post-modernism, 220
preadaptation, 79–80, 171
PREDICT (USAID), 228–29, 232
President's Council of Advisors on Science and Technology (PCAST), 230
protistan pathogens and disease: *Anaplasma phagocytophilum*, 247; *Babesia bigemina* (Texas cattle fever), 37; *Myxobolus cerebralis* (whirling disease),

stock, deer and relatives): Camelidae, 158, *158*; camels, 158, *158*, 236–37, *237* vertebrate hosts (non-human), Mammalia (mammals) – Ursidae (bears), 151–52, *152* vertebrate hosts (non-human), Osteichthyes (bony fishes): Cypriniformes, 125; Esociformes, 125; Gasterosteiformes, 125; Perciformes, 125; Salmoniformes, 125; *Salmo trutta*, 54; *Onchorhynchus mykiss*, 54; trout, 2 viral pathogens and diseases: Andean virus, 123; avian influenza, 228; chikungunya, 3, 155; Crimean-Congo hemorrhagic fever, 3; dengue fever, 3, 15–16; eastern equine encephalitis, 120; Ebola (EBOV), 2–3, 15, 17, 115, 118, 120, 128, 203, 213, 228; filoviruses, 126–28; foot-and-mouth disease, 44–45; hantavirus pulmonary syndrome (HFRS), 123; Hendra virus disease, 115; herpes B, 115; human immunodeficiency virus (HIV-AIDS), 115, 131; influenza A, 128–29; Marburg hemorrhagic fever, 115; MERS coronavirus, 228; monkeypox virus, 115, 213, 231, 289n23; Prospect Hill vi-

rus, 240; rabies virus, 213; rhinovirus, 3; rinderpest, 37, 53, 69, 86, 226; rubella, 53; SARS coronavirus, 15, 115, 117–18, 228; simian immunodeficiency virus (SIV), 131; smallpox, 27, 29, 34, 43, 52, 69, 159–60, 214, 226, 231; Texas cattle fever, 36–37, 45; West Nile virus (WNV), 155, 213; yellow fever, 2–3, 35, 43, 45–46, 52, 165, 167, 203, 213; Zika virus (ZIKV) 2–3, 15, 17, 115, 117, 124–25, 155 Virchow, Rudolf, 36, 251 virulence, 119 von Ihering, Hermann, 278n11, 279n26 von Kölliker, Albert, 278n13 von Nägeli, Carl, 32 Vrba, Elisabeth, 80

Ward, Henry Baldwin, 284n39 water sustainability and climate change, 273n4 Wegener, Alfred, 279n27 West-Eberhard, Mary Jane, 177 Whitman, Charles Otis, 41 Wilson, Edward O., 139–40, 202

zoonoses, 28–29, 150, 155–59, 291n26